International Architectural Design

国际新建筑

《国际新建筑》编写组 编 常文心 译

辽宁科学技术出版社

图书在版编目（CIP）数据

国际新建筑 / 《国际新建筑》编写组编 ；常文心译. -- 沈阳 ：

辽宁科学技术出版社，2013.9

　　ISBN 978-7-5381-8258-3

　　Ⅰ．①国… Ⅱ．①国… ②常… Ⅲ．①建筑设计

－作品集－世界－现代 Ⅳ．①TU206

　　中国版本图书馆CIP数据核字（2013）第206214号

出版发行：辽宁科学技术出版社
（地址：沈阳市和平区十一纬路29号 邮编：110003）
印　刷　者：利丰雅高印刷（深圳）有限公司
经　销　者：各地新华书店
幅面尺寸：240mm×330mm
印　　张：64
插　　页：4
字　　数：50千字
出版时间：2013年 9 月第 1 版
印刷时间：2013年 9 月第 1 次印刷
责任编辑：陈慈良
封面设计：段娉婷
版式设计：吴　杨
责任校对：周　文
书　　号：ISBN 978-7-5381-8258-3
定　　价：520.00元

联系电话：024-23284360
邮购热线：024-23284502
E-mail: lnkjc@126.com
http://www.lnkj.com.cn

前言/Preface

当今的建筑与从前已经大不相同。一方面，新材料和新技术正在重塑建筑的设计方式。另一方面，许多建筑师和设计师正尝试以现代方式来运用传统材料和技术。在经济高速发展、社会生活多样化、环境敏感度提高的作用下，建筑设计摒弃了传统的设计风格，正朝向舒适、经济和简约美观发展。

以专业化的视角放眼全球，深度剖析建筑设计经典案例。本书精选全球6大洲50多个国家建筑设计的精品之作250个,以其不同的角度展现了设计师对其当地历史文化在建筑设计中的创新性应用。

本书为读者提供了全球最为经典的建筑设计视觉盛宴。内容涉及文化建筑、商业建筑、医疗建筑、教育建筑、办公建筑、居住建筑、酒店建筑、交通建筑、娱乐建筑、综合建筑10个类别。每个项目都配以实景图片、平面图及标注、项目信息、图注、设计理念以及材料的使用等相关信息，为读者准确详实的提供了高品质图书保证。

本书突出时效性、全球性、地域性、专业性，有助于全球的读者寻找设计灵感，了解新材料，改造旧项目，传承建筑设计的地域文化。

Buildings are now on the drawing board and they are nothing like the places we may recall from our childhood. New materials and new technologies are reshaping the way we build. At the same time, many architects and designers are also drawing upon ancient materials and building techniques but interpret them in modern ways. With the advanced development of economy and diversity of social life as well as increased sensitivity to the environment, architecture design is far from the primitive forms and styles; it is on the way to be comfortable, economical and rustically beautiful.

The book *Atlas of World Architecture*, with 250 projects selected, is a detailed and comprehensive portrayal of the best and latest architecture projects from six continents of more than fifty countries. Designers can be inspired a lot to search a balance between the overwhelmingly globalised trend and the increasingly personalised feature.

It offers readers a visual feast with the collection of world's most classic architecture projects and is categorised into ten parts, including Cultural, Commercial, Hospital, Educational, Corporate, Residential, Hotel, Transportation, Recreational and Complex architecture. Each project is illustrated with real photos, plans and a text. In addition, each geographic region is distinguished by a different colour-code. We firmly believe and hope it will serve as a source of pleasure and inspiration to all its readers.

Featured with its timeliness, globalisation, regionalisation, and professionalisation, it will help readers from all over the world to find inspiration and approach new materials and the cultural heritage.

Location of the selected projects of *Atlas of World Architecture*
《新世纪建筑地图》项目分布

1. Canada 加拿大
2. USA 美国
3. Mexico 墨西哥
4. Colombia 哥伦比亚
5. Chile 智利
6. Brazil 巴西
7. Iceland 冰岛

8. Norway 挪威
9. Finland 芬兰
10. UK 英国
11. Denmark 丹麦
12. Germany 德国
13. Poland 波兰
14. Portugal 葡萄牙

15. Spain 西班牙
16. France 法国
17. The Netherlands 荷兰
18. Luxembourg 卢森堡
19. Switzerland 瑞士
20. Italy 意大利
21. Austria 奥地利

22. Slovenia 斯洛文尼亚
23. Hungary 匈牙利
24. Greece 希腊
25. Angola 安哥拉
26. Libya 利比亚
27. Turkey 土耳其
28. Georgia 格鲁吉亚

29. Cyprus 塞浦路斯
30. UAE 阿联酋
31. India 印度
32. China 中国
33. South Korea 韩国
34. Japan 日本
35. Malaysia 马来西亚

36. Singapore 新加坡
37. Indonesia 印度尼西亚
38. Australia 澳大利亚

Contents /目录

Daytime view 日间景色

The Belleville Public Library and John M Parrott Art Gallery

贝尔维尔公共图书馆和约翰·M·帕洛特美术馆

The new Belleville Public Library not only provides resources for research and recreation; it is also a cultural and community destination. At 38,000 sf, the building includes a library, art galleries, meeting rooms, and a café as well as a significant outdoor public space. A large plaza frames the rotunda building, welcoming people from Campbell and Pinnacle Streets.

Interpretative and flexible spaces are at the heart of the architectural design of the building's programmatic elements. A rectangular element houses the galleries, library stacks, lounges and study spaces while the circular element – the rotunda – is the public hub of the library and plaza that includes the entrance, gift shop and street café. The third floor gallery entered from the rotunda connects the building activities vertically and increases the diversity of the building programme. The library provides both quiet spaces for contemplation and study as well as dynamic light–filled open spaces for other social activities.

新建的贝尔维尔公共图书馆不仅为科研、消遣提供了资料，而且是一个文化交流中心。建筑总面积3,530平方米，包含图书馆、美术馆、会议室、咖啡厅以及一个优美的户外公共空间。圆形大楼的前面是一个巨大的广场，迎接着来自坎贝尔大街和尖峰街的人群。

灵活和便捷的空间是建筑设计的核心。矩形建筑内设置着美术馆、图书馆的藏书室、休息大厅和学习空间；圆形建筑内则设置着图书馆的公共大厅和入口广场（包含入口、礼品商店和街角咖啡馆）。从圆形建筑可以进入三楼的美术馆，美术馆将两个建筑连接起来，增加了建筑的多样性。图书馆提供了一个安静思考和学习的空间，光线充足的公共空间十分适合社交活动。

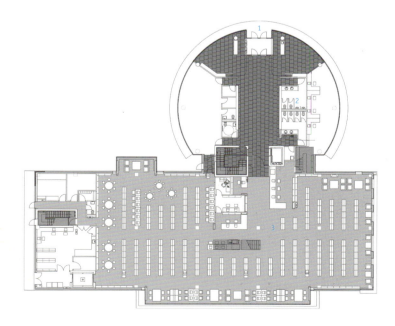

1. entrance	1. 入口
2. WC	2. 洗手间
3. reading room	3. 阅览室

夜景　Night view

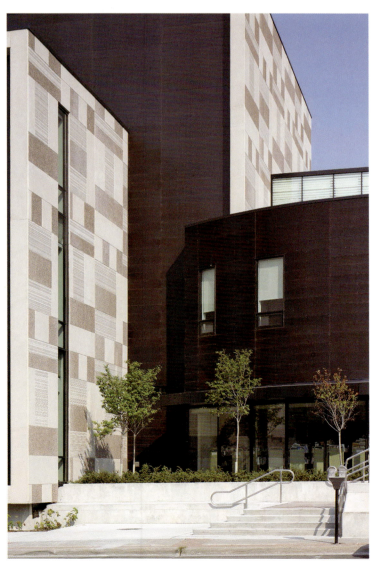

建筑入口 Entrance

阅览室 Reading room

Photo: Tom Arban

Completion Date: 2006

Architect: Zeidler Partnership Architects

Educational

Hilton Baltimore Convention Centre Hotel

The hotel is poised to play a key role in the continued success of the Pratt Street and Inner Harbour Entertainment and Convention Centre District. To take advantage of this unique position, the design team aimed to create and enhance the pedestrian experience that flows from the convention centre to Camden Yards. Civic spaces and defined urban edges are critical components to defining the area, which long lacked cohesive commercial activity and animation.

The hotel's exterior skin was designed to embody Baltimore's complex personality, hinting both backward and forward. Red brick façades wrap the building's lower floors and establish visual connection with the historic brick warehouse across the street that serves as a backdrop to the Ballpark at Camden Yards and with the traditional row houses that line the residential neighbourhoods to the west. If brick serves as a nod to the past, the metal cladding makes a more overt nod to the future, calling to mind Baltimore's industrial bulwark while offering a modern edge that relates to the sleek high-rises bordering the site.

The interior continues the sense of openness and visibility that drives the public spaces. Arranged to limit barriers between interior and exterior, the lobby and public areas provide constant but unobtrusive visual interest and activity.

希尔顿酒店

希尔顿酒店的建成对于普拉特街以及会议中心区的持续发展起到了至关重要的作用，为充分利用其地理位置优势，设计师致力于营造并强化行人体验。酒店外观设计体现了巴尔的摩市典型的综合性特色——建筑底层采用红色砖石"包裹"，与周围的古老砖石结构形成视觉连续感。如果说砖石的采用是对历史的延续，那么金属覆层则代表了对未来的向往，同时与周围的时尚高楼交相呼应。

室内设计同样突出开阔感及可视性，尤其在公共空间中更为明显地展现。为缩小室内外的障碍，大厅及公共区极为通透，让人一览无余。

Details 建筑细部

General view 全景图

室内 Interior

大堂 Lobby

入口 Entrance

1. Eutaw Street
2. Paca Street
3. Howard Street

1. 尤托街
2. 帕卡街
3. 霍华德街

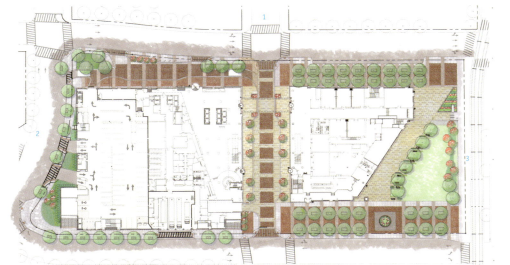

建筑景观 Landscape

通道 Passageway

Photo: RTKL/David Whitcomb

Hotel

Completion Date: 2008

Architect: RTKL Associates Inc./Mckissack & Mckissack; Interiors: Daroff Design
Landscape Architects: Mahan Rykiel Associates Inc.
Lighting Design: Brandston Partnership Inc.

Experimental Media and Performing Arts Centre

The building incorporates a wide variety of venues designed to the highest professional standards, which accommodate both the traditional performing arts and new, experimental media. Also provided are artist-in-residence studios, audiovisual production and postproduction suites, audience amenities and student and support facilities.

By taking advantage of the slope of the hillside site, the design solves one of the persistent challenges of performing arts projects: concealing the windowless mass of a very large hall and fly tower. This use of the topography also creates vistas over Troy towards the Hudson River, as seen from the campus approach and from major visitor spaces within the building.

The entire north façade of the building is a glass curtain wall, providing transparency between the EMPAC interior and the city of Troy. The glass wall allows daylight to flood the atrium, augmented by a halo skylight around the top of the concert hall that washes the cedar hull with the changing light of the day. By night, the wood hull is lit up from within the building and creates an iconic external identity that can be seen from distance.

体验媒体与表演艺术中心

这一建筑内集合了不同的场馆，其设计完全符合最高专业标准，包括古老的表演中心、新建的媒体中心、艺术家工作室、影音制片室、后期制作室以及学生活动中心。

建筑设计充分利用其所处地点的山坡地形，解决了长久以来一直困扰着设计师们的问题——将无窗户设计的大型空间"遮蔽"起来。巧妙地利用地形特色可以堪称完美，同时将哈德森河的美丽景致引入进来。

建筑北立面完全采用玻璃材质打造，在室内和特罗伊城之间建立了一道无形的屏障与纽带。光线透过玻璃涌入到中庭处，经由音乐大厅顶部周围的天井之后照射到木质表皮处。夜晚，木质外壳在灯光的照耀下格外吸引眼球，使得整幢建筑如同灯塔一般。

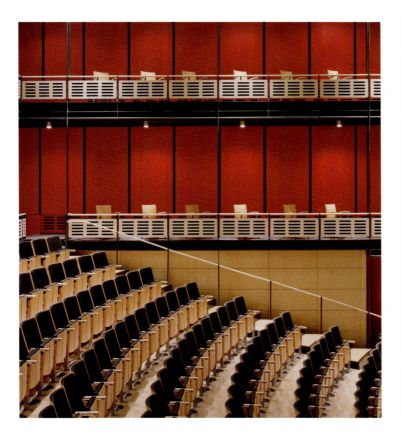

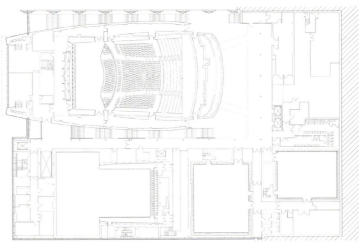

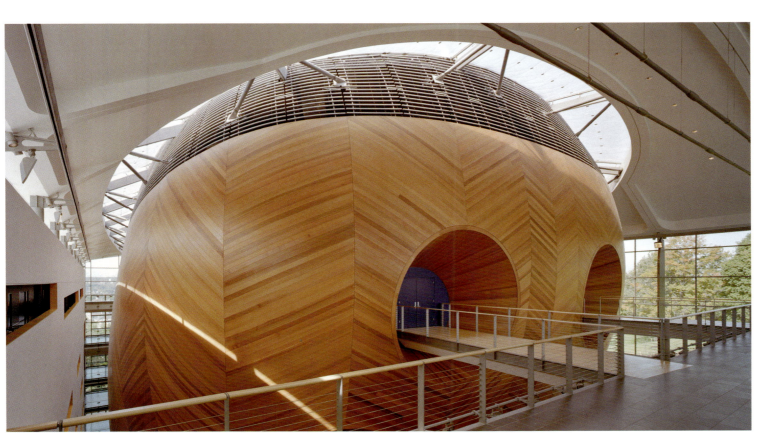

Photo: Aaron Esto, Paul Rivera

Cultural

Completion Date: 2008

Architect: Grimshaw Architects

General view 建筑外观

Evans House

Here is a residential building designed by Bittoni Design Studio (architects Mark Bittoni, Ross Jeffries and Salomé Reeves). This project is actually a redesign of a private residence located in the Crestwood Hills, near Los Angeles. This house has a special site, located on the hill with a panoramic view all around. Inside the house there is also a luxurious interior, spacious rooms, large windows, polished floors, comfortable beds, swimming pool and kitchen and adequate dining room. It is correct to say that this house is a real dream.

埃文斯别墅

埃文斯别墅由一幢私人住宅重新改造而来，位于克雷斯特伍德山区，洛杉矶附近。别墅所处地理位置独特，环绕于壮丽的景致中。住宅内部彰显奢华——宽阔的房间、高大的窗户、抛光地面、舒适大床、游泳池、厨房及餐厅无一不展现着主题。这里真的可以称为"梦幻之家"！

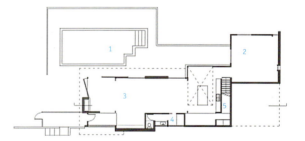

1. entrance   1. 入口
2. kitchen   2. 厨房
3. living room 3. 客厅
4. toilet   4. 洗手间
5. staircase   5. 楼梯

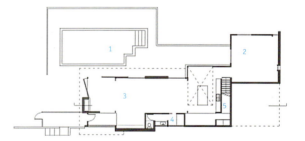

Exterior 建筑外观

起居室 Living room

厨房 Kitchen

窗外和厨房 Outdoor view and kitchen

Photo: Bittoni Design Studio

Completion Date: 2009

Architect: Bittoni Design Studio

20th Street Offices

20号街办公

Environmental sensitivity went into all aspects of the design and construction of the 20th Street Offices. The initial concept began with an open linear tube-like form sitting atop a series of moment frames. This concept allowed the occupiable space to be lifted above the at-grade parking, maximising opportunities for open green space, natural ventilation and daylight. With the open ends oriented to the east and west, the natural flow of air coming off the Pacific Ocean circulates through the tube, maximising fresh air and minimising the need for mechanical systems. The building envelope of the tube element consists of custom-designed diamond-patterned cladding, fabricated out of sheet metal. This cladding combined with recycled content insulation of high R-values, minimises heat gain and puts less stress on the mechanical systems as well.

Broken up into different multifunctional spaces the building allows occupants, visitors and clients to congregate for discussions and events, hold visual presentations, share a meal, watch a film or even hold a yoga class on the green roof. The 20th Street Offices strives to create a lifestyle, an office culture and a connection to the community synonymous with its environmentally conscious informed design. The building functions as a laboratory and gallery to explore ideas, test products, promote green initiatives and market "building responsibly" to its clients and the surrounding community.

环保理念贯穿设计及建筑的各个方面。最初的理念是打造一个开放式管状结构，并通过一系列柱状结构支撑，使其脱离地面，以便预留出更多的绿色空间，同时增强自然通风及光照。整个结构东西两侧开敞，来自太平洋的凉爽空气便可通过所有空间。建筑表面采用专门设计的钻石形状覆层（由金属片铸造而成）包裹，加之高性能可回收绝缘材料，极大地减少了热能吸收，进而减轻了机械系统的负担。

整个空间被分割成不同的多功能区，方便员工及来访者相互交流以及娱乐休闲。可以说，20号街办公营造了一种全新的生活方式——独特的办公文化。这里更是一个实验室或画室，让客户尽情探索，并将环保理念传达出去。

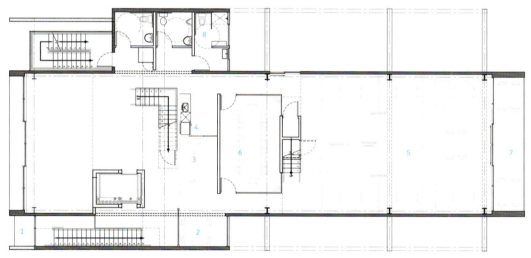

1. 下层入口
2. 前台
3. 经理办公室
4. 厨房
5. 工作室
6. 会议室
7. 阳台
8. 洗手间

1. entrance below
2. reception
3. manager office
4. kitchen
5. work studio
6. conference room
7. balcony
8. restroom

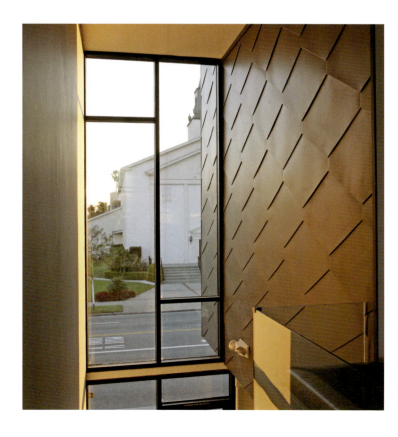

Photo: Belzberg Architects

Completion Date: 2009

Architect: Belzberg Architects

Façade 建筑外观

Arizona State University Walter Cronkite School of Journalism & Mass Communication

亚利桑那州立大学新闻大众传媒学院

Located in downtown Phoenix, the new six-storey, 22,500-square-metre building has become an integral part of the fabric of ASU's energising downtown campus and a harbinger of Phoenix's redevelopment. As truth and honesty are guiding principles to journalism, so are they to the design of the building. The architecture is specifically expressive of function and materiality. The design is based on an economical 30-foot-square exposed structural concrete column grid with post-tensioned concrete floor slabs. The exterior is clad with glass, masonry and multi-coloured metal panels – the pattern of the panels is inspired by U.S. broadcast frequency spectrum allocations (the Radio Spectrum). The composition is kinetic and dynamic – symbolic of journalism and media's role in our society. The building's massing incorporates appropriate sun screens on each of the four façades; their specific architectural treatment reduces the heat loads and is one of many of the LEED Silver building's sustainable strategies. Burnished concrete block walls, ground and polished concrete floors and warm wood ceilings further express the forthright and direct nature of news delivery.

The Cronkite School occupies all of the second and third floors and a portion of the fourth and sixth floors. The airy, multi-tiered First Amendment Forum is the heart of the school. By day, students gather spontaneously between classes, and in the evenings, the grand hall transforms into a public forum where students and industry leaders discuss the most critical issues facing today's news media.

亚利桑那州立大学位于菲尼克斯市中心,新闻大众传媒学院新建6层建筑,面积达22500平方米,已完全融入到校园中,同时引领了菲尼克斯市的发展规划。真实、诚实作为新闻专业的重要指导原则,因此也被运用到建筑设计中。建筑以边长为30英尺的正方形水泥结构为基础,采用水泥板装饰,并拼接成格子造型,突显功能性与物质性。外观采用玻璃板、石板及多色金属板饰面,其灵感来源于美国广播电台频谱分布图案,增添动感的同时,更强调了新闻传媒的重要作用。此外,建筑外观大量运用遮阳板,以减少表面热量。

新闻及大众传媒学院占据二三两层以及四层和六层的部分空间。其中通透而层次感分明的第一修正案论坛广场构成了中心结构。白天,学生们聚集在一起探讨;夜晚则转变成了大型公共论坛区,学生和行业领袖们可以尽情探讨当今传媒领域的重大课题。

Front view 建筑正面

夜景 Night view

1. 门厅
2. 服务大厅
3. 展示大厅
4. 教室

1. lobby
2. service centre
3. presentation hall
4. classroom

展示大厅 Presentation hall

服务中心 Service centre

Photo: Bill Timmerman

Completion Date: 2008

Architect: Ehrlich Architects (www.ehrlicharchitects.com)

Educational

Exterior view 建筑外观

Northern Arizona University

The building forms a long arc oriented to the south to capture the winter sun in a glass-enclosed three-storey gallery that serves as a thermal buffer space for the offices behind. Louvres and blinds shade the gallery from the hot summer sun, yet allow sun penetration to warm the building in the winter. The south façade of the building is expressed in brick, wood, glass and aluminium. Sustainable strategies integrated into the building include:
a long and thin shape to maximise daylight and minimise electric lighting needs;
concrete structural frame stores heat in the winter and cool in the summer to reduce energy required for heating and air-conditioning;
low-pressure under-floor air distribution reduces fan sizes and energy requirements;
nearby field of photovoltaic panels (donated by APS) produces 160 kilowatts and produces more than 20% of the building's electricity;
triple-glazed windows on the building's north side minimise unwanted energy loss and gain;
automated shade controls regulate solar gain to maintain a comfortable gallery environment.

The combined impact of all these strategies is to reduce the energy consumption by 89% compared to a typical building. Ninety percent of construction waste materials were recycled, and 30 percent of the materials used in construction were made from recycled materials, including insulation made from recycled denim jeans. Water conservation measures include the use of indigenous landscaping, low-pressure faucets, waterless urinals and dual-flush toilets.

北亚利桑那大学

建筑南侧外观由砖块、木材、玻璃及铝材打造，呈现出弧形结构，办公区前面三层高的全玻璃结构行使着热缓冲器的功能。百叶窗的运用同时使得建筑免受夏日强光的灼晒。设计中格外注重环保理念，包括：
细长的造型便于充分接受自然光线，进而减少电灯照明的需求；
混凝土材质框架利于调节温度，确保冬暖夏凉；
地下低压配风系统的运用减少了风扇规模及需求；
光电板的运用可提供建筑20%的电能所需；
南侧三层玻璃窗结构减少了不必要的能源消耗及获得；
光线调控设备可自动调节温度，营造舒适的环境。

以上这些方式的运用使得建筑同其他普通建筑相比减少了89%的能源消耗。此外，90%的建筑材质都是可循环利用的，而30%的材质都是经循环而来。节水措施包括低压水龙头、无水便池、双按水马桶等。

1. 入口
2. 中庭
3. 会议厅
4. 机房
5. 办公室
6. 阳台
7. 咖啡厅
8. 泄洪区
9. 入口广场

1. entrance
2. atrium
3. conference pod
4. plant
5. offices
6. terrace
7. café
8. detention basin
9. entrance plaza

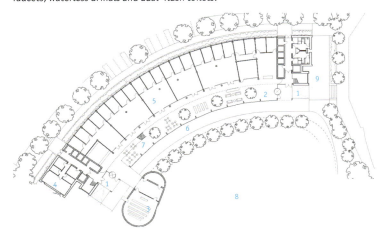

建筑外观 Exterior view

建筑外观　Exterior view

走廊　Passageway

Photo: Timothey Hursley

Educational

Completion Date: 2008

Architect: Hopkins Architects with Burns Wald and Executive Architects

21

Exterior view of north façade 北立面外观

Taubman Museum of Art

Located at one of Roanoke's most visible and historic downtown intersections, the new Museum is the first major purpose-built museum ever constructed in the city. The building, with forms and materials chosen to pay homage to the famed Blue Ridge and Appalachian Mountain surroundings, quadrupled the size of the Art Museum's previous facilities at Centre in the Square. The building features flexible exhibition galleries for the Art Museum's important permanent collection of 19th and early 20th century American art, contemporary art and regional crafts; education facilities with a library, studio and study Centre; a multi-purpose auditorium; a café; a book and gift shop; a black-box theatre; and outdoor terraces providing unique vistas of the city.

The finish on undulating, stainless steel roof forms reflects the rich variety of colour found in the sky and the seasonal landscape. Inspired by mountain streams, translucent glass surfaces emerge from the building's mass to create canopies of softly-diffused light over the public spaces and gallery level. As it rises to support the stainless steel roof, a layered pattern of angular exterior walls is surfaced in shingled patinated zinc to give an earthen and aged quality to the façade.

陶布曼艺术博物馆

陶布曼艺术博物馆位于历史古城罗阿诺克一个十字路口处，格外引人注目，这也是该地区第一个为特定目的而建的博物馆。建筑在造型及选材上突出了布鲁里奇以及阿巴拉契亚山脉的特色，其规模较之原来的博物馆扩大了四倍。内部包括画廊，用于展示美国19及20世纪的永久性收藏品、当代艺术品以及当地手工艺品；教育空间，如图书馆、工作室及研究中心；多功能礼堂；咖啡厅；书店及礼品店；黑盒子剧院；室外露台。

建筑屋顶由不锈钢材质打造，弯曲的造型极为引人注目，色彩装饰更突出了蓝天以及四季景观，五彩斑斓。通透的玻璃表面设计灵感来源于山间溪流，将建筑上方包裹起来并在公共区及画廊一层之上形成顶篷，用于调节光线的强度。层叠的角状外墙采用锈迹斑斑的锌片饰面，营造出年代久远的特色。

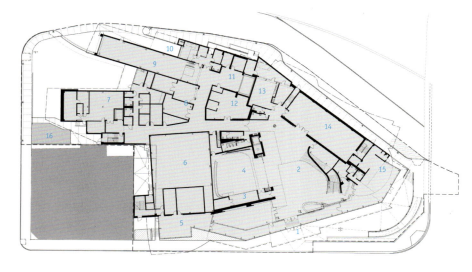

1. Salem Avenue entrance	1. 赛伦街入口
2. museum lobby	2. 博物馆大厅
3. museum shop	3. 博物馆礼品店
4. auditorium	4. 礼堂
5. theatre foyer	5. 剧院大厅
6. theatre	6. 剧院
7. mechanical	7. 机械室
8. museum services	8. 博物馆服务中心
9. art handling	9. 艺术品处理室
10. loading dock	10. 装货平台
11. protective services	11. 防护设备
12. catering kitchen	12. 餐厅和厨房
13. e & o studio	13. e & o工作室
14. art venture gallery	14. 艺术美术馆
15. museum café	15. 博物馆咖啡厅
16. electrical equipment	16. 配电室

塞伦街外景 Exterior view from Salem Avenue

从美术馆看中庭 Atrium view from gallery level

诺福克街外景 Exterior view from Norfolk Avenue

Photo: Timothy Hursley

Completion Date: 2007

Architect: Randall Stout Architects, Inc.

41 Cooper Square - The New Academic Building for The Cooper Union

41 Cooper Square, the New Academic Building for The Cooper Union, aspires to manifest the character, culture and vibrancy of both the 150-year-old institution and of the city in which it was founded. 41 Cooper Square aspires to reflect the institution's stated goal to create an iconic building – one that reflects its values and aspirations as a centre for advanced and innovative education in Art, Architecture and Engineering.

In the spirit of the institution's dedication to free, open and accessible education, the building itself is symbolically open to the city. Visual transparencies and accessible public spaces connect the institution to the physical, social and cultural fabric of its urban context. At street level, the transparent façade invites the neighborhood to observe and to take part in the intensity of activity contained within. Many of the public functions – an exhibition gallery, board room and a two-hundred-seat auditorium – are easily accessible on one level below grade.

The building reverberates with light, shadow and transparency via a high performance exterior double skin whose semi-transparent layer of perforated stainless steel wraps the building's glazed envelope to provide critical interior environmental control, while also allowing for transparencies to reveal the creative activity occurring within. Responding to its urban context, the sculpted façade establishes a distinctive identity for Cooper Square. The building's corner entrance lifts up to draw people into the lobby in a deferential gesture towards the institution's historic Foundation Building. The façade registers the iconic, curving profile of the central atrium as a glazed figure that appears to be carved out of the Third Avenue façade, connecting the creative and social heart of the building to the street.

Built to LEED Gold standards and likely to achieve a Platinum rating, 41 Cooper Square will be the first LEED-certified academic laboratory building in New York City.

1. entrance
2. loading bay
3. retail
4. storge
5. office
6. classroom
7. main lobby

1. 入口
2. 装卸台
3. 零售区
4. 仓库
5. 办公区
6. 教室
7. 主大厅

41柯柏广场——柯柏高等科学艺术联盟学院

41柯柏广场——柯柏高等科学艺术联盟学院，旨在展现学院及其所处城市150年的特色、文化及活力。设计目标即为打造一个标识性建筑，能够展现其作为艺术、建筑及机械设计中心的价值及理想。

学院一向崇尚自由、开放及通俗的教育方式，这从设计中展现出来——学院大楼面向整座城市开放。视觉的通透感以及开放式的公共空间使得学院完全融入其所处城市的自然、社会及文化结构。

建筑外观为双层结构，半透明的带孔不锈钢材质"包裹"在玻璃表层，在视觉上调控着室内的环境，同时营造了通透感。独特的外观将中央心房的蜿蜒造型清晰展现，仿若雕刻出来的一般。角落处的入口稍微抬起一些，引领来访者走进大厅。此外，41柯柏广场是纽约市第一个通过LEED（能源与环境建筑）认证的学术性建筑。

街景 Street view

远景 View from distance

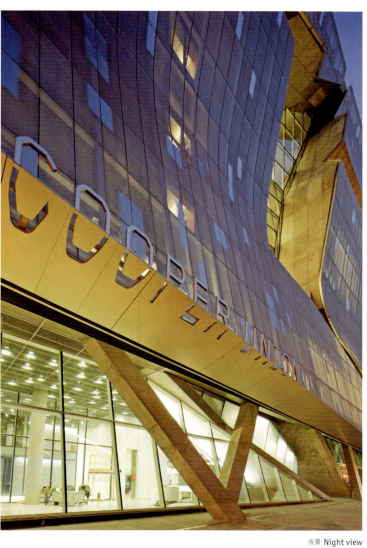

夜景 Night view

楼梯井和屋顶 Stairwell and roof

楼梯 Staircase

View from west; manufacturing building at night 西立面；右侧是制造装备楼

322 A Street Office and Manufacturing Facility

The building is clearly split to reflect the two parts of the programme: a single-storey metal fabrication workshop and two levels of office area. The office building is elevated to create a covered parking area underneath, raise the occupied spaces above the 100-year flood plane, and assure that the presence at the street intersection is given due prominence. The lower-level offices house the metal fabrication administration connected through a bridge to a mezzanine in the shop. The second level is the headquarters of the real estate development company. The building is book-ended by two high spaces: a lobby atrium that functions as exhibition space for company products and a common space oriented towards the river and the city.

The material palette consists of two primary materials: zinc flat locks panels are used to highlight people spaces, and pre-manufactured, field-assembled corrugated metal panels are used for the workshop. Glass curtain wall assemblies present the multi-storey gathering areas to the public and provide near and distant views to the occupants.

The office floor plan is only 60 feet (18 metres) wide. Offices, located on the perimeter, and all conference rooms are enclosed by client-made translucent shoji screens or transparent glass walls fabricated with painted steel angles and exposed fasteners. The desks, tables and credenzas within were all produced by the client's shop and continue the integration of design and manufacturing to showcase metalwork as a craft.

322A街办公和加工大楼

建筑被清晰地划分为两个部分：单层的金属零件加工楼和两层的办公楼。办公楼整体被架高，下面的空间形成了一个停车场，这让建筑主体高于近百年最高的洪水水位，也保证了建筑在十字路口更加显眼。办公楼的底楼是金属零件加工厂的行政部门，通过一个廊桥与加工楼相连。二楼是房地产开发公司总部。办公楼的两端都是较高的空间：一个展示公司产品的中庭大厅和一个朝向河流和市中心的公共区域。

建筑主要使用了两种材料：办公楼采用了镀锌面板，而加工楼则采用了预加工波纹金属板。公共区域的玻璃幕墙为人们呈现了远近的风景。

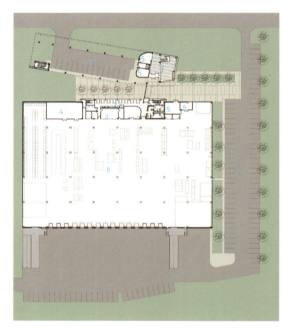

1. lobby	1. 大厅
2. lunchroom	2. 餐厅
3. take-off	3. 卸货区
4. storage	4. 仓库
5. mechanical	5. 机械室
6. manufacturing	6. 加工区
7. parking	7. 停车场

办公楼的楼面只有18米宽。办公室设在建筑的四周，它们和会议室都由公司自己生产的透明隔断和透明玻璃墙隔开，隔断和玻璃墙上镶嵌着彩钢边角和金属配件。楼内的办公桌椅、书架都是由公司自己生产的，将设计和金属制造业连接成一个整体。

办公楼西侧的公共区和平台 View from west towards office building common area and terrace

办公楼和制造装备楼的连廊 View of connecting corridor between office and manufacturing building

二楼会议室 First floor conference room

Photo: Woodruff/Brown Architectural Photography

Corporate

Completion Date: 2005

Architect: KlingStubbins

A generously-sized, colourful urban oasis   一个开阔、多彩的城市绿洲

Lausd William J. Clinton Middle School

Located just south of downtown Los Angeles in an isolated pocket of light manufacturing and vacant buildings with minimal community life, this middle school links two city blocks and provides a safe learning haven for more than 1,500 students. Within a tight budget, a clear and simple design was employed to allow for quality materials including corrugated steel and concrete masonry.

The 150,000-square-foot campus (plus 65,000 square feet of structured parking, canopies and bridges) includes a three-acre academic quad and, across the street, a 6-acre athletic quad. The two-storey academic building houses administration, library, lecture space and shared classrooms on the first floor, with individual classrooms on the upper level. Each side of the U-shaped plan, colour-coded in blue, green or yellow, forms a community of about 500 students; each individual classroom within its cluster is painted a different shade of the principal colour.

The two quads are joined by a pedestrian bridge over a wide, busy street that links classrooms with a gymnasium, playing fields, a track, and faculty parking for 142 cars. All shade canopies are tilted at an optimal angle to the sun, to enhance the performance of photovoltaics.

威廉姆•J•克林顿中学

这所中学位于洛杉矶南部一处独立的场地，周围环绕着轻工业工厂和空置的建筑物，极少有社区居民活动。学校连接了两个街区，为1,500名学生提供了一个安全的学习环境。由于预算紧张，建筑师采用了一个简洁的设计方案，运用了如波纹钢和混凝土砌块等高品质材料。

校园占地13,935平方米（附加6,038平方米的停车场、华盖和天桥），共有12,140平方米的教学空间和24,280平方米的体育空间。教学楼的一楼是行政办公、图书馆、讲堂和公共教室，二楼是独立教室。U形大楼的三面分别是蓝、绿、黄三种颜色，每一面大约能容纳500名学生。每个独立教室都采用了与所在楼面颜色同色系的颜色。

教学空间和体育空间中间是一个过街天桥，将教室和体育馆、操场、跑道和停车场连接起来。学校的华盖按照太阳直射的角度而设计，充分利用了太阳能电池板。

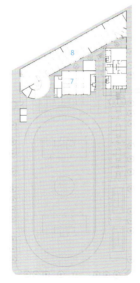

1. main administration	1. 主行政区
2. library	2. 图书馆
3. multi-purpose	3. 多功能教室
4. general classroom	4. 普通教室
5. art room	5. 艺术教室
6. kitchen	6. 餐厅
7. gym	7. 体育馆
8. parking	8. 停车场

场地设计极具城市特色 The site design treats the school in an urban way

设计中采用了经久耐用的材料 Durable, low-maintenance materials were used throughout the design

彩色波纹钢和混凝土石砌墙 Coloured corrugated steel and concrete masonry

巨大的露天华盖 Large open-air canopies

Photo: Tom Bonner

Educational

Completion Date: 2007

Architect: Ehrlich Architects (www.ehrlicharchitects.com)

View of the Margot and Bill Winspear Opera   温斯皮尔歌剧院全景

Bill Winspear Opera House

The Margot and Bill Winspear Opera House is engineered specifically for performances of opera and musical theatre with its stages equipped for performances of ballet and other forms of dance. The opera house's principal entrance features the 60-foot Annette and Harold Simmons Signature Glass Façade that wraps three quarters of the way around the building, creating a transparency between the opera house and the surrounding Performance Park. The transparent façade provides dramatic views of the Margaret McDermott Performance Hall, which will be clad in vibrant red glass panels. From within the Winspear Opera House, the Simmons Glass Façade provides a sweeping view of the skyscrapers of downtown Dallas that line the northern edge of the Performance Park.

Radiating from the Winspear Opera House on all sides, the sky canopy will provide shade over three acres of the Sammons Park, creating new outdoor spaces for visitors to gather and relax. The glass solar canopy's louvres will be arranged at fixed angles following the path of the sun, calculated to provide optimal shade for the outdoor spaces throughout the day, as well as preventing direct sunlight from hitting the Simmons Glass Façade during the warmest months of summer.

温斯皮尔歌剧院

温斯皮尔歌剧院主要包括歌剧院及音乐厅，其舞台设计专门满足了芭蕾舞及其他各种舞蹈表演的需求。入口处的特色玻璃外观占据了大楼四周近四分之三的通道，在剧院及周围的背景中营造了一道透明的纽带。透过玻璃外观，可以看见剧院内的表演大厅，同时在剧院内部也可以看到达拉斯城区内巍然耸立的高楼。

为进一步增强剧院与公园的联系，84英尺宽的玻璃幕墙被改造成23英尺高的垂直外观，这样，大厅、咖啡厅及餐厅的景色便可在室外一览无余。顶棚的设计既减少了玻璃外观的强光照射程度，同时也营造了凉爽的室外活动区。

1. theatre
2. entrance
3. restroom

1. 歌剧院
2. 入口
3. 洗手间

Front view   建筑正面

建筑外观 Exterior view

楼梯和大厅 Stair and lobby

剧院顶部的吊灯 Chandelier

Photo: Iwan Baan, Tim Hursley

Cultural

Completion Date: 2009

Architect: Foster + Partners

Public gallery flanked by social lounges and student entrance 公共美术馆两侧是休闲楼和学生入口

Hassayampa Academic Village, Arizona State University

亚利桑那大学Hassayampa教学村

Located at the southeast corner of Arizona State University's Tempe campus, The Hassayampa Academic Village interlaces 1,900 beds with classroom, computing, dining and retail components. The buildings are organised as a series of four-storey courtyard buildings sharing a public gallery space with a seven-storey tower. The towers flank the primary east-west and north-south connections to campus and serve as thresholds to the gallery spaces with their entrance to the residential buildings. Each of these buildings is composed of four floors of forty student communities sharing a social lounge with the adjoining floor. Together, the four floors of student suites gain a shared identity through the colour of their respective courtyard elevations, thereby promoting an individual identity for each building within the life of the academic village.

The project is designed to respect the demands of the climate and environment through its orientation, building envelope, mechanical systems and harnessing of breezes. Devices such as canopies will shade outdoor public spaces, which in turn temper the environment around the buildings. Coupled with material selection and efficiencies of the building, these strategies to reduce heat gain have achieved an LEED Silver rating for the complex.

Hassayampa教学村位于亚利桑那大学坦佩校区的东南角，为学生提供了1,900个床位和教室、计算机房、餐厅、零售等配套设施。教学村里的建筑以一系列的四层庭院楼为主，它们共同使用一个七层高的公共设施塔楼。塔楼将宿舍楼和校园连接在一起。

每座宿舍楼里有40个学生社团，他们共用一个休息大厅。每座宿舍楼都有一个独特的颜色，象征着自己的独特地位。建筑的朝向、外墙设计、机械系统和通风设备都反映了环境和气候的需求。天棚等设施遮蔽着户外公共空间，调节了建筑周围的环境。建筑所使用的环保材料和节能策略为教学村赢得了LEED银奖认证。

1. Chuparosa Court
2. Arroyo Court
3. Jojoba Court
4. Verbana Court
5. Acourtia Court

1. 丘帕罗萨庭院
2. 旱地庭院
3. 霍霍巴庭院
4. 威尔巴纳庭院
5. 菊类庭院

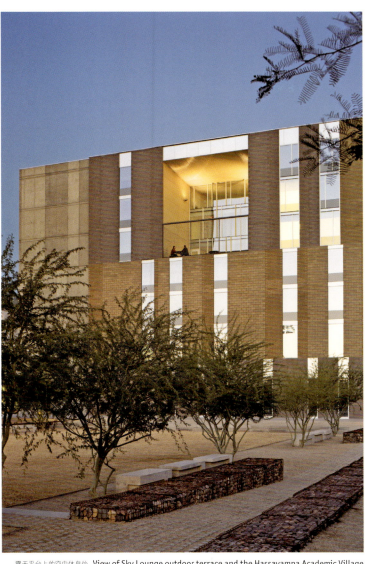

露天平台上的空中休息处 View of Sky Lounge outdoor terrace and the Hassayampa Academic Village

处于休闲楼和学生入口中间的公共美术馆 Public gallery flanked by social lounges and student entrance

休闲楼前的旱地庭院 Arroyo Courtyard looking towards social lounge

Photo: Farshid Assassi, Anton Grassi/Esto

Educational

Completion Date: 2007

Architect: Machado and Silvetti Associates with Gould Evans, LLC.

Front with pool (west) 别墅正面和游泳池（西侧）

VilLA NM

VilLA NM is not a regular house; it is not meant for everyday living. It is a house for summers, for weekends, for stolen time. This is a house that you share with your immediate family, with your most intimate friends. The house is compact as vacation homes often are: like the dacha and lake-side cabin of Russia and Scandinavia, the house offers a simple, private, family-and-nature orientated retreat from urban life.

The conceptual model for Villa NM is a box with a blob-like moment in the middle; a twist in both plan and section that causes a simple shoebox to bifurcate into two separate, split-level volumes. One side clings to the northern slope of the hill; the other detaches itself from the ground, leaving room underneath for a covered parking space.

All the internal spaces maximise the potential for wraparound views. The kitchen and dining area on the ground floor are connected by a ramp to the living space above, the 1.5-metre (5 feet) height change allowing for a sweeping outlook over the surrounding woodland and meadows. A similar ramp connects the living area to the master and the children's bedrooms on the first floor. Facilities such as the bathroom, kitchen and fireplace are clustered in the vertical axis of the house, leaving the outer walls free. Large glazed windows feature in all but the most private rooms.

NM别墅

NM别墅不是一座普通的住宅。它并不是为日常生活所设计的，而是为避暑、度假而设计的。在此你可以和亲朋好友共度美好时光。别墅的设计简洁紧凑，正如俄罗斯和斯堪的纳维亚半岛的乡间别墅或是湖滨小屋一样，别墅提供了一个远离喧嚣都市的简洁、私密、居家、自然的世外桃源。

NM别墅的概念模型是一个放着杂乱结构的盒子。一个在平面和剖面上的扭曲造型将简单的鞋盒结构分割成两个独立的错层空间：一侧贴近北面的山坡；另一侧抬离地面，在下面形成了一个停车场。

室内空间的设计保证了从窗口可以看到最多的风景。一楼的厨房和餐厅通过一个坡道和上面的起居空间相连，1.5米的高度差让室内可以欣赏到周边树林和草地的全景。另一个类似的坡道连接了起居区和二楼的主卧和儿童房。浴室、厨房、壁炉等设施聚集在别墅内的一个垂直轴线上，其他的外侧墙壁都是空白的。除了私密的房间之外，几乎所有空间都装有大型玻璃窗。

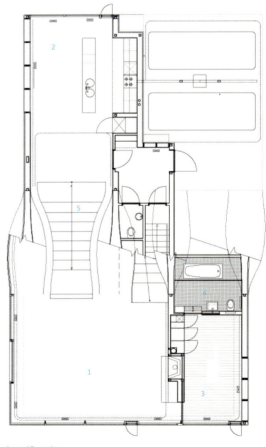

Ground floor plan

一楼平面图

1. 客厅	1. living
2. 厨房	2. kitchen
3. 卧室	3. bedroom
4. 浴室	4. bathroom
5. 楼梯	5. stairs

远景 Distance view

入口 Entrance

室内设计和家具摆设 Interior with furniture

Photo: Christian Richters

Completion Date: 2007

Architect: UNStudio

Green Circle Shopping Centre

The Green Circle Shopping Centre located in Springfield, Missouri, is one of the most sustainably developed retail spaces in the United States. Slated to achieve an LEED Platinum rating, the highest rating possible, the 23,000–square-foot centre incorporates recycled materials, utilises sources of renewable energy, and maximises energy efficiency.

Site location was treated with great sensitivity. The conventional shopping centre would clear the site of trees and maximise parking and retail space. As a sustainable alternative, Green Circle preserved over forty existing trees on site and in doing so provided building tenants and customers with green space for recreation and visual relief.

A geothermal system with forty wells located under the parking lot utilises the earth's heat energy for heating and cooling 100% of the spaces. Paired with both an ERV (Energy and Heat Recovery Ventilators) and heavily insulated walls and floors, the geothermal system provides a 50% improvement in efficiency and decreased utility demands when compared with the baseline case of a typical shopping centre. Increased efficiency translates to lower electrical bills for the tenants and less air pollution for the environment. Lighting controls, efficient light fixtures, photovoltaic panels and extensive daylighting by the strategic placement of windows play a large part as well. The roof and the south façade have photovoltaic panels producing several kilowatts of electrical energy for building use. Interior spaces are capable of having almost no artificial lighting during daylight hours. Furthermore, all of the glass used for daylighting is high performance which minimises solar heat gain where necessary and transmits a high percentage of visible light.

绿环购物中心

绿环购物中心位于密苏里州斯普林菲尔德市，是美国最具可持续特征的零售空间之一。这个2,136平方米的购物中心以LEED白金标准建造，采用了再生材料和可再生能源，最大化了能源效率。

项目对场地的处理具有高度可持续性特征。传统的购物中心一般会将场地上的树木清除，以最大化停车和零售空间。而绿环购物中心保留了场地上原有的40棵树木，让这些绿树为建筑和消费者洒下一片绿荫。

位于停车场下方的地热系统有40口热井，为整个购物中心提供空调服务。地热系统与能源和热回收通风系统、隔热墙面和地板一起，增加了50%的能效，减少了购物中心的能量需求。这些高效设施既节省了租户的电费，又减少了对环境的污染。照明控制系统、节能灯具、太阳能光电板、大范围自然采光和窗户的朝向等策略同样为建筑节能做出了巨大贡献。屋顶和南面外墙上的太阳能光电板为建筑提供了电能。白天，室内几乎不需要人工照明。此外，天窗上的高性能玻璃可以减少热能吸收，同时最大限度地过滤可见光。

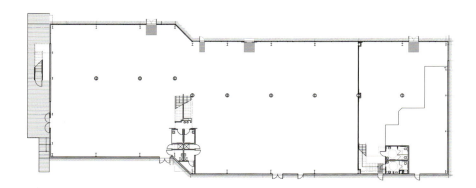

Aerial perspective © Bob Linder 鸟瞰图

鸟瞰图——屋顶 Aerial perspective with roof details © Bob Linder

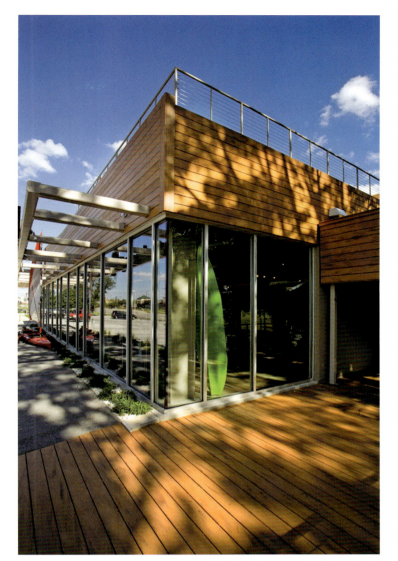

外观一角 Exterior corner details © Bob Linder

户外楼梯 Exteiror stairs © Sinclair

室内 Interior © Bob Linder

Commercial

Photo: Bob Linder, Mike Sinclair

Completion Date: 2008

Architect: Hufft Projects, LLC / Matthew Hufft

Exterior night view ©Travis Fullerton 外观夜景

Virginia Museum of Fine Arts

The design of the new $151 million building, plaza and sculpture garden opens up the Virginia Museum of Fine Arts (VMFA) to the city. The E. Claiborne and Lora Robins Sculpture Garden combined with the Mary Morton Parsons Plaza act as a square onto the Boulevard, and under a hill at the end of the gardens a new 600 car parking deck completes the redevelopment of the campus which also includes three other historic buildings.

The new 15,330-square-metre McGlothlin Wing provides extensive new space for the museum's collections and study centre. The museum reports that this expansion makes them one of the ten largest encyclopaedic museums in the United States.

VMFA announces itself to Richmond with the atrium's twelve-metre-high window facing the Boulevard. Natural light floods into the heart of the museum and a wall of glass opens the atrium, café and restaurant onto the pools, fountain and sculpture garden. Five bridges connect the new to the original museum, across all floors of the building. It adds fifty percent additional exhibition space to the existing building. There is 139 square metres of changing exhibition space for major touring exhibitions. The building also includes the Art Education Centre, the Freeman library, a gift shop, state-of-the-art object and painting conservation facilities, a 150-seat lecture hall, the "Best" café with garden terrace and pool, and a restaurant overlooking the Sculpture Garden.

1. existing museum
2. James W and Frances G. Mcglothlin Wing
3. Robinson Farm House
4. E. Claiborne and Lora Robins Sculpture Garden
5. parking deck
6. Vmfa Centre for Education and Outreach
7. Confederate Chapel
8. United Daughters of the Confederacy

1. 原有的博物馆
2. McGlothlin翼楼
3. 农场屋
4. 雕塑花园
5. 停车场
6. VMFA教育中心
7. 同盟礼拜堂
8. 南部联邦的女儿纪念堂

弗吉尼亚美术馆

造价为1.5亿美元的美术馆大楼、广场和雕塑花园将弗吉尼亚美术馆展示在民众面前。E. Claiborne and Lora Robins雕塑花园和Mary Morton Parsons广场直通林荫大道，山脚下花园旁的停车场共有600个停车位。整个园区还包括三幢历史性建筑。

新建的15,330平方米的McGlothlin翼楼为美术馆提供了宽敞的展览、研究空间。这一扩建工程使弗吉尼亚美术馆跻身于美国十大综合博物馆。

美术馆中庭12米高的落地窗正对里士满林荫大道。阳光透过中庭洒入博物馆。通过中庭的玻璃门，中庭、咖啡厅、餐厅和户外的泳池、喷泉、雕塑园联系在了一起。翼楼的各层共有五个廊桥将新旧建筑连接在一起。项目为博物馆增添了50%的展览空间，其中139平方米可以作为巡回展览空间。新建建筑还包括艺术教育中心、自由人图书馆、礼品商店、绘画存储设施、150个席位的演讲厅、"Best"咖啡馆、花园平台、游泳池和俯瞰雕塑花园的餐厅。

建筑外观 Exterior ©Travis Fullerton

东侧落地窗 East window ©Travis Fullerton

中庭廊桥 The atrium bridges ©Tippy Tippens

Photo: Travis Fullerton, Tippy Tippens

Cultural

Completion Date: 2010

Architect: Rick Mather Architects

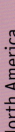

Social Condenser for Superior

The Social Condenser project is located at the base of the Superstition Mountain Range in the Town of Superior, Arizona which was founded in 1882 and has strong ties to mining of copper, silver and gold.

The project is uniquely positioned between historic Main Street and Queen Creek. The site consists of two parcels, the project parcel to the north and an open landscaped parcel to be developed into future outdoor dining and music pavilion, and is bisected by an access path from the upper street level and a lower wooden footbridge that spans across the creek.

The project is a renovation and expansion of an existing two-storey block building and addition of an exterior dining terrace. The lower level is developed into kitchen, mechanical and storage spaces and the upper level is designed as an open gathering space. The south-facing wall of the upper level of the existing building is removed to expose the volume within. The remaining form is rendered to closely match the shadow tones of the surrounding hills and acts as both backdrop and anchor for the new addition.

The project was informed by the concept of the "public house". Classically an obscured, introverted diagram, the Social Condenser conversely aims to balance concealment with exuberant exposure of the internal activities to the streetscape, the pedestrian walking path and the adjacent landscaped parcel.

The project is envisioned to be the living room of the community; a place to congregate, socialise, view work of provincial artists and enjoy the breathtaking landscape vistas that envelop the region.

苏泊利尔社交中心

该项目位于亚利桑那州苏泊利尔镇，地处迷信山山脚。苏泊利尔镇建于1882年，以铜、银和金矿而闻名。

苏泊利尔社交中心夹在镇上的大街和王后溪之间，一面朝北，一面朝向一片开阔的景观区域，可用于未来露天餐厅和音乐凉亭的开发，两面由较高的街面道路和小溪上地势较低的木板桥分割开来。

项目是对原有的两层楼高的建筑的翻新和扩建工程，将为社交中心添加一个露天就餐平台。建筑的一楼是厨房、机电室和仓储空间，二楼则被设计为开放式集会场所。原有建筑朝南的墙壁被移除了，行人可以透过落地窗看到内部的情景。其他的建筑结构得以保留，封闭的造型和周边的山脉十分相称。

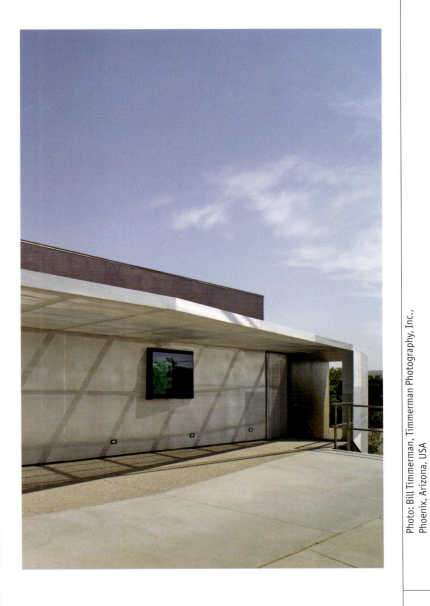

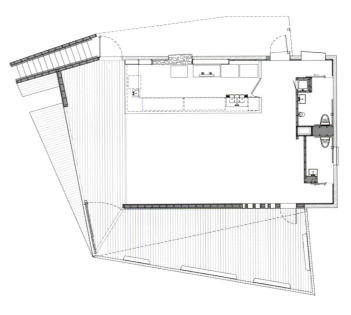

Photo: Bill Timmerman, Timmerman Photography, Inc.,
Phoenix, Arizona, USA

Recreational

Completion Date: 2007

Architect: Blank Studio

Building 建筑全景

Sustainable Residence – 3716 Springfield

The residence at 3716 Springfield in Kansas City is an environmentally conscious, modern home performing completely "off-the-grid". Being the first LEED platinum home in the Kansas City Metropolitan area, the building serves as an example of sustainable practice and living for buyers considering a life in the city. The combination of the passive glazing with the louvres for shading and active systems integrated with the roof plane flush to the siding calls out visually the environmental building standards. Also, the residence is dedicated to teaching the community and provides tours for all interested parties in order to effectively encourage the neighbourhood to become knowledgeable in sustainable architecture.

A broad south exposure was purposely sought after to support the passive solar effort. Additionally, operable windows along the lower south glazing and roof-top skylight allow for stack ventilation throughout the interior. The entire site was planted with drought-resistant landscaping and the south was intentionally left open to encourage the homeowner to plant a native garden. The hardscape surrounding the exterior of the home use pervious concrete which permits rainwater to seep through and into the water table.

春田街3716号可持续住宅

堪萨斯城春田街3716号住宅具有现代环保特征，完全实现了"自给自足"。作为堪萨斯城区第一座经过LEED绿色建筑认证的住宅，它是城市可持续建筑工程的典范。带有百叶窗的被动式玻璃窗和屋顶侧面排水系统形象地体现了其环保建筑标准。此外，这座住宅还在社区中起到了一定的教育意义，鼓励社区建造更多的可持续建筑。

住宅向南的朝向支持了被动式太阳能系统，南侧的可控式窗户和屋顶的天窗保证了室内的自然通风。住宅的周围种满了抗旱景观植物，南面的庭院特别被空出来供住宅的主人自己打造花园。

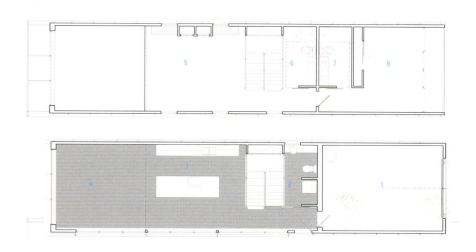

1. 车库
2. 浴室
3. 厨房
4. 客厅
5. 伸缩空间
6. 浴室
7. 主浴室
8. 主卧

1. garage
2. bath
3. kitchen
4. living room
5. flex space
6. bath
7. master bath
8. master bedroom

建筑东侧 East view

厨房和餐厅 Kitchen and dining space

一楼浴室 Bath on the ground floor

Photo: Courtesy of Studio 804, Inc.

Residential

Completion Date: 2009

Architect: Studio 804, Inc.

Two additions in relation to the existing structure 扩建结构与原有结构相连

Cohen Levine Residence

Reconsidering the expression "the whole is greater than the sum of its parts", a series of modern interventions were introduced to a traditional home. Responding to a number of paradigm shifts in the clients' lives, each project was contemplated as an independent entity yet considerate of the assemblage as a whole.

Initially, the entrance sequence was rethought, moving the front door out of the living room and creating an intimate yet articulated space. A flat roof stitches together the new and old entrance while introducing the modern idiom to the traditional fabric. The living room was expanded into the side porch, adding volume and light to the tired dark environment.

The largest intervention is a new building equalling the footprint of the existing home. Capped with a butterfly roof, the solution was a direct response to not overwhelm the small scaled colonial house. Inside, light shafts link the open-plan living, dining and cooking space to the first floor and inverted roof above. Cabinetry elements were inserted into the various living spaces to provide storage and act as a bridge between the interior spaces.

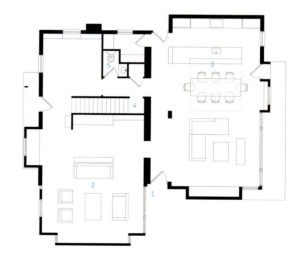

1. entrance	1. 入口
2. living room	2. 客厅
3. kitchen	3. 厨房
4. stairs	4. 楼梯
5. bath	5. 浴室

科恩·莱文住宅

建筑师以"整体大于部分总和"的理念为一座传统住宅添加了一系列的现代元素。在设计过程中，建筑师进行了换位思考，从住户的角度为住宅进行了整体性规划和改造。

首先，建筑师重新排列了入口的空间排列，前门被移到客厅之外，形成了一个私密而连贯的空间。一个平顶遮阳板将新旧入口连接起来，在传统建筑中引入了现代元素。客厅扩张到了侧门廊，扩大了空间，为昏暗的室内引入了光亮。

最大的改造工程是建造一座与原有住宅等大的新住宅。这座蝶形屋顶的新住宅并不会抢走旧住宅的风头。内部的轻质立柱将开放式客厅、餐厅和厨房与二楼和屋顶连接在一起。室内空间的木质家具为住户提供了储藏空间，也在其间搭建了一座桥梁。

扩建结构中的客厅 Living room of the side addition

餐厅可以仰视上方的蝶形屋顶 Dining area, looking up to underside of butterfly roof

后立面 View of back elevation

Photo: Paul Warchol Photography

Residential

Completion Date: 2010

Architect: David Jameson Architect

45

Spertus Institute of Jewish Studies

斯珀特斯犹太研究所

Acclaimed as the finest cultural addition to Chicago after Millennium Park, the Spertus Institute of Jewish Studies is one of the city's most celebrated and discussed new projects.

The façade is an innovative, 21st-century approach to the use of glass and the expression of identity. Folding glass planes on the façade give drama and presence to this ten-storey building located in a city of skyscrapers. The folds also scale and modulate the Spertus façade and relate it to the mostly brick and stone neighbouring buildings in the Michigan Boulevard Historic District.

The project includes an asymmetrical, fan-shaped auditorium with seating for 400 people, and full projection, lighting and sound mixing capabilities. The balcony is uniquely multi-tiered so that smaller groups can attain a level of intimacy with the onstage performance even when the auditorium is partially filled. The back of the auditorium has a faceted form, revealed within the main lobby of the building, that gives a dynamic expression to the activities inside.

Moving up into the building, visitors follow a sequence of light-filled spaces which connect and overlap Spertus's mixed-use programme of exhibition galleries, library, auditorium, college classrooms and administrative offices. A great hall at the top floor looks back across Grant Park to the skyline and the lake, connecting both the institution and visitors back to city and nature.

斯珀特斯犹太研究所被誉为自千禧公园以来，芝加哥最出色的文化扩建工程，也是芝加哥最炙手可热的新项目。

建筑的外墙极具个性，创新性地使用了玻璃材料。建筑共有十层楼，折叠拼接的玻璃面板为其增添了戏剧化的效果。这些折叠结构调节着建筑的立面结构，在它与密歇根大道历史街区上的砖石建筑之间建立了联系。

斯珀特斯犹太研究所的礼堂采用了不对称的扇形结构，可容纳400人，具有全套的投影、照明和音响设备。多层结构的看台也可以让小部分的群体和台上进行互动。礼堂后部的反射结构可以让外面大厅里的人清楚地看见礼堂内部的活动。

建筑的上层是光线充足的展览馆、图书馆、礼堂、学院教室和行政办公室。顶楼的大厅可以看到建筑后方的格兰特公园和湖上的景色，将人们从研究院带回了城市和自然景色之中。

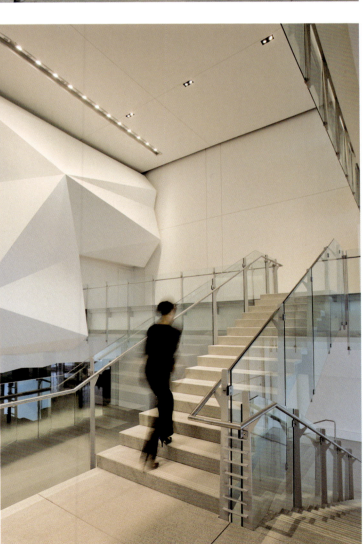

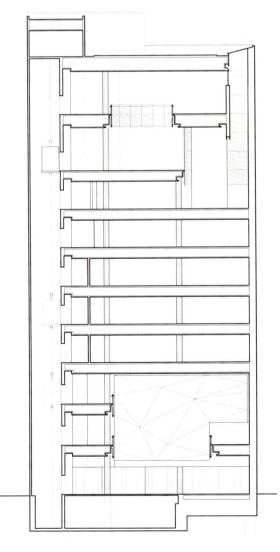

Photo: William Zbaren

Educational

Completion Date: 2008

Architect: Krueck+Sexton Architects

General view 建筑全景

Crockett Residence

It was decided to locate the living level on the upper floor to capture views and allow privacy without curtains. Within the lower level, a flex space of sorts was created which could act as a family room, a room for mother-in-law, or be closed off to act as a separate (legal) rental unit.

The surrounding area is an eclectic mix of charming craftsman homes and apartment buildings. Early on the decision was made to go with surface (though mostly covered) parking rather than a garage to both allow more space for the home and avoid the "garage door dominated" front façade look. This also speaks to the clients, lifestyle as they commute more on bicycles than they do in cars.

The final design was realised through the creation of three simple elements: a horizontal volume, a vertical volume and a folded plane. The private programme elements (bedrooms & bathrooms) were located in the two volumes while the public areas were contained within the interstitial spaces created when the three elements were combined. The entrance and stairs occur in the space between the folded plane and the metal clad volume; the living level within the upper fold of the Minaret plane. In both cases, the simple massing allows the spaces to seem as if they flow out from the home to connect with the neighbourhood, adding energy and size.

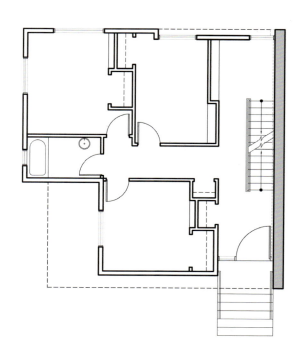

克罗克特住宅

住宅的起居空间被设在二楼，以便将更多的风景收入眼中，而且不用窗帘也能保证隐私。一楼的空间相对灵活，可以作为家庭娱乐室、给家里的老人居住，或是作为单独的出租空间。

建筑的设计由三个简单的元素组成：一个水平结构、一个垂直结构和一个叠层楼面。卧室和浴室等私密空间被设在水平和垂直两个结构里，而公共区域则设在三个结构的叠加空间里。入口和楼梯设在叠层楼面和金属壳结构之间，而起居空间设在二楼。在这个简单的模式里，所有的空间都仿佛溢出了住宅，与周边的空间相连，增添了住宅的空间。

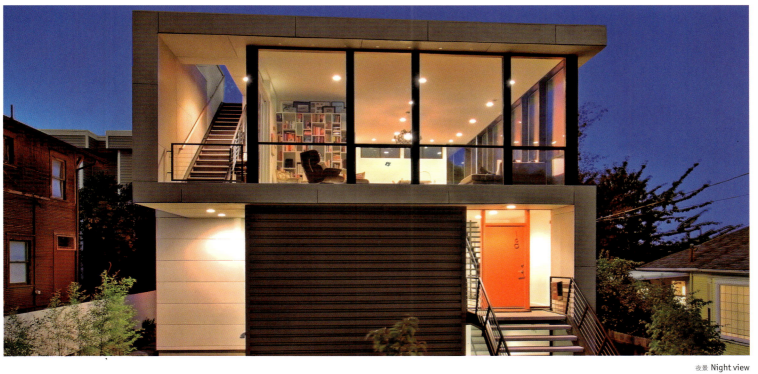

夜景 Night view

屋顶 Roof

室内 Interior

Photo: Pb Elemental Architecture

Completion Date: 2008

Architect: Pb Elemental Architecture, Dave Biddle, Chris Pardo

Building in the evening 夕照中的建筑

Strauchaus

This home rests on the edge of a wooded area. The diagramming process analysed the programme and needs of the two owners in effort to resolve a cohesive experiential sequencing of spaces, suggest massing arrangements and uncover elevation compositions. The abstraction of the programme is made in effort to make the two paths: the movement, and the connections become an extension of the landscape.

As the building takes form, the graphic maps and modeled paths begin to prescribe how the structure relates to the surrounding context. A patterning of thin floor-to-ceiling windows connects the structure to the neighbouring forestry by continuing the event of a body passing through the forest.

Strauchaus住宅

住宅坐落在一片树林的边缘。建筑师在设计阶段综合考虑了项目和两位户主的需求，确定了空间的排列、模块的构建以及里面的组合。项目规划被抽象成两个方面：移动路径和与外部景观相连接的路径。

图纸和既定路径在建筑建造过程中预先描绘了建筑与景观的联系。细长的落地窗模仿了树林里树木的造型，在结构上将建筑和周边的林地联系了起来。

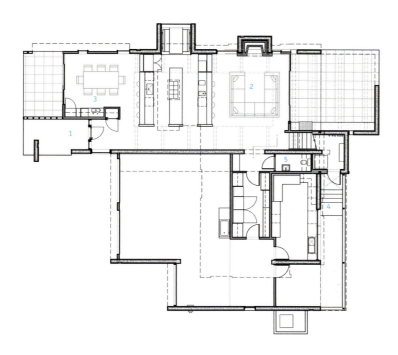

1. 入口
2. 客厅
3. 厨房
4. 楼梯
5. 浴室

1. entrance
2. living room
3. kitchen
4. stairs
5. bath

傍晚的中庭 Patio in the evening

走廊 Hallway

Photo: Peter Jahnke

Residential

Completion Date: 2008

Architect: PIQUE llc.

傍晚，建筑正面 Front side in the evening

Mt. Crested Butte Residence

王冠峰住宅

This is a special house built in a ski resort community by two brother carpenters who happen to own a truckload of redwood siding. The house was designed to be attractive to two families who might want to be in a resort at 3,100-metre elevation. From the front door one could ski down to the ski lift, take the lift and ski back down to the house. A central shaft that runs vertically through the house supports the house. In the interior the central shaft is a fireplace for each of the house's three floors. The house is designed with two garages on each side of the entrance. One enters the house on a split-level. A half level up has two master bedroom suites separated by the central shaft with a fireplace and whirlpool in it. A half level down from the entrance is the community family areas, living, kitchen, etc. The lowest level is a children's playroom/dormitory. The undulating roof reflects the mountaintop behind the house. Local people refer to the house as the "snow clam".

这座特别的住宅建在滑雪胜地，由一对木匠兄弟用他们的红木木板建造。住宅建在海拔3,100米的度假村里，适合两个家庭居住。从前门可以直接滑向滑雪缆车，乘坐缆车到达山顶，然后再从山上直接滑到门口。住宅的中心轴柱支撑着整个建筑结构，轴柱在室内则被设计成三个壁炉，每层各一个。大门的两侧各有一个车库。住宅内部采用叠层结构，上半层是由轴柱隔开的两间主卧室，下半层是家庭生活区，如客厅、厨房等。最底层是儿童游乐室和寝室。起伏的屋顶对应着屋后的山顶，当地人形象地称其为"雪蛤"。

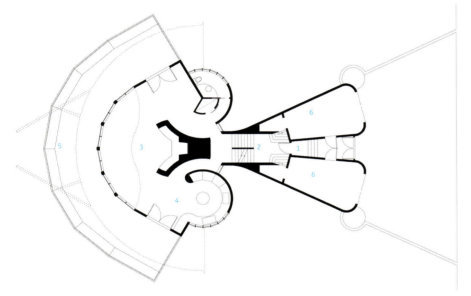

1. 外部入口
2. 内部入口
3. 大厅
4. 厨房
5. 平台
6. 车库

1. outer entrance
2. inner entrance
3. great room
4. kitchen
5. deck
6. garage

Completion Date: 2008

Architect: Robert Oshatz

Driveway 入口车道

Whitten House

This residence sits on a remote site composed of Sage & Juniper trees in Central Oregon. Conceived as two simple cubes in the landscape, one box for sleeping and one for living, the structure offers two distinct means of interaction with the landscape. The larger sleeping box is low & burrowed into earth, while the living box floats above, hovering just at treetop level. East & South orientations are exploited for views as well as passive solar orientation of the home.

Exterior materials were chosen for durability and fire resistance. All rainwater will be harvested & stored for landscaping or fire fighting purposes. The future pool will provide an additional margin of wildfire safety as a usable body of water on the remote site.

Evacuated tube solar water heaters will efficiently provide most of the heating for the home through in-floor radiant tubing. The building's narrow profile and extensive glazing combined with the region's low humidity allow for passive cooling of the home.

维顿住宅

住宅坐落在俄勒冈州中部一处偏远的松树林中，外部造型极为简单，一个盒状机构用于睡眠，另一个是起居空间。大的睡眠盒在下面，嵌入地面；而起居盒则悬在上面，和树顶同高。

建筑的外部材料具有耐久性且防火。所有的雨水都将被收集起来用作景观灌溉和消防用水。未来建成的水池将进一步保障建筑远离森林火灾，同时也可作为储备水。

太阳能加热器将通过室内热辐射管为住宅提供大部分的热能。建筑狭长的造型和大面积的玻璃窗与当地低湿的气候相结合，保证了住宅的被动式制冷。

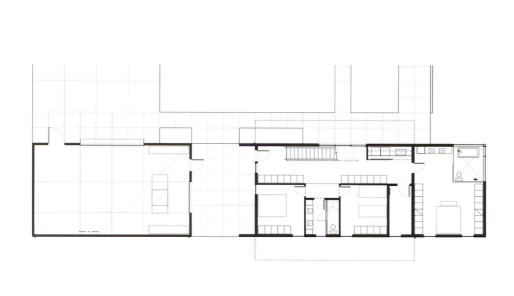

建筑背面 Back view

客厅 Living room

书房 Library

Photo: Peter Jahnke

Residential

Completion Date: 2009

Design: PIQUE llc.

Weiss Residence

维斯住宅

Located on a sloping site in hills of West Linn, the Weiss Residence provides an exclamation mark in the streetscape. It's strikingly rigid geometric forms and unusual detailing which accentuate its shape make for an intriguing composition.

The 179-square-metre building is spread over three floors giving the house a vertical emphasis and making the east and west elevations somewhat tower-like, yet the building is by no means menacing from the street. The entrance to the home is made onto the middle level after descending a short concrete driveway, while the upper floor is almost absorbed as the eaves slope downward to form the diamond shapes that can be seen on the side elevations. The building opens out to a large rectangular deck through sliding doors. The deck is intended to act in the same way as a traditional porch would help to merge the lives of the home's occupants with the rest of the neighbourhood, which helps to soften the building from the street. The deck is also used to roof the garage and reduces the scale of the building by extending out towards the sidewalk. The façade is further broken down by the planter box, adorned with small shrubs which help to reduce the scale of the building.

维斯住宅坐落在西谷山的坡地上，外形令人惊艳。它引人注目的几何造型和与众不同的细节设计共同构成了它迷人的结构。

这座总面积为179平方米的建筑共分三层，东西两侧的里面类似一座塔，从街道上看来十分抢眼。住宅的主入口设在二楼，经过一条较短的水泥车道便可到达。三楼几乎完全隐蔽在屋檐里，从侧面看形成了一个钻石形。建筑的平台上设有拉门，相当于传统建筑的门廊，拉近了住宅里的住户和周围的邻里之间的关系。平台也是车库的屋顶，减少了建筑占用人行道的面积。花池进一步分割了建筑的外墙结构，旁边点缀的小型灌木弱化了建筑的规模。

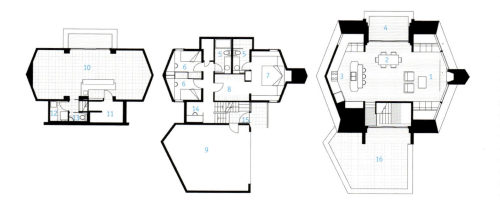

1. living	1. 起居室
2. dining	2. 餐厅
3. kitchen	3. 厨房
4. deck	4. 平台
5. bathroom	5. 浴室
6. bedroom	6. 卧室
7. master bedroom	7. 主卧室
8. closet	8. 壁橱
9. garage	9. 车库
10. reception room	10. 接待室
11. wine cellar	11. 酒窖
12. laundry	12. 洗衣房
13. powder room	13. 化妆间
14. computer alcove	14. 电脑室
15. entrance	15. 入口
16. stairs	16. 楼梯

Architect: Robert Harvey Oshatz   Completion Date: 2007   Photo: Meredith Brower   Residential

Aerial view of Ford Assembly Building and San Francisco Bay 福特装配大楼和旧金山湾鸟瞰图

Ford Assembly Building

This waterfront project rejuvenated the formerly abandoned and dilapidated Ford Assembly Building. The historical factory was transformed into a vibrant centre of 21st-century building uses, including entertainment, dining, office, and a visitor centre. Today it has a lively mix of public/private uses and accommodates a range of commercial tenants with offices, Research & Development facilities, light industrial, retail functions, and the NPS Visitor Centre celebrating WWII's "Rosie the Riveter". The project also incorporates significant sustainability features.

The designers' vision for the rebirth of this magnificent edifice was to retain yet enhance the architectural aspects of the original building's awe-inspiring shell, continuous bands of steel sash windows and floods of daylight, while maintaining its original waterfront relationship. This goal to renew the building was driven by an impetus to salvage and restore features inherent to the building's architectural spirit. "Intervention elements" of our century: lighting, furnishings, free standing buildings within the building, rooms, stairs, ramps, platforms, walls, etc, placed and designed to work with existing 1930's industrial architectural features are most apparent in the results for the Boilerhouse Restaurant/Café, SunPower Corporation and Mountain Hardwear projects. Low-water usage landscaping was designed on the building's west side to reflect the more public and formal façade. Lighting internally and externally was a way to highlight the building, particularly at night. The red-lit stack of the Boilerhouse, itself an icon of the project, is especially arresting at night.

福特装配大楼

这个滨海项目翻新了被荒废的福特装配大楼，将这座老工厂打造成充满活力的21世纪新建筑，使其集娱乐、餐饮、办公和游客中心为一身。改造后的大楼兼具公私两种功能，其商业租户包括办公、研发机构、轻工业、零售店和纪念二战时期"铆工路斯"的NPS游客中心。此外，项目还极具可持续设计特征。

设计师对于这座宏伟建筑的复兴具有以下想法：改进原有建筑宏大的外壳、连续的钢铁框格窗以及泛滥的自然采光，同时保持建筑和海滨的良好互动。这个目标的原动力来自保持和维护大楼的建筑精神。在Boilerhouse餐厅/咖啡厅、SunPower公司和Mountain Hardwear项目的设计中，21世纪的元素（灯光、陈设、楼中楼、房间、楼梯、坡道、平台、墙面等）体现尤为明显。大楼的西侧采用了低用水量的景观美化工程。室内外灯光设计进一步突出了建筑，尤其是在夜晚。红灯闪烁的Boilerhouse餐厅是项目的标志，在夜晚特别引人注目。

1. parking		1. 停车场
2. Mountain Hardwear		2. 山景硬壳
3. loading dock		3. 装卸码头
4. other tenants		4. 其他租户
5. SunPower Corporation		5. 太阳能公司
6. electric vehicle parking		6. 电器停车场
7. Vetrazzo		7. Vetrazzo公司
8. Rosie the Riveter Visitor Centre		8. 铆工路斯游客中心
9. Boilerhouse Restaurant		9. Boilerhouse餐厅
10. the craneway pavilion		10. 起重机轨道房
11. San Francisco Bay		11. 旧金山湾
12. sf bay trail		12. 海滨小路

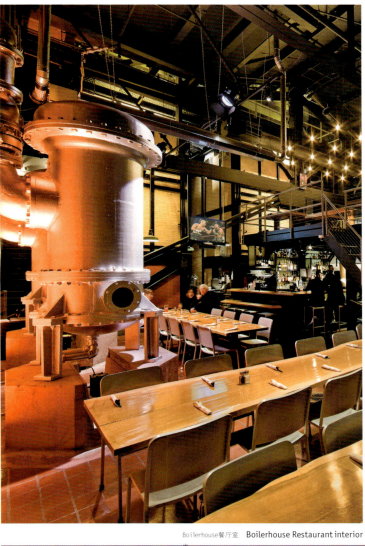

Boilerhouse餐厅室 Boilerhouse Restaurant interior
内

起重机轨道房室内 Craneway interior

Boilerhouse餐厅和起重机轨道房外景　Craneway and Boilerhouse exterior

West façade elevation – entrance approach to Recreation Building 西立面，娱乐中心入口

Orange Memorial Park Recreation Centre

Orange Memorial Park is the most important public recreation venue for the citizens of South San Francisco and is the context for the new recreation building which is encircled by soccer, picnic, basketball and other outdoor amenities. The building's most significant element is an airy, light-filled Activity Pavilion for cultural, recreational, celebratory, and educational activities. The architects chose wood flooring and an exposed wood truss roof to add warmth and grace to this important room.

The recreation building is conceived as a focal point of the park and an icon for the community. Towards that goal, a juxtaposition of two distinct rectangular masses was created – one large, light and largely transparent, housing the Activity Pavilion, with large areas of glass in concert with red and yellow cedar, and another mass that is by contrast a smaller, nearly solid box of basalt stone. The interior use of cedar specifically in the Activity Pavilion creates a dynamic and inviting environment for a central meeting place for the community. Moreover, the horizontality of the building is accentuated by the roof of the Pavilion whose paired glu-lam wood trusses span the room; these trusses cantilever beyond the enclosed footprint to provide covered outdoor patio areas.

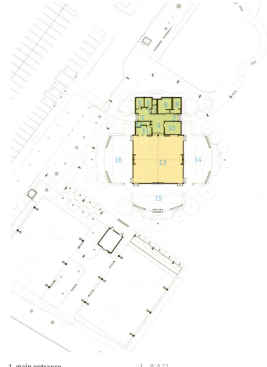

橘子公园娱乐中心

橘子公园是南旧金山居民最重要的公共休闲场所，人们可以在公园里进行足球、野餐、篮球和许多其他的户外活动。新娱乐中心最重要的组成部分是轻盈而阳光充足的活动中心，里面可以进行文化、娱乐和教育活动。建筑师选择木地板和清水木桁架为这一重要的空间增添了暖意和优雅。

娱乐中心是公园的焦点，也是社区的标志性建筑。为了实现这一目标，建筑师设计了两座并列的矩形结构——其中大而轻盈的透明结构是活动中心，其混凝土结构上装有大片的玻璃和红色、黄色的杉木；而另一个结构则是较小的由玄武岩组成的盒结构。活动中心室内所使用的杉木为社区集会营造了一个动感而吸引人的空间。此外，活动中心的成对胶合木桁架水平方向装点了建筑，扩展了空间。这些桁架延伸到建筑覆盖区之外，在下面形成了一个户外露台区。

1. main entrance	1. 主入口
2. entrance from fields	2. 场地入口
3. reception counter	3. 接待处
4. foyer	4. 门厅
5. office	5. 办公室
6. office	6. 办公室
7. utility room	7. 机械室
8. men's restroom	8. 男洗手间
9. women's restroom	9. 女洗手间
10. storage	10. 仓库
11. kitchen	11. 餐厅
12. pantry	12. 备餐间
13. multi-purpose activity pavilion	13. 多功能厅
14. east patio	14. 东平台
15. south patio	15. 南平台
16. west patio	16. 西平台

东立面，黄昏的景色 East façade elevation – view from fields at dusk

Detailed exterior cedar and glass façade of multi–purpose pavilion on west, south and east sides

多功能厅的雪松和玻璃外墙细部

多功能厅室内 Interior of multi–purpose pavilion

Library and Media Centre of the University of Guadalajara

瓜达拉哈拉大学图书馆和媒体中心

This building is a pioneer in the implementation of standardised norms for accessibility for people with disabilities; it has a set of ramps and aisles specially designed to make it 100% accessible. It will have a collection of 120,000 books, DVDs, and videos in a total surface of 5,346 square metres, making it the biggest public library in the western region of Mexico and the second one after the recently opened Central Library Jose Vasconcelos in Mexico.

The programme was met within a very narrow margin, and the goal was of course to have first of all a very functional building. The building is organised into three different prisms, each built with different materials in order to make each volume's programme recognisable. The reading volume is built with red brick and is mainly opened to the north so the interior receives the best illumination for reading without any direct sunlight. The volume that contains the books is built in concrete and completely closed to the exterior, so this volume is read from outside as a closed, protected box. The media centre is allocated inside a metallic volume; this volume makes use of the latest technology in metal cladding and isolation.

这一项目是无障碍建筑领域的标准开拓者，它的走廊和坡道的设计完全达到了100%无障碍的标准。图书馆总面积为5,346平方米，共有120,000册藏书和DVD等影像资料，是墨西哥西部最大的公共图书馆，也是墨西哥第二大图书馆。

项目计划以极少的预算建造一座功能齐全的建筑。大楼共分为三个部分，每一部分都采用了不同的材料，以增加各个部分的辨识度。阅读区采用了红砖结构，朝向北面开放，既可以获得最佳照明，又避免了阳光直射。藏书区采用了混凝土结构，从外面看，是一个全封闭的"盒子"。媒体中心采用了金属结构，运用了最先进的金属外墙和隔离技术。

General view 建筑外观

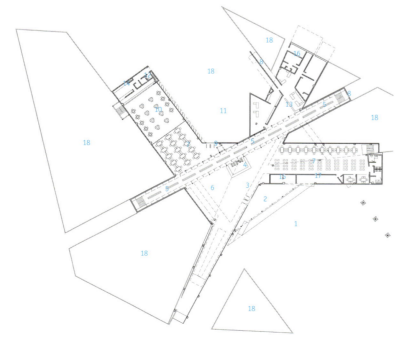

1. plaza
2. entrance
3. main lobby
4. front desk
5. book shelves
6. casual reading area
7. reading area
8. ramps and stairs
9. cubicles
10. snack area
11. reading plaza
12. toilets
13. administration
14. kitchen
15. copy centre
16. service
17. lockers
18. green areas

1. 广场
2. 入口
3. 主大厅
4. 前平台
5. 书架
6. 休闲阅览区
7. 阅览区
8. 坡道和台阶
9. 小隔间
10. 零食区
11. 阅读广场
12. 洗手间
13. 行政区
14. 厨房
15. 影印中心
16. 服务区
17. 更衣室
18. 绿化区

从广场看建筑正面 **Main façade from plaza**

媒体中心 **Media volume**

廊桥 Bridge

It is an almost 100-metre-long building, with a series of vertical strips used for ventilation by tip-up windows 建筑总长度近100米，竖条结构的窗户保证了通风

A.M. Celaya

The A.M. newspaper in the state of Guanajuato, Mexico, is subsidiary to one of the main national journals of Grupo Reforma. Its local contents are produced in Leon where it is printed, and from where it is distributed throughout the state, once it has received the news generated in Celaya. Therefore, printed without any industrial facilities, the A.M. Celaya had been operating in a rented unit with no urban presence whatsoever. So, in order to create an appropriate projection according to the social importance and financial viability of the newspaper, the owners destined an outstandingly located terrain. This was a long and slim land, facing one of the main avenues in the city, to be used for three purposes: the newspaper headquarters, shopping units and an additional rental business block in the upper level.

This is how the architectural programme comes out, bringing a triple-height cube as main volume for the newspaper brand image. The central stairway is set here, communicating the main entrance to the public attention areas as well as meeting rooms, whereas on the first floor comes the editors and directors office, as well as the newspaper offices. On the other end of the building is the main entrance for the rental business units, having underneath the shop units, all of them with direct access from the street.

A.M.塞拉亚报业大楼

A.M.报业位于墨西哥的瓜纳华托州，是改革报业集团的附属报业。A.M.报在塞拉亚编辑新闻，到利昂印刷，最后发放到州内各城市。由于不需要印刷设施，A.M.报业一直在一座出租楼内办公。为了提升报纸的社会影响力，A.M.报业的经营者决定为它建造一座新的办公楼。办公楼位于一片细长的场地上，正对城市的主要街道，共有三项功能：报业总部、商业网点和上层的出租办公区。

建筑的主体是三层高的立方体，代表着报业的品牌形象。立方体结构里的中心楼梯连接着主入口、公共区域和会议室；二楼是编辑、总编的办公室和报纸编辑中心。建筑的另一侧是出租楼层的入口，它与下方商业网点的入口都可以直接通向街道。

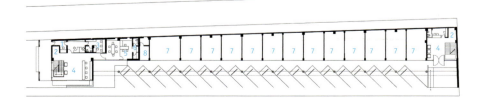

1. security	1. 保安室
2. lifts	2. 电梯
3. classified section	3. 收发室
4. lobby	4. 大厅
5. sala de juntas	5. 会议厅
6. files	6. 档案室
7. local commercial	7. 本地商业设施
8. services	8. 服务中心

The programme comes out of a triple-height cube as main volume for the newspaper brand image
三层高的立方体结构是建筑的主体

The long and slim terrain, facing one of the main avenues in the city, dictated the programme
细长的建筑主体面向城市的主要街道

The volumetric statement is the triple-height lobby hall with a big granite plate that holds the
楼面的花岗岩石板上写着报业的标识 newspaper's logo

The stairs lead to the first floor where the editors and directors office, as well as the newspaper
楼梯通往二楼的编辑办公室和报业工作室 offices come

Manantiales de Espejo Apartment Building

The building is located in a regeneration zone of Mexico City that is growing in popularity, housing forty-seven apartments, fifty-five square metres each. The building has a green rooftop that stands in contrast with the grey surroundings of the area. The structure is "U" shaped, with all the apartments having views both on the inside and outside façades. The windows on the outside are designed as a reference to barcodes used to tag commercial products. The inside windows bring light and movement together. The architecture of the building reflects the site, located next to a speedy avenue that connects the northern side of the city and surrounded by small streets with low circulation and speed, with a very dynamic expression in façades which contrasts the regular form of the plan. In the lower level there is the parking. The next six levels are apartments and the upper level is the roof garden. The materials used are mostly concrete for the main core of the building and steel for the circulations, making the building look both strong and modern in contrast to the building and houses in the immediate surroundings that are much older.

马南蒂阿莱斯公寓大楼

马南蒂阿莱斯公寓大楼位于墨西哥城的一个迅速发展的重建区域，共有47间55平方米的公寓。建筑的绿色屋顶和周边的灰色建筑形成了鲜明的对比。建筑呈U形，每间公寓在内外两侧都设有窗户。外侧的窗户造型和商品的条形码相似；内侧的窗户则负责室内采光和空气流通。公寓大楼靠近一条连接城市北部的快速干道，四周环绕着低车流量的小街道。建筑的外墙和常规设计不同，充满了动感和活力。建筑的底层是停车场，中间六层是公寓，而顶层则是屋顶花园。建筑主要采用了钢筋混凝土结构，看起来坚固而现代，与周边老旧的建筑和住宅形成了鲜明对比。

1. main bedroom	1. 主卧室
2. bedroom	2. 卧室
3. living room	3. 客厅
4. dining room	4. 餐厅
5. kitchen	5. 厨房
6. utility	6. 杂物间
7. toilet	7. 浴室
8. WC	8. 洗手间
9. study	9. 书房
10. balcony	10. 阳台
11. lift	11. 电梯
12. corridor	12. 走廊

Photo: Gustavo Slovik

Residential

Completion Date: 2008

Architect: SLVK: Gustavo Slovik, Daniel Dickter, Emanuel Teohua

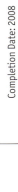

Omegablock, Colegio Anglo Colombiano (Omega Block, Anglo Colombiano College)

安德鲁·哥伦比亚学院欧米茄综合楼

The Omega Block Building is located in the Anglo Colombiano School in the Bogota, capital of Colombia. This building is the summary of three buildings in progression, which are adapted to the morphology of the triangular area disposed for its development.

The three volumes are combined with low to high mass or vice versa. Greater Volume houses general classrooms, specialised classrooms medium volume music and art rooms and smaller volume of media. The master volume is structured through the subtraction of the mass, the average volume level through the envelope, and the smaller volume is a sieve or perforated in the mass sequence. The dominant materiality of the project, large format brick and slender pre-stressed concrete elements (beige in colour), binds the outside spaces with that of the large internal atrium space. Further tonal and material compliment is found in the vibrant green special divisions (doors and divisions), with smooth Formica finish.

A grand staircase, or tiered seating, ensures a fluid connection between the base plane (ground floor) with level above, while offering a congregational space. With this element, the scale of the building changes, and the use of the atrium space is re-interpreted.

欧米茄综合楼位于哥伦比亚首都波哥大的安德鲁·哥伦比亚学院内，共有三座楼组成，三座楼呈有机结构完美地组合在一起。

三座楼由高到低依次排列，最大的楼里面是普通教室，中等大小的楼里是专科音乐和艺术教室，最小的楼里是媒体教室。主楼门窗向内凹，中楼的外墙像一个包层，小楼的外墙上则布满了小孔。项目中庭和外墙所采用的主要材料是大块红砖和米黄色的预制混凝土长条。门窗上一些活泼的绿色让建筑极具生气。

巨大的楼梯（也可作为阶梯座椅）保证了楼内各个楼层流畅的连接，也提供了集会的场所。楼梯的设置改变了建筑的规模，也重新定义了中庭的空间。

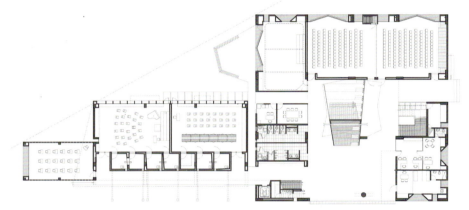

General view 建筑全景

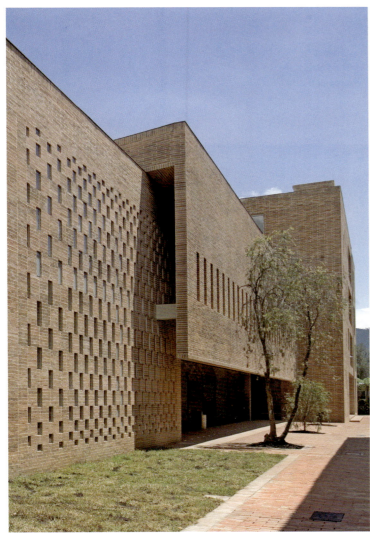

室内细部 Details

Photo: Rodrigo Davila

Educational

Completion Date: 2009

Architect: Daniel Bonilla

Front View 建筑正面

Museum of Memory and Human Rights

记忆与人权博物馆

The Museum is conceptually organised in two moments: the Exposition Beam and the Base. The first, elevated, housing the history, the information, and the act of living memory, is open on both ends, like someone who lets life pass naturally. The other, the Base, in a first step deep as a mine, where the study, the production, the invention, the seminars, the knowledge of the land and the territory are located; in another step is the necessary support by the administration sectors. The Beam is a specific museographical space and the Base is a museological one, a place of study and support.

Along both sides of the Exposition Beam are the circulations, bathrooms and support systems, lightened by the translucency of the perforated copper plates and the second external skin, made of transparent glass, thus creating a totally controlled inner environment. On a second effect, the light also penetrates all the exposition space through the semi-opaque glass walls.

博物馆由两部分组成：展览横梁和底座。展览横梁架空而建，里面是历史、信息以及生命记忆活动展览。底座下层像一座金矿，着重于地质和地形形成过程以及地貌的研究；底座里还包含其他的行政辅助设施。横梁注重图形展示，而底座注重理性研究。

展览横梁的两侧是走廊、洗手间和辅助设施，半透明穿孔铜板和透明玻璃外墙令室内空间显得十分明亮，阳光透过半透明玻璃墙渗透到展览空间。

Exterior 建筑外观

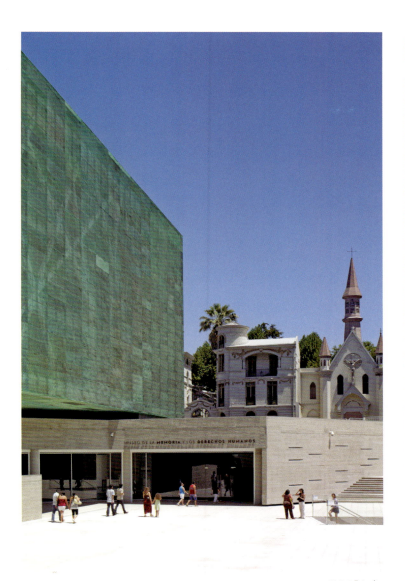

建筑外观 Exterior

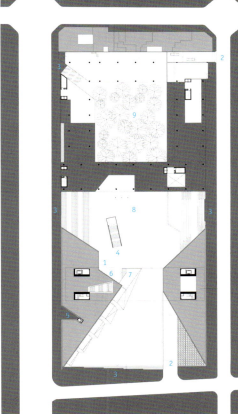
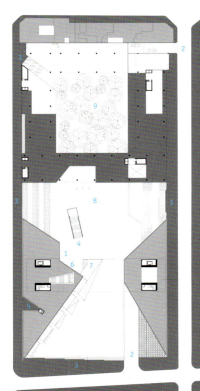

1. museum access
2. parking access
3. pedestrian access to the square
4. memorial alfredo jaar
5. lift access
6. subway connection
7. parlatorio
8. square memory
9. garden desires

1. 博物馆入口
2. 停车场入口
3. 广场入口
4. 记忆博物馆通道
5. 电梯
6. 地铁连接口
7. 天井
8. 记忆广场
9. 欲望花园

建筑内部 Interior

室内楼梯 Interior stairs

Façade 建筑外墙

Benavides Drugstore and Warehouse

An architecture of simple lines, contrast of transparencies and closed walls answers to two complementary functions: the buildings (distribution centre and offices) and the usability (services in the pharmaceutical area, incorporated with technologies of the last generation in the production and passive ventilation systems for the buildings).

The lot is located in a new expansion of an industrial park, made by the firm Kalos in the north of Monterrey (international airport highway). The new pharmaceutical complex is the gateway to the urbanisation, on one side the great park, on the other the central axis with green spaces.

The architecture of the complex is simple and outright. A vertical and diagonal texture of relief lines on a big closed white volume produces a game of light and shadows while giving movement to the façades. It contrasts with the administrative building, which is designed like a cube of blue glossy glass, subdivided by aluminium blades, giving the building a decisive character of transparency, both during day and night. The second skin is suspended from and offset to the volume of the administrative building, above a multiuse water pond. The water pond permits the thermal control of the building through evaporation produced during the hot months and the effect of transparencies, reflex, light and shadows.

1. access
2. hall
3. water
4. cantina
5. kitchen
6. bathroom

1. 入口
2. 大厅
3. 水景
4. 餐厅
5. 厨房
6. 洗手间

贝纳维德斯制药厂

建筑的线条简洁明快，透明的玻璃外墙结构和封闭的混凝土结构形成了鲜明的对比。建筑分为两部分：配送中心和办公楼。制药区服务设施十分人性化，产品融合了传统生产技术，整座大楼都采用了被动通风系统。

项目位于蒙特雷北部卡洛斯公司的一个新工业园区，紧邻机场高速。这座制药楼一侧是工业园区，另一侧是绿地。

建筑的造型干净简单，配送中心是封闭的白色混凝土结构，外墙上有一种奇妙的光影效果；而行政楼则是一个包裹着蓝色玻璃的立方体，其间镶嵌着铝片，使得建筑在夜间和白天看起来都十分透明。

建筑主外墙 Main façade

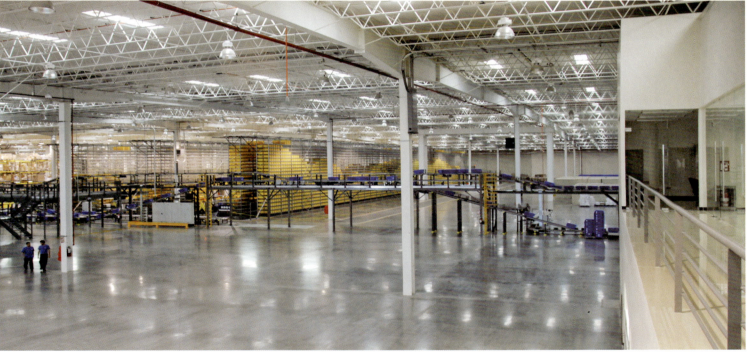

工厂大厅 Factory hall

仓库 Warehouse

Photo: Guillermo Hevia

Completion Date: 2009

Architect: GH+A Arquitectos

Far view 远景

ALSACIA

The volumes level the architecture to the lines of the perimeter, while fitting itself to the urban context and the counterfort of the cordillera, using a roundly colour scheme of red and black. The project defines itself by the use of simple lines, extreme colours and pure volumetry.

The use of the buildings (distribution and classification of auto parts) and its complexities define the architecture to a big, closed volume which terraces itself to follow the urban boundaries. The contrast between the red and black, draws attention to accent the simplicity of the volumes, transforming the architectural ensemble to a unique icon of the access to the ENEA business site. The main building contrasts with its black metal to the administrative building and the service areas due to its morphology and materiality. The second volume is made out of insight concrete, glass and steel. The suspended roof crosses the volume, bearing itself on both sides on metal columns of double height, generating covered and protected areas. An access atrium, preceded by a water pond (obey the function of chilling the main façade by evaporation) and the square in the subsequent façade assigned for social activities, contains the big stairs which cross the diagonal, dynamising the space and giving access to the educational and staff rooms.

ALSACIA公司总部

这座以红黑为主的建筑各个部分错落有致，与山脉和城市都十分相称。简洁的线条、极端的色彩和纯粹的造型使它十分显眼。

建筑的主要功能是汽车零件的配送和分类空间，其巨大而封闭的建筑结构稳稳地坐落在城市的边缘。红黑两色的鲜明对比进一步强调了建筑的简单性，使其成为ENEA商业区入口处的一处显著的地标。由黑色金属板包裹的主楼以其独特的材料和造型与行政楼和服务区区分开来。二号楼由混凝土、玻璃和钢铁建成。高悬的屋顶横跨整个楼面，在建筑两侧形成了遮蔽和保护区域。入口中庭前的水塘通过蒸发来为建筑外表降温。中庭处巨大的楼梯连通了各个楼层的空间和员工办公室。

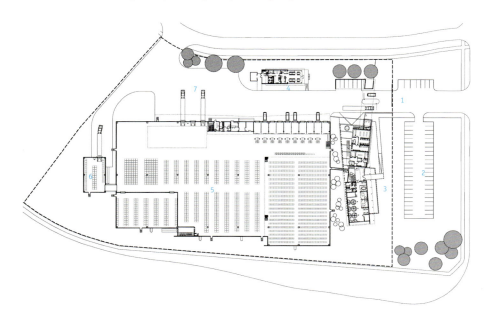

1. 入口
2. 停车场
3. 办公和服务区
4. 餐厅
5. 配送中心
6. 危险品仓库
7. 货车操作台

1. access
2. parking
3. offices and services
4. cafeteria
5. distribution centre
6. storage for dangerous elements
7. truck operation patio

黄昏景色 Sunset view

楼梯 Stairs

走廊 Corridor

廊桥 Bridge

Photo: Cristián Barahona, Guillermo Hevia H., Guillermo Hevia García

Corporate

Completion Date: 2008

Architect: GH+A Arquitectos

Entrance 入口

SESC

The library evidently has a great influence of the modern architecture from São Paulo. It is possible to perceive the influences from Vilanova Artigas and Paulo Mendes da Rocha. The influence of Artigas' building, the FAU–USP building, is also visible, almost as a tribute from the alumni to the building in which they studied architecture.

This construction is composed by a great entrance atrium surrounded by a concrete beam one storey high. This beam, which has the role of structure and sealing to the superior portion, organises a long footbridge which gives access to the library's general collection. Therefore, this structure not only serves as sealing, but it also shelters the collection and creates the footbridge for visitors. Under the concrete beam, the glass sealing produces a contrast between the weight of the concrete on the superior portion versus the transparency and lightness of the inferior sealing. It is at this tension point that the visitor is invited to enter the library.

Part of the library is suspended over SESC's great central water mirror, and the access is made through a footbridge which symbolises the passage to a study place. When over the water mirror, it is possible to have very interesting views that may change during the day as daylight itself changes.

SESC图书馆

建筑采用以入口中庭为中心的单层混凝土横梁结构。这个横梁既起到了框架作用，又封闭了上层结构的一部分。横梁下方是通往图书馆各个藏书区的天桥。因此，横梁起到了构造天桥结构的作用。混凝土梁下的玻璃墙与上层厚重的混凝土结构形成了鲜明的对比，这种张力吸引着游客进入图书馆。

图书馆的一部分浮在SESC的中心倒影池之中，随着日光的变换，倒影池会倒影出形形色色的有趣景象。

建筑全景 General view

建筑细部 Details

室内 Interior

室内 Interior

Front building 建筑正面

Laranjeiras House

This is a lovely beach house. Ultra wide openings and imperceptibly transition between indoors and outdoors that make this place very comfortable. Minimalism softened through wood surfaces and stone walls make the house design's modern but not cold. Six bedrooms with balconies are located on first floor creating great places for rest and catch sunrise views. Ground floor have kitchen and maids quarters at the one wing and dining room, living areas, bar and games room at the other.

Elements of the colonial architecture of the historic city of Paraty were incorporated into the modern design of the House of Laranjeiras. The ceramic tiles and the use of rustic material offering great thermal isolation, remit back directly to the large old houses of the neighbouring city.

Likewise, ecological concerns were incorporated into the architecture through the use of these materials (both the ceramic and wood were recycled) and an integration with the surroundings, a beautiful beach in the state of Rio de Janeiro.

拉兰热拉斯住宅

这是一座可爱的海滨住宅。超大的开窗和室内外精妙的转换让这个空间十分舒适。木质平面和石墙柔化了极简主义设计,让住宅看起来现代而又不冰冷。二楼6间带阳台的卧室营造了极佳的休息空间,还可以欣赏落日的景象。一楼的一侧是厨房和佣人房,另一侧则是餐厅、客厅、吧台和游戏房。

帕拉蒂城的建筑元素被融入了拉兰热拉斯住宅的现代设计之中,瓷砖和粗手工材料的运用保证了建筑的隔热性,与帕拉蒂城古老的住宅极其相似。

此外,住宅的设计也融入了生态元素——瓷砖和木材都是再生材料;住宅与其周边环境——里约热内卢一处美丽的海滨完美地融合在了一起。

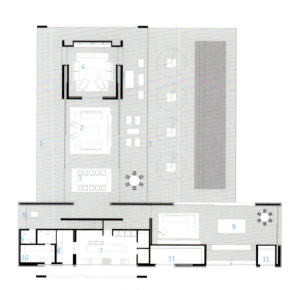

1. terrace	1. 露台
2. living room	2. 起居室
3. dinning room	3. 餐厅
4. TV room	4. 电视房
5. main entrance	5. 主入口
6. bath	6. 浴室
7. kitchem	7. 厨房
8. laundry	8. 洗衣房
9. cardroom	9. 棋牌室
10. bedroom	10. 卧室
11. storage	11. 仓储室

庭院，游泳池　Swimming pool in the yard

游泳池和平台 Pool and deck

起居空间 Living space

Photo: Nelson Kon

Completion Date: 2008

Architect: Studio MK27

Front façade 建筑正面

House 53

The House 53 volumetry was defined following São Paulo city building laws and the site's peculiar shape, which is just over ten metres in front and approximately thirty metres in length. According to the legislation one can build in the neighbourhood up to a two–floor building, settled upon the site's lateral limits. A third floor is allowed as long as the lateral setbacks are respected.

The house was conceived as a wood–and–mortar monolithic block with another concrete and glass volume upon it. Due to the ground's small front and volumetry, the box's two edges had to make the most of light's entrance, which explains the large windows. It was also desirable that these windows would make it possible to darken the internal environment whenever needed.

The house's interior volume, which comprises the living room on the ground floor and the bedrooms on the first floor, is a glass box with wooden brises that open as folding doors. The rooms' front and back façades were designed to be completely closed or opened.

53号住宅

53号住宅的建造遵循了圣保罗市建筑法规，适应了场地的特殊情况（场地正面仅有10米宽，共计30米长）。根据建筑法规，该地区的住宅不能超过两层楼高，如需建造第三层楼，其侧面必须呈阶梯状缩进。

住宅的主体是木材和灰泥组成的盒结构，上面还有一个混凝土和玻璃框架结构。鉴于场地十分狭小，盒状结构的两侧必须要摄入足够的日光，所以采用了巨大的窗户。而室内的窗帘可以令室内环境在必要的时候变得昏暗。

住宅一楼是客厅，二楼是卧室。建筑侧面看起来像是一个包裹着木质遮光罩的玻璃盒子，这些遮光罩可以打开成为折叠门。住宅的前后两面外墙设计成或是全封闭的，或是全开放的。

1. living room 1. 客厅
2. bedroom 2. 卧室
3. lounge room 3. 休息室

平台夜景 **Deck at night**

建筑正面 Front

餐厅 Dining room

Photo: Rômulo Fialdini

Residential

Completion Date: 2009

Architect: Studio MK27

Star Place

The city of Kaohsiung is the second largest city on the island of Taiwan and has a population of some 1.5 million. Its geographical position along the Taiwan Strait has allowed it, over time, to become a maritime hub. After the Second World War, the city grew rapidly, transforming itself from an undeveloped fishing village into a booming, heavily industrialised port city. Today, like other places in this position, Kaohsiung is in the process of re-envisioning the industrial identity that has brought it so far. For a city such as Kaohsiung, which takes a secondary role on the world stage in comparison to global cities such as Hong Kong or Beijing, this question of reassessing the identity can be problematic to address. The kind of urban environment that is produced by rapid industrialisation does not provide a platform that is conducive to effortless regeneration. Moreover, the absence of symbolic markers such as those found in leading cultural or administrative cities makes it difficult to identify obvious starting points for the desired regeneration.

What seems to be happening in Kaohsiung is that the city has sought to find a focus for its urban renewal in the residential experience, that is to say, in what the city means in the everyday life of its citizens. The overall goal is to enhance the attractiveness and the comfort level of the city as a place to live. In order to fulfill the goal of establishing a sustainable urban environment, the planning focus has been directed towards strengthening both the natural aspects of the city and its public infrastructure. The SHE concept (safe, healthy, ecological) is the driver behind the development plans, which include cleaning up the river and creating parks and wetland regions.

Within this context, Star Place fulfills a role in the lively and resident-orientated urban landscape that is more on the urban than on the natural scale of the spectrum. As a shopping centre, the project typifies a contemporary public-private form of architectural space. In summary, while the project does not find itself directly within the scope of the nature-enhancing urban regeneration schemes, we still feel that there is a link with the overall improvement goals of Kaohsiung, because thoroughly urban facilities with mass-appeal also form a contributing factor to the livability of a city. Moreover, the project is completely in synch with a resident-orientated approach to urbanism, in which the experience of everyday space by the citizen is the central consideration.

大立精品馆

高雄是台湾第二大城市，总人口约为150万。它地处台湾海峡，是一处主要港口。第二次世界大战以后，高雄从一个小渔村迅速发展为一个高度工业化的港口城市。现在，与其他港口城市一样，高雄市正试图重塑自己的工业形象。与香港、北京等城市相比，高雄在世界舞台上只是一个二流城市，因此，重塑自身形象对其至关重要。高速工业化所带来的城市环境并没有为重建带来有力的平台。此外，城市标志性形象（如先锋文化或行政中心等）的缺失也让形象重塑工程步履维艰。

对于高雄来说，找到一个体现城市居住环境变革的理念已成为重中之重。整体目标即为增强本身的吸引力和舒适度，为实现这一点，营造可持续发展城市环境已被确立为首要方法。

大立精品馆便"诞生"于这样一个大环境中，在活力十足的"宜居"氛围中实现了自身的作用。作为一家购物中心，这一项目在建筑领域中所展现的是现代化的公共空间与个体生活相结合的形式，与整体环境打造理念相一致。

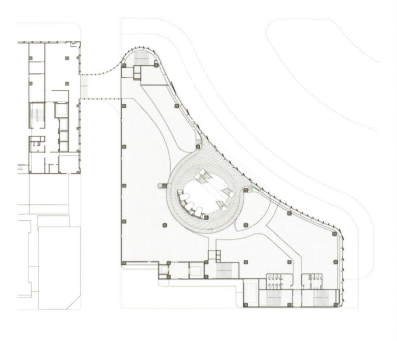

Photo: Christian Richters

Completion Date: 2009

Architect: UN Studio

Exterior view 建筑外观

Kerry Plaza, Futian, Shenzhen

Located on the southern part of the Futian CBD between the Exhibition Centre and Civic Centre, this comprehensive development houses a six-star flagship Shangri-la Hotel complex and the Futian Kerry Plaza.

The Futian Kerry Plaza calls for a pair of connected towers. The two 23-storey office towers soar up to 99.8 metres high. Symmetry is forgone for a more tectonic interplay between two rectangles. The two towers are connected on grade with a podium containing the hotel lobby and retail facilities. The entrance is framed with an iconic portal and matching canopy in the form of sweeping curves.

嘉里建设广场

嘉里建设广场选址在深圳福田区商业中心南部,介于展览中心和市政中心之间,整体开发工程包括六星级香格里拉酒店大楼及嘉里广场写字间。

嘉里广场写字间由两幢相连通的塔楼结构组成,共为23层,高达99.8米。设计摒弃了对称性原则,特别强调两个长方形结构之间的相互作用。塔楼之间通过基座连通,这里包括酒店大堂及零售店。入口突出曲线结构,带有顶篷,大气十足。

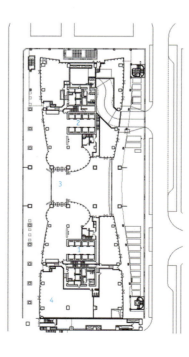

1. office areas
2. office areas
3. central passage way
4. activity space

1. 办公区
2. 办公区
3. 中心走廊
4. 活动空间

建筑外观 **Exterior view**

建筑外观 **Exterior view**

Complex

Completion Date: 2007

Architect: Wong & Tung International Ltd.

入口通道 **Entrance way**

Full view 全景图

Nokia China Campus

Nokia China Campus in Beijing is China's first LEED – NC (Leadership in Energy and Environmental Design – New Construction) Gold certified building. The facility includes a research and development laboratory, office space, canteen, auditorium, formal and informal meeting areas, gymnasium and bike shed.

From concept to detailed design, ARUP was aware of the sustainability issues and the building boasts energy-efficient features such as a temperature-controlled cavity between the panes of the glass façade which balances the sun's natural heat and the building air-conditioning so the intense summer and winter climates of Beijing do not affect the internal climate. The building also incorporates water conservation techniques, methods to reduce air pollution and improved air ventilation. Thirty design techniques altogether result in 37% of water saving and up to 20% energy saving. Ninety-seven percent of the interior is afforded views from the glass façades, and skylights and a large communal atrium provide natural light and ventilation throughout the building.

This six-storey facility is the realisation of an integrated multidisciplinary design services. The building is the product of cutting-edge engineering design services – ARUP employed Computational Fluid Dynamics, thermal and energy modelling, structural optimisation and building sustainability tool kit to design the energy efficient building.

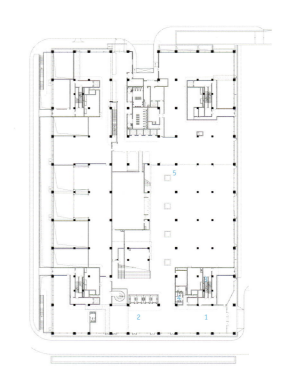

诺基亚中国园区

北京的诺基亚中国园区是中国第一座获得LEED-NC（能源与环境设计先锋奖——新建建筑类）金奖认证的建筑。园区包括研发实验室、办公空间、餐厅、礼堂、正式与非正式会议室、健身房和自行车棚。

从总体设计理念到细节设计，ARUP始终注重设计的可持续性。玻璃外墙之间的控温空心夹层平衡了太阳热能和室内的空调系统，因此，无论是酷暑还是寒冬，室内气候始终保持不变。建筑还应用了节水技术、低污染气体排放技术以及优化通风技术。30种可持续技术的结合减少了37%的用水量和20%的能耗。97%的室内空间都可以透过玻璃外墙看到外面的风景；天窗和巨大的中庭为建筑提供了自然采光和通风。

这座六层的建筑是多学科设计服务相结合的产物。建筑采用了多种先锋工程设计——为了设计这座节能建筑，ARUP应用了计算流体力学、热能建模、结构优化、可持续建筑工具箱等手段。

1. hall	1. 大厅
2. reception	2. 前台
3. stairs	3. 楼梯
4. WC	4. 洗手间
5. dining	5. 餐厅

建筑外墙 Exterior façade

室内大厅 Interior hall

结构细部 Structural details

Photo: Jerry Lee, M Moser, Ben McMillan

Completion Date: 2007

Architect: ARUP

Exterior view 建筑外观

Bird's-eye view 鸟瞰图

Zhongguancun Christian Church

With its free curved shape, the building forms a solitaire in the park-like open space between Zhongguancun Cultural Tower and the "City of Books". On the upper floors it houses China's largest Christian church, while on the ground floor commercial spaces are situated. The shape of the church body does not only allow for a sightline to the south media façade of the Cultural Tower but also emphasises its special function in contrast to the surrounding commercially used buildings. The rod system of the façade forms a homogeneous skin, which however lets in sufficient daylight.

The entrance to the main church hall faces northeast and opens up to the street as well as to Zhongguancun Cultural Tower. A cross, clearly identifying the building as a Christian church, develops from the façade rod system. Through a large portal, worshippers mount a stairway to enter the main church hall on the first floor. With its alternation of open and massive wall sections, the façade rod system creates a special lighting atmosphere inside the church hall, matching the ecclesiastical function of the space. The entrance for clergy and church employees is placed on the northwest side and gives access to a side chapel as well as to stairways and a lift to all the floors of the church building. The parish offices and community spaces are on the second and third floors of the south and west wings. Some of the upper-floor spaces open onto a roof terrace – a substitute for a churchyard – that offers parishioners an attractive outdoor space.

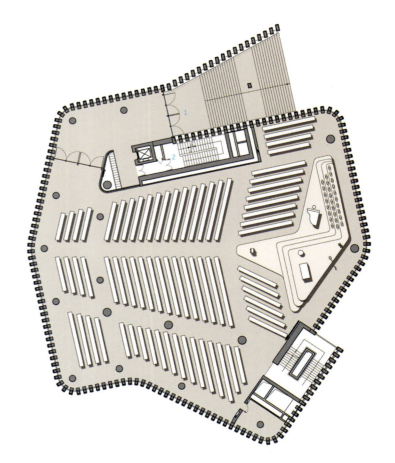

1. entrance
2. stairs

1. 入口
2. 楼梯

中关村基督教教堂

建筑位于中关村文化大厦和图书城之间，外部造型像一副摊开的纸牌。建筑的上层是中国最大的基督教堂，一楼有一些商业空间。从教堂可以看到文化大厦南侧的媒体墙。它的外观与周围的商业建筑截然不同，外墙的杆形结构形成了均质的外壳，其间的缝隙可以满足室内自然采光。教堂大厅的入口朝北，正对街道和中关村文化大厦。入口的十字架与外墙的杆形结构异曲同工，标志着教堂的身份。礼拜者从大门经过楼梯进入二楼的教堂大厅。外墙的杆形结构在开阔的大厅里营造了特别的光影效果，与教堂的宗教氛围十分相称。神职人员和工作人员的入口位于建筑的西北侧，他们可以由此进入一个小礼拜堂或是通过楼梯和电梯进入教堂大楼。教区办公室和社区空间位于西、南两翼的三四楼。

建筑室内 Interior view

建筑室内 Interior view

夜景 Night view

Photo: Bildnachweis

Completion Date: 2007

Architect: Gmp Architekten

Shanghai–Pudong Museum

Opposite the historically grown city centre of Shanghai a new "Manhattan" comes into being on the other side of the river: the district Shanghai–Pudong with the highest office – and hotel building at present in China. The Shanghai–Pudong Museum is one of the most important urban projects in this new district. It is supposed to document and archive the district's history and development comprehensively. At the same time modern multifunctional and open exhibition spaces are created to inform the public with a permanent exhibition and special exhibitions about selected topics.

Three elements form the building complex: the square-shaped horizontally orientated glass body with exhibition halls, a much broader, four-metre-high base with surrounding stairs, which accommodates the archives, and a bar-shaped building on the eastern side for the administration. The base as one of the main architectural features of the museum lifts the main building with the exhibition halls above the level of the surrounding streets and emphasises the central importance of the complex. Simplicity and reduction of the materials dominate the clear cube. The façade of the upper, closed part of the main building not only serves as weather protection but also as a communication surface. It is made of two parallel façade-layers. The outer layer consists of glass and the inner one of room-high closed wall panels. These elements can be rotated along their longitudinal axis and can be opened or closed, according to the particular requirements of the exhibition concept, so that views from the inside to the outside and vice versa are generated.

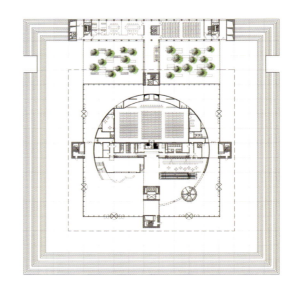

上海浦东博物馆

曼哈顿大厦巍然耸立在上海老城中心的对岸，至此浦东新区再添一幢新建筑。浦东博物馆同样位于全新城区内，用于记载这一区的历史及发展，同时也设置着多功能现代展览空间，以永久展览及专项展览的方式诠释着各种不同的话题。博物馆大厦主要由三部分构成：方形的玻璃结构（用作展厅）、4米高的基座（四周设计有楼梯，用于贮藏文件）以及东侧的行政管理大楼。其中基座作为整幢建筑的特色将展厅"托起"，简约而功能性十足。主建筑上层外观共为两层，外层为玻璃，内层为封闭的墙板，既可阻挡风雨的侵蚀又可用作媒体幕墙。

General view 全景图

建筑正面 Front view

夜景 Night view

室内 Interior view

Photo: Christian Gahl, Berlin

Cultural

Completion Date: 2005

Architect: Gmp – Von Gerkan, Marg and Partners Architects

Night view 夜景

SOHO Shangdu

The faceted façades of the two 32-level towers and smaller landmark building onto Dongqiaolu are clad in a random pattern of grey glass and aluminum panels. The dynamic qualities of the pattern readily allowed the proportion of glass orientating to the south and west to be reduced, thus lowering the heat gain to units in those directions. Inscribed within the faced facets are a large-scale geometrical parametric network of lines which at night create continuous light-lines and the project's distinctive nocturnal image. The façade pattern was directly generated on the same principles and visual image of the traditional Chinese ice-ray pattern, a decorative motif used primarily for joinery and paving, which is parametric in its geometric properties.

The organisation of the retail was designed on the principle of a vertical hutong with a series of internal streets and passages which vary in their position, width, height from floor to floor, thus generating differentiated circulation patterns which in turn, create the potential for localised and specialist retail zones.

Two large internal courtyards, which spiral vertically and link all the floors provide a navigational and activity focus to the retail area's "east" and "west-ends", facilitating a range of events from fashion parades, commercial launches to concerts and talks activities.

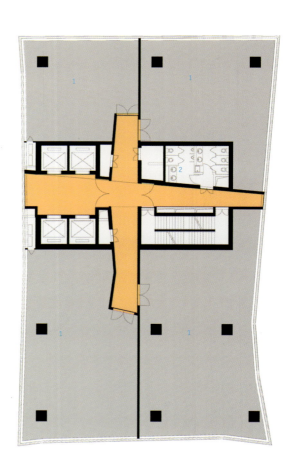

尚都

尚都由两栋32层大厦及一幢小规模地标性建筑构成，外观采用灰色玻璃和铝板镶嵌而成。随意的拼贴突出动感，同时减少了西、南两侧的玻璃材质数量，进而降低了强光的照射。尤为重要的一点是，表层下布满了各种电线，夜晚光线闪耀，并投射出不同的图案，这一设计源于中式传统的木工制品及铺石路面的图样模型。

商业区的设计以"垂直的胡同"为理念，内部街道及通道不断变换着宽度和高度，构成不同的样式。两个开阔的室内庭院蜿蜒向上，将不同的楼层连结起来，同时将大厦内部的景象尽收眼底。与周围的同类建筑相比，尚都可以说是别具特色。

1. office area
2. WC

1. 办公区
2. 洗手间

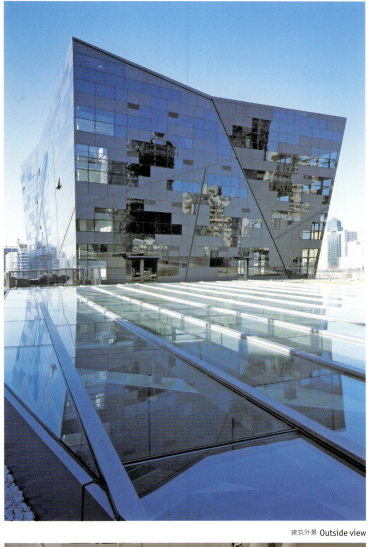

建筑外景 Outside view

建筑外景 Outside view

大厅 Hall

室内 Interior view

Bird's-eye view 鸟瞰图

Jishou University Research and Education Building and Huang Yongyu Museum

The project is concerned mainly with two important issues relating to site: the first is the relationship between the architecture and surrounding environment, and the second is how to establish a relationship between local architectural tradition and local culture.

The building sits on development-ready levelled land that once was part of the hillside on the university campus. The Research Education Building and the Museum form a wedge-shaped composite section that juts into the land. The building mass, multiple roofs and integrated windows blur the vertical and horizontal forms of the walls and roofs, which in turn, contribute to rebuilding and reestablishing the physical presence of the site.

Respect for cultural tradition of architecture evolves into two types in Jishou: protection of the typical traditional architecture in old town; copy of the traditional residence regardless of the difference in structure, material, function, and size. Conceiving maintaining a modern, new architectural logic as precondition, they try to introduce the style of traditional residential building into the building, so as to build relationship between new structure and local building visually. Therefore, in concept, this building is a "mountain" as well as a "village".

吉首大学综合科研教学楼及黄永玉博物馆

建筑位于湖南省吉首市吉首大学校园内，由两部分组成——综合科研教学楼和黄永玉博物馆。设计主要关注两个问题：一个是建筑如何重组建筑与周边物理环境的关系；二是建筑如何与当地原有的建筑文化传统建立积极的联系。

校园建在山地上，几乎所有的建筑都是依山而建。建筑的基地位于校园中心的人工湖南侧，原是坡地，后被削平。教学楼与博物馆形成的整体以楔状的剖面形态插入基地，用建筑的手段恢复了基地物理环境的秩序。裙房部分的屋顶与高层部分的北侧外墙在剖面形态上构成两个不同斜率的连续的表面，从而模糊了屋顶与外墙两种不同功能的建筑构件在形态上的差异。屋顶与外墙上相似的开窗方式进一步加强了这种混淆，使建筑整体加入"造山运动"。

对建筑文化传统的尊重在吉首地区分化为两种模式：一种是对老城区内建筑物"标本"式的保护；另一种是新建的建筑物不顾结构、材料、功能、尺度等方面的巨大差异，对传统民居单体形式的"戏仿"。我们以保持当代建造逻辑并接受新建筑的大尺度为前提，尝试将传统民居村镇聚落的肌理带入建筑的形式系统，从而在视觉上建立起新建筑与当地建筑文化传统的呼应。因此，在概念上，这栋建筑既是"山"又是"村"。

入口广场 Entrance plaza

教学楼立面的一部分 Part of the elevation of the academic building

博物馆大厅 Entrance hall of gallery

Photo: Shu He

Completion Date: 2006

Architect: Atelier Feichang Jianzhu

Educational

Ninetree Village

A small valley, bordered by a dense bamboo forest, forms the site for this luxury housing development, situated near the Qian Tang River in Hangzhou, southeastern China. The particular charm and beauty of the place are the determining factors. Twelve individual volumes are arranged in a chessboard pattern to create the maximum amount of open space for each building. Through planting new vegetation, each apartment building is set in its own clearing in the forest. The buildings adapt to the topography, creating a flowing landscape through a slight turning of the blocks.

The grounds will be accessed from the southern entrance via a network of lanes. All buildings are linked to an underground car park, enabling the site to be free from vehicles above ground. Within the development there are six types of building differing in size and floor plan depending on the location, view and light conditions.

The individual apartment buildings contain five generously proportioned apartments, each accommodating a full floor of approximately 450 square metres. The floor plan concept creates a flowing interior space defined by solid elements which accommodate auxiliary functions.

九树村高端公寓

九树村高端公寓坐落在杭州钱塘江边的一座小山谷里，四周环绕着茂密的竹林，风景优美，十分宜人。12座公寓楼呈棋盘图案分布，为每座楼都营造出最开阔的空间。通过景观园艺设计，每座公寓楼都仿佛融入了森林之中。公寓楼依地势而建，组成了一幅流动的景观山水画。

住户可以通过南面的入口进入到公寓区。所有的公寓楼都和一个地下停车场相连，保证了小区里没有机动车通行。公寓区内共有六种楼型，根据地点、景观和日光条件的不同，楼面的大小和布局也不尽相同。

每座公寓楼里有五套比例适当的豪华公寓，每套公寓约为450平方米。公寓的室内布局简洁流畅，内部设施功能全。

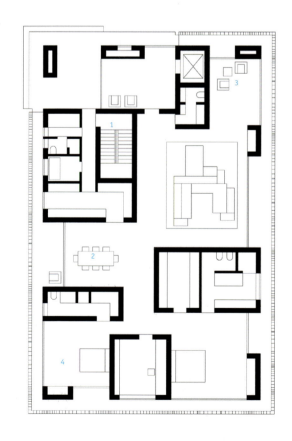

1. stairs	1. 楼梯
2. dining	2. 餐厅
3. living room	3. 客厅
4. bedroom	4. 卧室

Photo: Christian Richters, Shu He

Residential

Architect: David Chipperfield Architects

Completion Date: 2008

Liberal Art Department, Dongguan Institute of Technology

The ground floor of the Liberal Art Department building is plugged in the hill, thus the building is melted in the terrain. The ground floor's two sides are opened to the outside and own the nice scenery, while the other part of the building get the daylight from the L-shaped courtyard. The roof of the ground floor become an open platform in rolling hills, while the elevated square building stands above the platform. The platform, courtyard and corridors together create a centralism of outdoor public square. In the main direction of walking, strong feeling of block building have clearly been faded away, which reduced the building's pressure to the environment. When eyesight crosses the elevated building and falls on the behind hills, people may recall the original image of this area.

东莞理工学院文科系馆

文科系馆首层的建筑被半插入山体，建筑的部分体量由此成为地形的一部分，因为地势的原因，首层建筑仍有两面是开敞的，并面对很好的池塘景观，其他部分则通过一个L形的内院采光。首层屋顶在起伏的地形中形成了一块开放的平台广场，上部建筑则呈正方形围合并架空在平台层上，这样内院、围廊、平台共同形成了一处具有中心感的公共开放空间。在人流的主要行进方向，建筑的体量被明显化解，有效地减轻了建筑对于环境的压力，当视线穿越架空层而停留在后面或远处隆起的山坡上时，人们大致可以想象或回忆起这处地形原本的模样。

General view 全景图

Southeast elevation 东南立面

二楼内庭　Inner court of the first floor

内庭和二楼　Inner court of the first floor

入口 Entrance

Photo: Deshaus Studio

Completion Date: 2004

Architect: Deshaus Studio

SIEEB (Sino–Italian Ecological and Energy Efficient Building)

Sino–Italian Ecological and Energy Efficient Building (SIEEB) is realised in the Tsinghua University Campus in Beijing. It is a 20,000-square-metre building, forty metres high and it will host a Sino–Italian education, training and research centre for environment protection and energy conservation.

The envelope components as well as the control systems and the other technologies are the expression of the most updated Italian production, within the framework of a design philosophy in which proven components are integrated in innovative systems.

The SIEEB building shape derives from the analysis of the site and of the specific climatic conditions of Beijing. Located in a dense urban context, surrounded by some high–rise buildings, the building optimises the need for solar energy in winter and for solar protection in summer.

Gas engines are the core of the energy system of the building. They are coupled to electric generators to produce most of the electricity required. The engines' waste heat is used for heating in winter, for cooling – by means of absorbtion chillers in summer and for hot water production all year round.

中意生态节能楼

中意生态节能楼位于清华大学的校园里，总面积为20,000平方米，有40米高，里面是中意合作开展的环保和节能教育、培训和研究中心。

建筑的外墙构成、控制系统和其他技术都采用了最先进的意大利技术，整体设计理念都以创新系统为主体。

中意生态节能楼的造型综合考虑了项目的地理位置和北京独特的气候特点。地处高楼林立的市区，建筑的设计优化了冬季的太阳能摄入量和夏季的日光防护措施。

内燃机是建筑节能系统的核心，它们与发电机相配合，提供了建筑所需的大部分电量。内燃机所产生的废热可用于冬季供暖，通过吸收式制冷机在夏季制冷，或是提供全年的热水供应。

Photo: Daniele Domenicali, Alessandro Digaetano, MCA Archivi

Educational

Completion Date: 2006

Architect: Mario Cucinella Architects

South façade 南立面

Ordos Protestant Church

Located on top of a hill of Ordos, planned as a green open space for the city, the project takes its inspiration from the topography of the land. The surrounding landscape offers a strong contrast of colours and depth, creating a rich changing background from day time to night time for the church's settlement. It sets framed views of the church in a characterised landscape of excavated rocks proper to a city like Ordos, in Inner Mongolia.

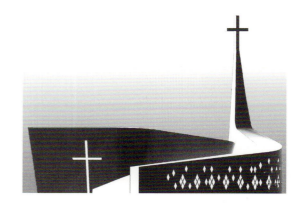

The scheme, named "Dove of Peace", gives its metaphor and poetry to the church by re-interpreting a contemporary and abstract silhouette of the bird caring a branch of Olive in its beak. The Church has a concrete structure and uses white crepi finish for its façades. Its dynamic shape follows the adjacent curved road that crosses the site. The dialogue between the outside and the inside space is emphasised by the play of shadows and light that creates complexity and depth in the reading of the space. Its elements are thought to reflect harmony and tranquility in this place for prays and celebration.

鄂尔多斯基督教堂

坐落于鄂尔多斯一座小山上，这座基督教堂的设计灵感之一来源于其地形。设计师首先从地形作为设计的出发点，结合鄂尔多斯实际的气候特质；鄂尔多斯总是阳光明媚，天总是比别的地方显得高一些，这些都被带入设计灵感中。周围的环境色彩对比强烈，蓝天和褐土对比产生出强烈的视觉效果，为教堂产生了完美的周边环境。营造出了丰富的昼夜背景变化。它形成了对于鄂尔多斯这座城市而言独特的教堂景观。

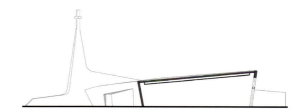

方案命名为"和平鸽"，通过重新诠释和平鸽用喙衔回橄榄枝的故事，赋予了教堂新的寓意和诗意。本教堂是混凝土结构，外立面为白色涂料表面。其流线的外形同时顺应了地块周边的道路走向。其内外空间关系是通过对光线和阴影的描写，深层次解读教堂的空间。其设计元素是为了表现出教堂文化的和谐和平静。

北立面 North façade

鸟瞰图 Birdeyeview

室内 Interior

Photo: Beijing Sunlay Architectural Design Co., Ltd.

Completion Date: 2009

Architect: Beijing Sunlay Architectural Design Co., Ltd.

Liangzhu Museum

The museum houses a collection of archaeological findings from the Liangzhu culture, also known as the Jade culture (~3000 BC). It forms the northern point of the Liangzhu Cultural Village, a newly created park town near Hangzhou. The building is set on a lake and connected via bridges to the park.

The sculptural quality of the building ensemble reveals itself gradually as the visitor approaches the museum through the park landscape. The museum is composed of four bar-formed volumes made of Iranian travertine stone, equal in width (eighteen metres) but differing in height. Each volume contains an interior courtyard. These landscaped spaces serve as a link between the exhibition halls and invite the visitor to linger and relax. Despite the linearity of the exhibition halls, they enable a variety of individual tour routes through the museum. To the south of the museum is an island with an exhibition area, linked to the main museum building via a bridge. The edge areas of the surrounding landscape, planted with dense woods, allow only a few directed views into the park.

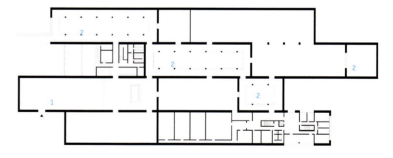

1. entrance
2. courtyard

1. 入口
2. 庭院

良渚博物院

良渚博物院内收藏着良渚文化（又称玉石文化，约公元前3000年前）的考古发现。博物院位于良渚文化村（一个靠近杭州的村落公园）的北角，建在湖上，通过栈桥与文化村相连。

当游客从公园向博物院接近时，建筑雕塑般的形象逐渐显露。博物院由四个长条形黄洞石建筑组成，宽均为18米，但高度不同。每座建筑都有个室内庭院。这些景观空间在展厅之间建立了联系，也为游客提供了休息空间。尽管展厅采用了线形结构，还是为游客提供了多样化的路径。博物院的南侧是一个展览岛，与主展馆通过小桥相连。博物院的四周种植着茂密的树木和植物，从公园内很难看到博物院的景象。

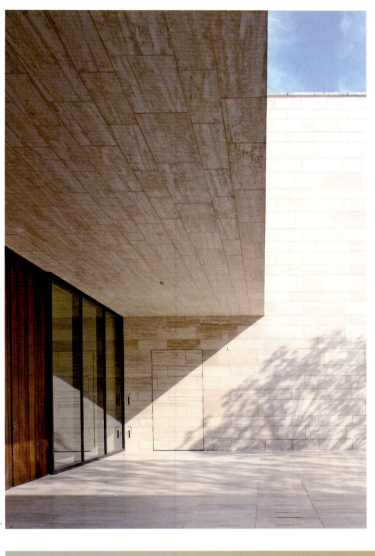

Tianjin Bridge Culture Museum

The Bridge Culture Museum is located at the centre of the QiaoYuan Park, Tianjin. Because of it special location, on the central axis facing to the main entrance, and because it is supposed to be the landmark of the Bridge Park, this area is designed as a leisure and exhibition space in order to meet people's needs of "pleasurable show space".

The origin of the design concept is "operable bridge". Steel bridge, folded surface, variable structural columns, U-shaped enclosed space and extended platform, all create a "Building for Strolling". This building is "accessible and passable". It brings people exciting and surprised experiences because of the conversion between interior and exterior spaces. The steel bridge between the buildings leads people to the exhibition space, which is all green; the bottom of the building is multidimensional surface, which is accessible for people to go through; the sixteen structural columns have been treated in different ways to distract people's sight; the Z-shaped building, U-shaped enclosed space, the coffee break deck and the combination of the interior and exterior spaces, merged the building together with the site. One can find attraction of the building through their own experience to it. The "accessible and passable" Bridge Culture Museum is interactive for people, both interior and exterior.

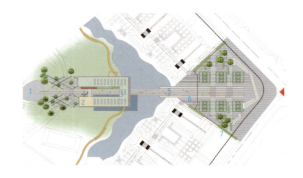

1. performance plaza
2. rooftop viewing terrace
3. green roof
4. water terrace
5. "Bridge Culture" steel plate
6. iron gate and fence
7. main entrance plaza

1. 演艺广场
2. 屋顶休闲观景平台
3. 屋顶绿化
4. 水台
5. "桥文化" 钢板
6. 铸铁大门+围栏
7. 主入口广场

天津桥园桥文化博物馆

桥文化博物馆位于天津桥园公园的中心位置，处于主入口相对应的中轴线上。它是桥园公园的 "点题" 主体建筑和区域标志。由于它所处的特殊场地，建筑师将它定义为 "休闲型展示空间"，满足人们 "看" 的欲望：在 "看展示" 的同时，也能 "看外面风景"。

最初的设计理念是设计一座 "可操控的桥梁"。钢桥、呈折线而起的建筑体、U形围合空间、延伸的咖啡平台，这些设计共同组成了一座 "可散步的建筑"。桥文化博物馆是可以被 "通过" 和 "穿越" 的，它扭曲的室内外空间为游客带来了奇妙的感受。建筑直接的钢桥将人们引入绿色的展览空间；建筑底部采取多维结构，人们可以自由进出；16根结构柱被多手法处理，以吸引人的眼球；Z字形建筑、U形围合空间、咖啡平台以及室内外空间的相互融合使建筑得以有机地契入场地。人们可以通过自己的体验发现建筑的迷人之处，这座 "可通过和穿越" 的桥文化博物馆的内部和外部同样诱人。

Architect: Sunlay Architecture Design Co., Ltd., Zhang Hua, Fan Li

Photo: Shu He

Completion Date: 2007

Cultural

Beijing Nexus Centre

The Nexus Centre is a structure of sophisticated form and superb function. Its column-free and unusually large 2,200-square-metre floor plate allows ultimate freedom for tenants to fully customise their space and maximise efficiency according to individual needs. Located on a long, linear site, the dual-tower complex slopes from south to north from its 551-foot apex in order to limit shadow casting on its northern residential neighbours. The eastern façade fronts the heavily-trafficked ring road and gives retailers prominence along the base. The western façade, by contrast, overlooks the serene setting of an urban park. On the south, the building curves in relation to its corner site, opening up to a large landscaped plaza that segregates vehicular and pedestrian traffic while creating an inviting amenity for tenants.

The design of the complex uses a combination of stainless steel and granite to define distinct fenestration that visually breaks down the overall mass into more slender elements. Expansive floor-to-ceiling glass is applied throughout in order to bring natural light deep into the office floors. Extensive vertical shading devices on the east and west façades control heat gain, and operable windows on all floors promote the introduction of fresh air throughout the work environments.

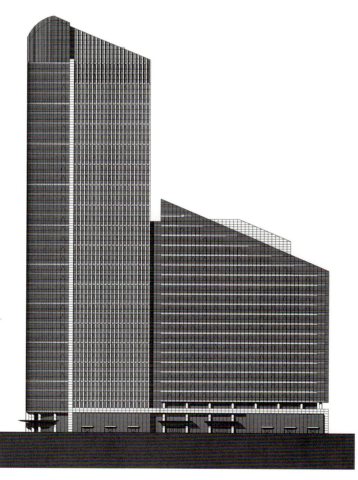

北京嘉盛中心

嘉盛中心造型独特，功能齐全。2,200平方米的开放式楼面让租户可以自由利用空间，根据自身需求将效率最大化。这座双重塔楼南高北低，位于南面168米高的最高端距离建筑的最北面有很长的距离，最大限度地避免了高楼遮光的问题。建筑的东立面朝向交通繁忙的东三环路，为零售商提供了优秀的商业空间。建筑的西立面朝向一个开阔的景观广场，与交通车道相隔绝，环境宜人。

建筑的设计采用了不锈钢和花岗岩的组合结构，独特的外窗设计覆盖了整个建筑的外墙，使建筑看起来更加细长，大片的落地窗为办公楼层提供了丰富的自然采光。建筑东侧和西侧的外墙采用了巨大的垂直遮阴设备，控制了建筑的热吸收。各个楼层的可调控窗户让新鲜空气能够顺利流入工作空间。

East elevation 东立面

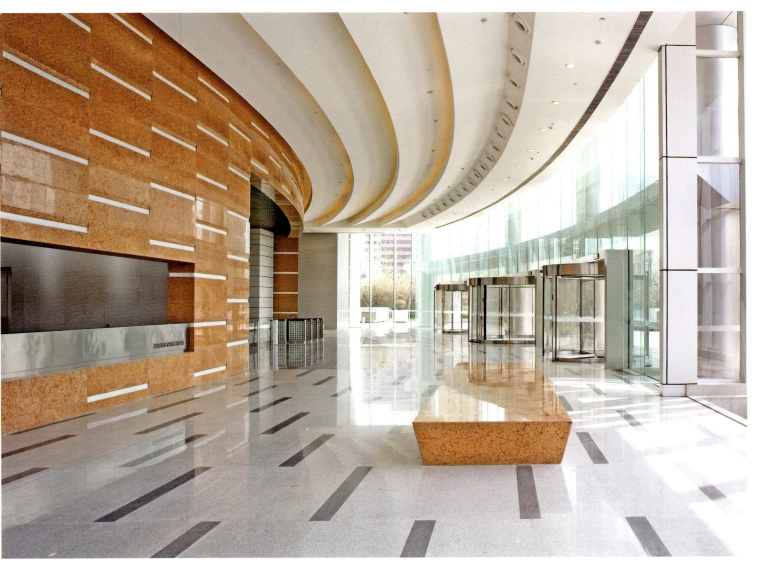

Photo: Doug Snower, Doug Snower Photography

Corporate

Completion Date: 2008

Architect: Goettsch Partners

UF Soft R & D Centre, Beijing, China

For the UF Soft R & D Centre, the architecture began the architectural design by studying the way computer programmers work and live. In order to create a healthier working environment, they borrowed the basic fabric of the old Beijing – the courtyard houses, which offer their residents the proximity of the nature – to mix indoor and outdoor spaces. To further reinforce the integration of the architecture and the landscape, the complex was limited to only three storeys and shallow depth that allows better natural ventilation and daylight. The end result is a mat building that could extend horizontally or a fine–grained fabric urban design that is the antithesis of treating buildings as objects. In the plan, the complex consists of three inter–connected buildings. Two major courtyards are situated in between the buildings with various smaller courts and terraces on different levels.

Besides the concern of health, making team working more spontaneous is another important aim of the design. The diverse public spaces indoor and outdoor have been proven to encourage people to interact, to communicate, and to socialise. While there are a number of materials and construction methods employed in the project, glass blocks embedded in the concrete masonry units is the main technological innovation, which gives the building a solid and transparent appearance at the same time.

用友软件园I号研发中心设计说明

在用友研发中心的设计中，建筑师首先研究了电脑程序员的工作与生活方式。为了创造一个更加健康的工作环境，建筑师借鉴了北京四合院的基本肌理：四合院通过室内外空间的融合，为居民提供了一个接近大自然的环境。为了更好地整合建筑与景观，该建筑仅建有三层，进深很浅，保证了较好的自然通风和日照条件。最终建成的是一座可以水平延展的地毯式建筑，或一个不把建筑物当作物体的纹理细密的城市设计。该建筑在平面上由三栋相互连通的建筑构成，两个主要的院落坐落在这三个建筑之间，还有些小院子和天台零散布置在不同的标高之上。

除了健康的因素以外，保证小组工作的自发性是设计的另一个目标，而多样性的室内外公共空间，最终被证实，能够激发员工们的交流、沟通与往来。该项目运用了多种材料和施工方法，其中最主要的技术创新是在混凝土块砌筑中嵌入玻璃砖，使整个建筑呈现出一种坚实通透的外观。

1. R & D Centre I
2. R & D Centre I, Phase 2
3. entrance & exit

1. I号研发中心
2. I号研发中心二期
3. 出入口

View of courtyard from bridge 从桥上看庭院

庭院夜景 Night view of courtyard

庭院 View of courtyard

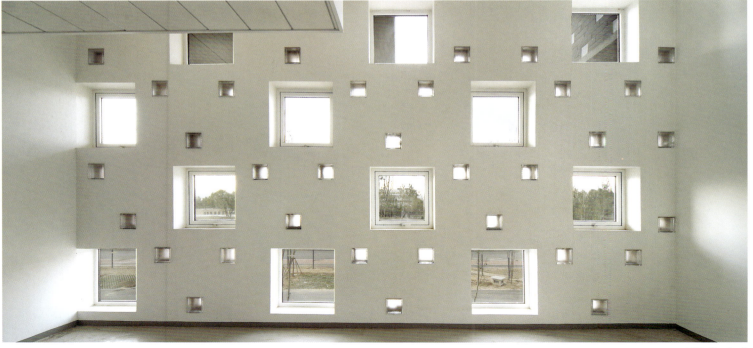

内墙 Inside of wall

Photo: Shu He

Completion Date: 2007

Architect: Jia Lianna, Chen Long, Liu Yang

View from far away 建筑远景

Raffles City Beijing

Raffles City Beijing is located at Dongzhimen with convenient access to the Subway, west to the East Second Ring Road and Southwest to the Airport Express Way. This building complex includes retails, offices, clubhouse, studio apartments, parking space and supporting facilities. The design concept is to create a unique impressive architecture in city centre with multiple function choices, clear and flexible functional spaces and smooth internal traffic.

Different buildings have been illustrated by different surface texture, for example, the surface of the podium is pixel–like glass pattern; the curtain wall of the office building is composed of aluminium frame and 1.4-metre–wide and full–height double glazing; the façade of the studio apartment is illustrated by panel curtain wall and French window system; the roof of the club employs coloured aluminium, to show the connecting of the pieces.

Around the buildings, the gardens are linearly distributed, with the plants and flowers connected to the open spaces with the buildings. The two gardening entrance can not only guides costumers but also emphasise the retail atmosphere. There is a courtyard on the roof of the podium, and the landscape of the courtyard creates a screen which separats the office building away. The central of the courtyard is also the glass roof of the atrium, whose diamond surface forms a special view of the yard.

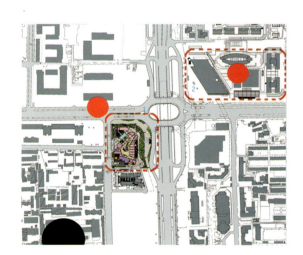

1. entrance
2. main part of the project
3. green plants and landscape

1. 入口
2. 建筑主体
3. 绿植和景观

北京来福士广场

北京来福士广场位于东直门，靠近地铁口，西邻东二环路，西南方向是机场高速。来福士广场集零售、办公、俱乐部、小型公寓、停车场和辅助设施于一体，旨在市中心打造一座独一无二的多功能建筑。来福士广场功能区的设置简洁灵活，内部交通四通八达。

建筑的不同部分采用了不同的表面纹理来诠释：底座结构的外墙像素块一般的玻璃图案；办公楼的幕墙由铝框架构成，采用1.4米宽的玻璃窗；小型公寓楼的外墙采用了玻璃幕墙和落地窗结构；

俱乐部的屋顶运用彩色铝板展示了不同面板的连接。

线形花园环绕着来福士广场，花草树木将开阔的空间和建筑连接起来。两个花园式入口不仅引导着顾客进入，而且烘托了商场的氛围。建筑底座的屋顶是一个庭院，庭院的景观形成了一道屏风，隔开了办公楼。

建筑全景 Full view

入口的灯光 Entrance with light

建筑细部 Details

自动扶梯 Escalator

Photo: Sunlay Architecture Design Co., Ltd.

Complex

Completion Date: 2008

Architect: Sunlay Architecture Design Co., Ltd.

View from southwest 从西南看建筑

2010 Shanghai EXPO – Shanghai Corporate Pavilion

In 2010, we have gone through a long period of rapid technological advancement and the amount of infrastructure in a building has dramatically increased to the point that technologies are today's basic building blocks. For Shanghai Corporate Pavilion at the World Expo, the architect would like to manifest this observation in their design: the interior spaces of the Shanghai Corporate Pavilion, which are shaped as a series of free, flowing forms, will be enclosed not only by walls of the static kind but also a dense, cubic volume of infrastructural network, including LED lights and mist–making system, which are capable of changing the appearance of the building from one moment to another as programmed through computer.

However, the design is not embracing technology for the technology's sake. Rather, they like to convey visually the spirit of the Shanghai Corporate Pavilion, the dream of a brighter future, through sophisticated technologies. Technology is about the enrichment of imagination and symbolic of the industry and industrialism of Shanghai. Also through technology, they like to address the pressing issue of energy and sustainability. A part of the architectural infrastructure is designated for the solar energy harvesting and rain water collecting, and the external façade will be made of recycled polycarbonate (PC) plastic.

2010年上海世博会　上海企业联合馆

2010年，建筑业经历了一段长时间的高速技术发展，建筑内部的大量技术组件将会成为建筑的基本元素。建筑师希望将这一观察应用到世博会上海企业联合馆的设计中去：上海企业联合馆是一个自由、流动的空间，它不只是由墙围合界定形成的，而更是由密集的技术网络立方体包裹而成。在这个技术网络立方体中，LED塑料管与喷雾系统可以依照电脑程序的控制，不断改变建筑的外观。

但是，这一设计并不单纯是为了应用技术，建筑师希望通过精密的技术在视觉上彰显上海企业联合馆的精神，即对光明未来的憧憬。技术是上海工业和工业化形象的浓缩。通过对技术的应用，建筑师实现了节能和可持续设计。上海企业联合馆具有太阳能采集和雨水收集系统。

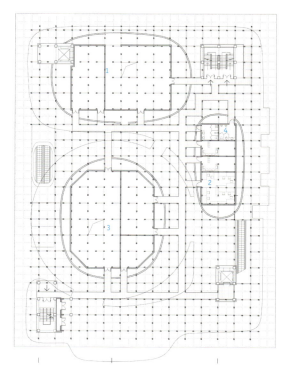

1. power distribution room
2. air condition room
3. office
4. WC

1. 配电室
2. 空调室
3. 办公室
4. 洗手间

西立面 West elevation

从西北看建筑 View from northwest

排队区 Queuing area

Photo: Shu He

Completion Date: 2010

Architect: Atelier Feichang Jianzhu

Qingcuiyuan Club House

Qingcuiyuan Club House is located on the west bank of Wenyu River, Chaoyang District, which has excellent condition of graceful natural landscape. The design concept is not only to create a unique building, but also to integrate it into this attractive landscape environment.

Qingcuiyuan Club House shows two topics, Collision and Blend, the collision of different cultures and styles, the blend of time and space. It is telling a story which is ancient and modern. It is describing an aesthetics encounter between east and west.

The architectural design scheme considers "axis and symmetry" which are emphasised by classical architectural language, and "fluxion and consistency" which are sought by modern architectural language. Red brick walls and transparent glazing are employed to create coagulated building mass and flowing spaces. Inside courtyard with comfortable scale is the continuation of greening, water scene and original natural landscape.

晴翠园别墅小区会所

晴翠园会所位于北京朝阳区温榆河西岸，风景优美。会所的设计理念不仅是建造一个独一无二的建筑，更重要的是将这个建筑融入周边迷人的风景之中。

晴翠园会所展示了两个主题——冲撞与交融：多元文化、风格的碰撞；划时空的交融。它讲述一个古老与现代的故事，是东西方美学世界之间的相遇。

建筑的设计将古典建筑语言所强调的轴线感和对称性与现代建筑语言所追求的空间流动感和内外一致性融合在一起。厚重的红砖墙和透明的玻璃幕墙凝聚了建筑主体和流动的空间。宜人尺度的内庭院是绿化景观、水景和自然景观的延伸。

Façade and entrance 入口和立面

室内景观 Interior landscape

走廊 Corridor

建筑夜景 Night view

Photo: Shu He

Completion Date: 2008

Architect: Sunlay Architecture Design Co., Ltd.

117

General view 全景图

Bridge School

Located at a remote village, Fujian Province in China, the project not only provides a physical function–a school + a bridge, but also presents a spiritual centre. The main concept of the design is to enliven an old community (the village) and to sustain a traditional culture (the castles and lifestyle) through a contemporary language which does not compete with the traditional, but presents and communicates with the traditional with respect. It is done by combining few different functions into one space – a bridge which connects two old castles across the creek, a school which also symbolically connects past, current with future, a playground (for the kids) and the stage (for the villagers).

A light–weight structure traverses a small creek in a single, supple bound, essentially, it is an intelligent contemporary take on the archetype of the inhabited bridge. Supported on concrete piers (which also have the function of a small shop), the simple steel structure acts like a giant box girder that's been slightly dislocated, so the building subtly twists, rises and falls as it spans the creek. Inside are a pair of almost identical wedge–shaped classrooms, each tapering towards the mid point of the structure (which holds a small public library). Although it's possible to use the building as a bridge, a narrow crossing suspended underneath the steel structure and anchored by tensile wires offers an alternative and more direct route.

桥上书屋

项目位于中国福建省一个偏远的小山村。它不仅仅是学校和桥的简单结合，还展示了一种内在的精神文化。现代设计并不会侵犯传统，相反，它将与传统对话，并将其传承下去。该设计的主要目的是通过现代建筑语言复兴古老的村落，传承传统土楼文化和生活方式。项目将多个功能区融为一体：一座穿过溪流连接两幢土楼的桥，一个在象征意义上连接过去、现在和未来的学校，一片操场（供学生使用），一个舞台（供村民使用）。

一个轻质结构横贯与小溪之上，基本上，它就是一个以可居住的桥梁为原型的巧妙现代设计。由混凝土基座（同时也是一个小商店）支撑的简单钢结构像一个被放错地点的巨大盒子。建筑在水面上微妙地扭曲着，一会儿上升，一会儿下降。内部结构和外部造型相似，桥内两间相同的楔形教室各自朝向桥的中央（中央是一个小图书馆）逐渐缩小。尽管也可以把建筑当成一座桥，但是悬在钢结构下面、以拉线固定的小道无疑是一条更方便的路径。

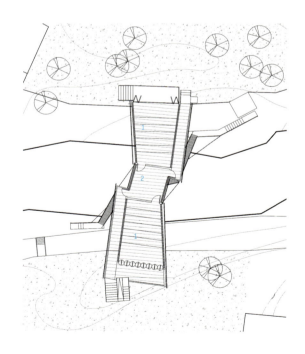

1. classroom
2. library

1. 教室
2. 图书室

建筑细部 Details

入口 Entrance

教室 Classroom

Photo: Li Xiaodong

Completion Date: 2009

Architect: Li Xiaodong

119

Luxehills Chengdu, China

Located in the beautiful foothills of Shuangliu County just south of Chengdu is the new 500–acre Luxehills international community. The project is designed as a world-class international destination for shopping, entertainment and sports, as well as a safe and desirable place to live, work, and raise a family. The entire development is inspired by a European lifestyle, complete with the usage of natural materials and rich landscaping. The 2,500 residents of Luxehills will enjoy an array of retail, restaurant, entertainment, and leisure amenities.

麓山国际社区

麓山国际社区位于成都城南双流县风景优美的山麓丘陵上，占地约200万平方米。项目旨在打造国际一流的购物、娱乐、休闲、居住、办公社区。整体规划采用欧式风格，充分利用了地区优美的自然风景并融入了多彩的景观设施。麓山国际的2,500位居民能够享受到高品质的零售、餐饮、娱乐、休闲设施。

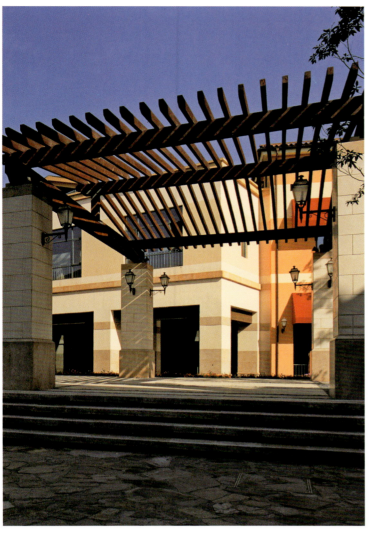

Photo: 5 Plus Design

Completion Date: 2007

Architect: 5 Plus Design

View from sea 海上看建筑

One Island East, TaiKoo Place

Design Concept
One Island East is the latest office development at TaiKoo Place, Quarry Bay, Hong Kong. It comprises fifty-nine office floors with a typical floor plate of 2,300 square metres, providing a total office area of 140,000 square metres. With a height of 308 metres, One Island East becomes the landmark of the district and the centre piece of this office campus.

Form / Façade of the Building
The basic form of the building is a square plan with central core. The four corners are rounded. Two corners facing north and south "open up" at the top floors to address the Harbour view. At the base, the two corners facing east and west "open up" to address the open space.

The edges of the four façades "sail" beyond, creating a "floating" effect, and giving lightness to the building. Lighting feature is incorporated into the edge of the four façade to enhance the floating effect. Architectural fins are introduced in a staggered pattern to add texture and scale to the façade.

Landscape Forecourt
Without a podium structure, the tower sits freely in front of a large landscaped open space to the east. The canopy at the porte cohere is specially designed as a piece of sculpture.

1. passenger lift lobby
2. service lift
3. service lift lobby
4. WC

1. 客用电梯大厅
2. 工作电梯
3. 工作电梯大厅
4. 洗手间

太古坊港岛东中心

设计理念
港岛东中心是位于太古坊的全新商业建筑，包括59层办公大楼。大厦高308米，面积达140万平方米，已俨然成为这一区的地标建筑以及核心结构。

建筑造型与外观
大厦基本造型为方形结构，并设计有中央大厅，四角呈现圆形。其中南北两侧在上层完全开敞，用于欣赏维多利亚港的美丽景致；东西两侧在底层以开放式

设计为主，借以突出开阔的空间。大厦外观边缘犹如帆船一般，营造悬浮感以及轻质感，而"嵌入"边缘的灯饰则更突出了这些特性。

景观庭院
整幢大厦并未设计基座结构，门廊经过特殊设计好似一座雕塑。

行人天桥 Footbridge to OIE

一楼自动扶梯 Escalators at ground floor

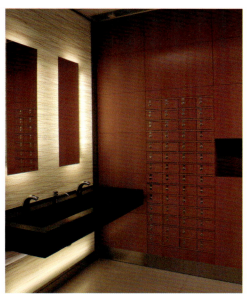

卫生间 Typical Lavatory

电梯厅大堂 Office lift lobby

主入口 Main entrance

外观 Exterior

Photo: Wong & Ouyang (HK) Limited

Commercial

Completion Date: 2008

Architect: Wong & Ouyang (HK) Limited

Perspective 全景图

Hong Kong Polytechnic University – Hong Kong Community College (West Kowloon Campus)

Innovation and Communal Spaces
Due to the limited land resources, the Hong Kong Community College (West Kowloon Campus) aims at experiencing the transformation of the campus communal space design, which is usually, provided in the form of garden on lower floors, into major sky decks similar to high-rise landscape gardens. With associating activities like canteen and student union facilities, campus atmosphere is provided to the sky communal decks. They form the nodal points in the upper campus, which are also proudly visible in the architectural form of the building. The two sky decks are provided to create a sense of community and a place for students' gatherings.

Material and Form
The twin towers and the connecting structures are carefully articulated. Solidity and transparency are the key architectural manipulation for such articulation. "Red-brick" tile is adopted for the solid twin towers while aluminium cladding with glass wall are employed for the transparent connecting bridge structures. This layering put a great emphasis on the connection between the two towers. The twin tower design can minimise the plan depth of each tower and so facilitate the penetration of natural ventilation and lighting into the tower campus.

Sustainability
The building design encourages embrace of natural environment and use of environmentally friendly materials. Innovative building technologies are used in the design with close collaboration with academics from the Hong Kong Polytechnic University in order for the building to achieve energy efficiency and a healthy community. Besides, the lower floors accommodate most mass teaching spaces to facilitate effective pedestrian circulation.

3F Podium Garden   空中花园
1. canteen   1. 食堂
2. podium garden   2. 平台花园
3. student union centre   3. 学生会中心

香港理工大学-香港专上学院(西九龙校园)

创意与公共空间
为了在有限的空间之上，争取最多的可用空间，香港专上学院(西九龙校园)重新设计空间布局，将原本的低层平台花园向高空发展，成为空中花园。并于高层空中花园附近加入食堂和学生会设施等聚集人气的活动，有效地将人气汇聚至校园的高层平台，符合垂直发展的建筑理念。另外，大楼上的两个空中花园不但能够营造理想校园环境，鼓励学生主动学习，增强归属感，更是师生交流及讨论的好地方。

选材与外观
校园的大楼采用双塔拱形设计，有效减少大楼的深度，一方面可以使室内空间摄取足够自然光线，加强对流通风，另一方面又可以避免景观受到旁边建筑物的遮挡。

校园于选材方面经过精心考虑，务求做到实体与透明的视觉对比。两座大楼主要运用红砖外墙以增加实体的感觉；而连接两座大楼的拱桥则以金属外层加上玻璃外墙覆盖，以加强透明感，与大楼的实体布局形成强烈对比，更加加强大楼之间的连贯性。

可持续性
香港专上学院(西九龙校园)的建筑设计尊重自然环境，采用环保物料，符合可持续发展的建筑理念。校方更与香港理工大学的学者教授紧密合作，运用创新的建筑技术，达到能源效益，共建环保校区。

东立面 East elevation

楼梯 Stairs

室内 Interior view

Photo: Wong Wingfai, Tang Kafai

Completion Date: 2008

Architect: AD+RG Architecture Design and Research Group Ltd.
In Collaboration with AGC Design Ltd and Wang Weijen Architecture

Hong Kong Polytechnic University – Hong Kong Community College (Hung Hom Bay Campus)

The form composes of various teaching blocks stacked spirally in the air, which are separated with sky gardens at different levels. The various blocks have different degree of opacity. The opacity controls the degrees of direct sunlight casting into the interior of the building with different degree of intensity. The interface of the low block and high block is spatially separated with a top sky-lighted atrium. It serves as a focal point to link up various facilities together and orientates the internal space and circulation.

The mass teaching facilities are arranged on the lower floors. They are shared and connected effectively with the escalators and lifts system. Main staircases are always provided next to the escalators and also link up all sky gardens together. It helps to bring the outdoor atmosphere to the interior.

Sky garden is one of the major design features. From the appearance of the building, a spiral-chain of sky gardens could be perceived and the conspicuous feature is distinctive from the surroundings. The sky gardens are conducive to students' discussions on projects and casual gathering for socialising. Among the intense urban fabric, the sky gardens provide good locations for viewing towards Hung Hom district and the relatively open Hung Hom Station and Coliseum areas. The choice of bamboo in the sky gardens suffices to let sun light shining into the interiors and some large trees are also planted to form a lively and pleasant atmosphere.

1. stage and backstage area
2. store room
3. multi-purposes hall
4. building line above
5. classroom
6. AHU
7. lecture theatre

1. 舞台和后台区域
2. 贮藏室
3. 多功能厅
4. 上方建筑界线
5. 教室
6. 空调机组
7. 阶梯教室

香港理工大学 – 香港专上学院（红磡湾校园）

建筑的外观由不同的教学楼块以螺旋状堆积而成，中间由不同楼层的空中花园隔开。各个楼块有不同的透明度。透明度控制着建筑内部的日光直射层级。高低建筑的接触面由一个顶部透光中庭隔开。中庭作为将不同设施连接在一个的焦点，为室内空间和交通确定了方位。

大规模的教学设施被安排在较低的楼层。它们由电梯和自动扶梯系统有效地连接起来。主楼梯设在自动扶梯的旁边，同时与空中花园相连。它将户外氛围带进了室内。

空中花园式设计的亮点之一。从建筑的外面就可以看到空中花园，这一特征使其独树一格。空中花园便于学生探讨项目，也便于他们休闲集会。在密集的城市网格中，空中花园是难得的观景佳处，可以俯瞰红磡区和相对开放的红磡车站和大剧院。空中花园上种植的竹子让阳光透过缝隙射入室内，而一些大型的树木则增添了生动愉悦的气氛。

Distant view 远景

东立面 East elevation

主入口 Main entrance

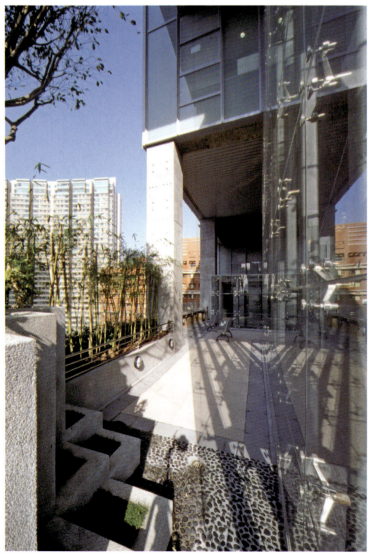

空中花园 Sky gardens

图书馆的自然采光 Natural daylight to library

Photo: Wong Wingfai, Tang Kafai

Architect: AD+RG Architecture Design and Research Group Ltd.
In Collaboration with AGC Design Ltd and Wang Weijen Architecture

Educational

Completion Date: 2007

Panoramic view 全景图

Kaohsiung Metro Station

Kaohsiung city decided a construction of the subway, and designated Shin Takamatsu to design a new subway station. The design range was the concourse (total floor area: 10,000 square metres) and the entrance. The station is located under an intersection in the central Kaohsiung city. Concretely, four main entrances on four corners and eight sub-entrances along the road are architecture on the ground. Therefore, through urban point of view, these designs influence the landscape of the city. As the result of researching the characteristics of the site with basic recognition, Shin Takamatsu found a historical fact that this intersection was the place where the Taiwan democracy movement began. The confusing design task because of the simplicity of functional request was rapidly solved by the symbolic information. An idea, which settles the four entrances located on four corners as an architectural synthesis and gives each entrance the form of joining hands while unifying them was generated in one sitting. At the presentation for the concept, these forms were named "Form of Prayer". Kaohsiung city gladly accepted it. An idea that the emitted laser beams from the top of the four entrances meet above the central intersection was proposed at the presentation.

高雄地铁站

建筑师高松伸受邀为高雄市设计一个新地铁站，设计包括站前广场（10,000平方米）和地铁入口。地铁站位于高雄市中心的一个十字路口，四个主入口分别位于四角，其他八个副入口设在沿街的建筑物地下。从城市规划的角度来看，这些设计影响着城市的景观。高松伸通过调研发现这个十字路口是台湾民主运动的发源地，这一发现促成了地铁站的设计。四角的四个入口被设计成合十的手掌形状，并被建筑师命名为"祈祷式"。高雄市政府对这一设计十分满意。四个入口的顶部还会发射出镭射光，并最终汇聚在十字路口的中央。

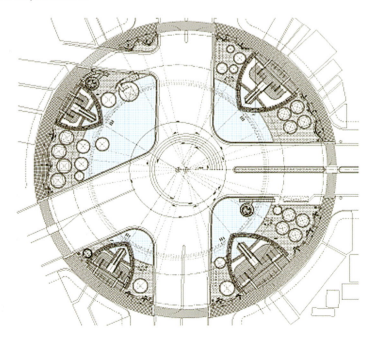

出口 Exit

地铁出口 Station exit

室内 Interior

Transportation

Photo: no credit

Completion Date: 2008

Architect: Shin Takamatsu +Shin Takamatsu Architect and Associates Co., Ltd.

Panoramic view 全景图

LV Taipei

Since Taipei is in the subtropical zone, the city is filled with lots of trees. The designers tried to use the landscape of trees for the motif of the façade. Tree does never relate to the brand image of Louis Vuitton, however they thought it was important. The main material of the façade is perforated stone. The designers used half–artificial stone, and made square holes with water jet machine. Then they filled up holes with synthetic resins. The pattern of the hole was based on Louis Vuitton's checker flag pattern called DAMIER pattern. Even though DAMIER pattern usually is homogeneous, the designers made each square in different sizes. There are more than thirty sizes. The smallest is seven millimetres, and the biggest is thirty-five millimetres. Using such difference of sizes, they made another pattern in bigger scale which is also the checker flag pattern.The bigger pattern is not homogeneous. The size of each square is different. Using such difference of sizes, they made one more pattern which is the silhouette of trees. What the designers really wanted to do is to weaken the symbolised image of the brand. Usually symbol does not relate to the context, which the designers really do not like. They want to make something linked to the surroundings delicately. If the designers can refine the commercial façade architecturally, this is one way to do it.

台北LV专卖店

由于地处亚热带地区，台北市里遍布各种各样的树木。设计师试图将树木景观融入建筑外墙的设计中。尽管LV的品牌形象与树木无关，设计师还是认为这很重要。外墙的主要材料穿孔石材。设计师在半人造石上通过喷水枪打出方形的孔，然后在孔内注入合成树脂。方孔所形成的图案以LV经典的DAMIER方格图案为基础。虽然DAMIER方格图案是均匀结构，设计师将每个方块赋予了30种不同的尺寸，最小的7毫米，最大的35毫米。设计师随后又用这些大小不一的方格拼成了大规模的DAMIER方格图案。大方格的尺寸也大小不一。设计师用这些方格拼成了树木的轮廓。设计师真正的意图是削弱品牌的既定形象。品牌形象通常与环境无关，但是设计师并不喜欢这种方式。他们想要刻意将品牌和周边的环境联系在一起。如果设计师能够从建筑上改善商店的外墙，这一设计就是一种可行的方式。

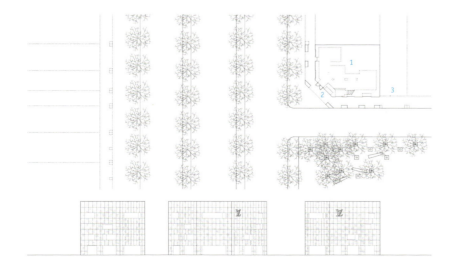

1. 零售空间
2. 入口
3. 拱廊

1. retail space
2. entrance
3. arcade

入口 Entrance

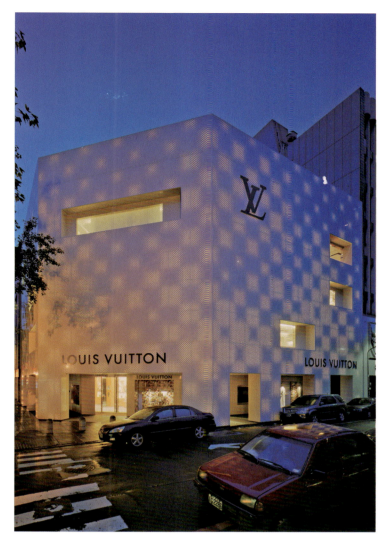

建筑外墙 Exterior façade

前厅 Lobby

Photo: Daichi Ano

Commercial

Completion Date: 2006

Architect: Kumiko Inui

Night view 夜景

Xi Gallery

Located in Yeonsan-dong, Pusan, this building was constructed for the purpose of promoting "Xi", a brand of apartments. In addition to the standard type of an apartment unit exhibition space, an even larger share of the floor area is allocated as a variable cultural space for the locals, which as a result creates a brand-new building typology: a Housing Cultural Centre. As economic forces and cultural activities seem to form complex interrelationships causing our private and public spheres to merge and invade each other, this building comes as a product of these current phenomena. The focus of the designers' investigation is to create a fluid space that can respond to the "continuously new" situations arising from the dynamic flux of economy and culture, and in the organisation of the movement system to correspond to such a space. This new movement organisation is necessary to maintain the existing individuality of the spaces, but at the same time be able to expand/unify them in diverse manners to suit future possible needs.

Xi展览馆

Xi展览馆位于釜山的莲山洞，是为了宣传"Xi"品牌公寓而建。除了标准的公寓样板房展示之外，更多的空间可以被用作举办当地的文化活动。这一举动创造了一个全新的建筑类型——住宅文化中心。经济力量和文化活动的相互关系十分复杂，让我们的个人和公共生活相互影响，而这座建筑就是这一现象的产物。设计师的重点在于设计一个能够适应不断更新的经济文化变化的流畅空间，并使建筑的室内交通符合空间的需求。这个室内交通系统既要维持不同空间的独立性，又要将它们依照未来需求合理地统一起来。

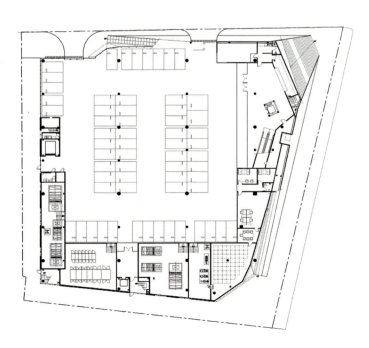

建筑正面 **Building front**

大厅 **Hall**

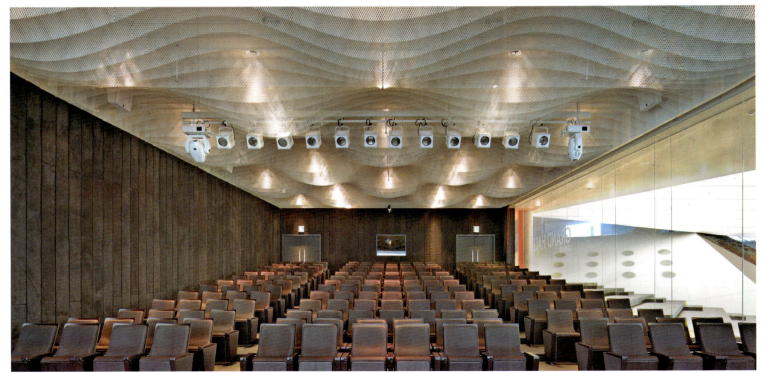

室内 **Interior**

Photo: Yong–Kwan Kim

Completion Date: 2007

Architect: Mass Studies

Building in normal environment 建筑外观

Model House

The Model House for I'Park City is a showroom to exhibit the residential urban and façade design for an eighty-eight–building development in Suwon. The principal design intention was to curate the visitor experience by elaborating on an implicit circulation strategy. The route of the visitor, from the approach to the building and throughout the tour within, is treated as an ongoing exhibition.

The building is therefore configured as a clear illustration of this strategy, which becomes the central feature both within the interior space and as a generator of the façade design. The resultant crossing of the circulation ribbon produces various viewpoints of the exhibited content, in this case, the façades of the show–units. In parallel, the same path oscillates between a focus on the interior exhibited content and the adjacent larger site and landscape, which is the location of the future development. The circulation route underscores the importance of the "coming home experience" which is the organisational driver of the entire I'Park City Urban development, and the core of the overall site branding strategy.

样板房

I'Park新城样板房是展示水原市一处88座住宅楼开发工程的陈列室。设计的主要目的是通过暗示性的路线策略向访客详细地展示样板间。从一进门到整个参观结束，访客的路线是一个不间断的展览过程。

建筑的配置清晰地展示了这一策略。无论是室内空间还是外墙设计都以此为中心。内部路线的交叉点为展览内容——样板间提供了多样化的视角。此外，展览路线还能够看到临近的大型场地和景观——未来开发工程的地点。展览路线着重强调了回家的归属感，这也是整个I'Park新城开放工程的原动力和品牌策略。

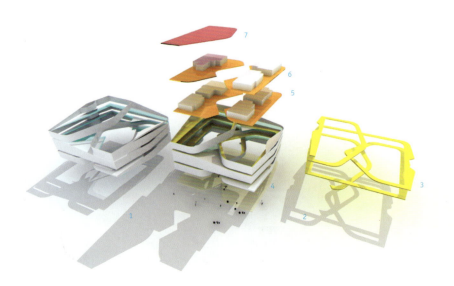

1. 建筑外壳
2. 建筑体量
3. 展览路线
4. 入口楼层
5. 1层
6. 2层
7. 屋顶平台

1. skin
2. volume
3. exhibition ribbon
4. entrance level
5. level 1
6. level 2
7. roof terrace observatory

外部细节图 **Façade details**

室内 **Interior**

建筑正面 **Building front**

Daytime view from the street 日间街景

Myeongdong Theatre

Due to the soaring land price in Myeongdong and increasing propensity to consume, the former Myeongdong National Theatre building was under threat of being torn down and replaced by a high-rise commercial building. However, the exterior wall was preserved thanks to the campaign for preserving the building and the purchase by Ministry of Culture, Sports and Tourism. The new cultural interface of the newborn Myeongdong Theatre holds the memories of the past and the vital energy of the district and has become an artistic beacon. The exterior wall, which keeps the memory of Myeongdong, was preserved and a vessel of regeneration (auditorium mass) was created for the culture of Myeongdong streets in the past, present and the future. The energy of the streets was brought inside the building and filled between the wall of the past and the auditorium mass.

For an urban space open to people, the auditorium mass was lifted above a floor to create a large lobby. The natural lighting pouring down through the skylight, which has been installed between the hall and the auditorium mass, runs along down the mass and reaches the lobby, making the narrow space into one enriched by light. As for the auditorium, the balconies of two floors are surrounded in a three-dimensional way and compose a friendly performance hall in the shape of a horseshoe.

明洞剧院

由于明洞飞涨的地价和人们不断攀升的消费倾向，前明洞国家剧院大楼存在着被推倒，并被高层商业楼替代的危险。好在韩国文体旅游部在历史建筑保护活动中将其买下，使得国家剧院大楼得以幸存。新的明洞剧院将传承历史，增添地区活力，成为一座艺术的灯塔。原有建筑的外墙被保留了下来，而表演厅大楼则进行了重建。明洞街道上的活力被引入了建筑内部，填补了过去与现在之间的鸿沟。

表演厅设在二楼，从而在一楼创造出一个巨大的门厅，作为公共活动空间。日光通过天窗洒在剧院内部，让里面充满了光线。表演厅呈马蹄形，三面环绕着三维立体包厢，环境宜人。

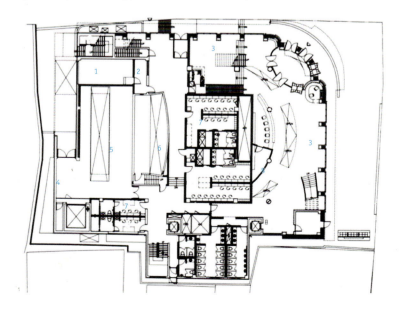

1. 避难室
2. 保安室
3. 大厅
4. 装卸区
5. 舞台下方
6. 乐池
7. 化妆室
8. 办公室

1. disaster preventing room
2. guard room
3. lobby
4. space for loading and unloading
5. under stage
6. orchestra pit
7. make-up room
8. office

屋顶 Roof

2层大厅，展览着保存下来的墙壁 2F Hall with showcase for preserved past wall.

礼堂 Auditorium

Photo: Park Young-chae

Architect: Myung-Gi Sohn, Jong-R. Hahn

Completion Date: 2009

Cultural

Daytime view 日间全景

Sungkyunkwan University's Samsung Library

成均馆大学三星图书馆

A resource centre where people can retrieve information systematically is an essential goal that the designers bear in mind for this information centre. The theme named the "Digital Library" has been the most basic concept to apply in this design. The Sungkyunkwan University's Samsung Library has an image of a ginkgo leaf, which is the symbol of Sungkyunkwan University as well as an image of an opened book. The Digital Library which has been embodied with transparent and metallic materials is set to reorganise the context of the campus as a hub of information exchange in the heart of the campus.

The centre is a multi-functional facility providing conventions and resting areas. The inner round-shaped void creates a hall, allowing sunlight to flow inside. Group study rooms are freely dispersed near the hall, creating dynamic rhythms. Besides, free community zones are a new type of multi-functional place, and are far different from the quiet and closed typical studying rooms. The Sungkyunkwan University Samsung Library represents a new type of library that is appropriate for the digital era where students can study while socialising with friends, surfing the internet, listening to music and drinking coffee.

在设计这个信息中心的过程中，设计师的基本目标是向人们提供一个可以系统检索信息的资源中心。"数字图书馆"是设计的基本理念。成均馆大学三星图书馆的造型像一片银杏树叶（成均馆大学的标志），又像一本打开的书。这个数字图书馆外部采用了透明的玻璃和金属材质，坐落在校园中心，是大学的信息交换中心。

图书馆是一个多功能设施，内部还提供会议和休息空间。室内圆形的区域形成了一个日光充足的大厅。小组研究室自由地分散在大厅的四周，产生了一种韵律感。而且，自由团体讨论区也是一个新型的多功能区域，与传统的封闭而安静的自习室大相径庭。成均馆大学三星图书馆展现了一个适应数字化时代的新型图书馆，在这里，学生可以在学习的同时交友、上网、听音乐，甚至是享用咖啡。

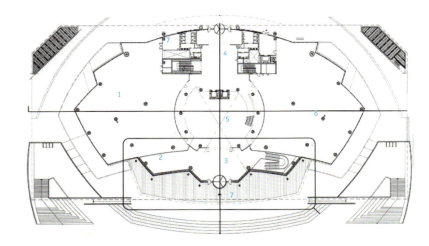

1. 查阅室
2. 展览厅
3. 门厅
4. 大厅
5. 上部开口
6. 自由阅览室
7. 数据室

1. search room
2. exhibition hall
3. lobby
4. hall
5. upper open
6. free reading room
7. data keeping room

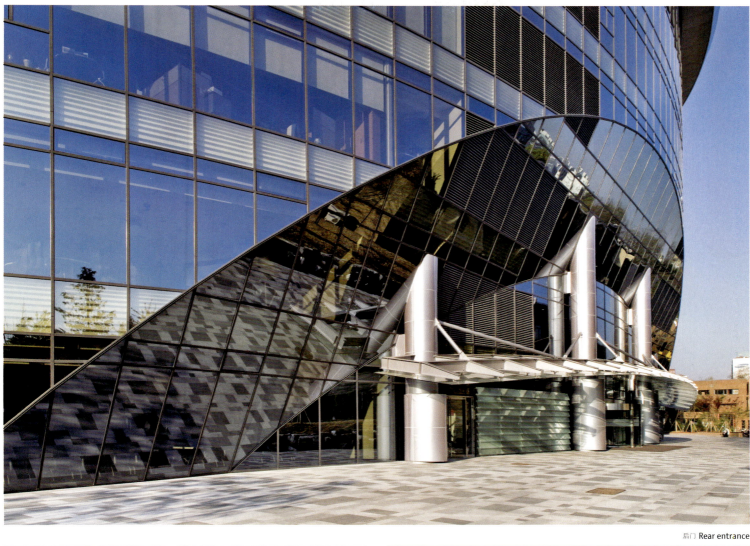

后门 **Rear entrance**

门厅 **Lobby**

会议室 **Meeting room**

Canadian Diplomatic Complex

The design of the Canadian Embassy in Seoul creates a dialogue between Korean and Canadian cultures, expressing common links and in particular a shared reverence for nature. This unique site shares a "place" with a 520-year-old tree, a living symbol of nature, called Hakjasu or "scholar" tree in the historic Jeong-dong district near Deoksoo Palace.

The composition pulls back and suspends the two main building masses creating an entrance plaza and gathering place with this tree at its focal point. The building base ties together these two main blocks. This undulating mass is wrapped with a continuous wooden screen composed of western red cedar. Its soft curves frame the public space around the tree adding to the rich composition of textures and materials at the entrance to the building. The wood will age to resemble the undulating walls surrounding Deoksoo Palace.

The historic Jeong-dong context plays an important part in the building materials and organisation of the embassy. Jeong-dong is built of stone, brick and wood in varying hues ranging from grey to red. Embassy materials have been chosen carefully to harmonise with these colours and textures and to extend the pedestrian walk that meanders along the undulating Deoksoog Palace wall.

A protection plan for the 520-year-old tree was critical. The massing was designed for minimum impact on the tree rootball. Using an existing urban site, directing drainage into landscaped areas, coordinating with local materials, minimising parking through a high-density automated system, providing open space, reducing light pollution by screening and curtain wall design, maximising energy efficiency, commissioning – these were all measures taken to promote sustainability.

1. reflective pool
2. garden
3. main public entrance
4. 520-year-old tree

1. 倒影池
2. 花园
3. 主公共入口
4. 古树

加拿大驻韩使馆

首尔的加拿大使馆搭建了韩国和加拿大文化交流的桥梁，尤其是在对自然的敬畏上达成了共识。使馆和一棵具有520年树龄的古槐树共享一片位于德寿宫附近的贞洞区。使馆的两座主楼略微后错，在门口古树的周围营造了一个小型广场。建筑的底座将两座主楼紧紧地连在一起。波纹形底座上包裹着连绵的红雪松木屏障。这条柔和的曲线框住了古树周围的公共空间，也丰富了建筑入口的

材质和纹理。日久天长，这些木屏障会变得与德寿宫起伏的围墙相似。

传统的贞洞文化在建筑材料和使馆结构方面起到了重要的作用。贞洞由石头、砖块和木头制成，颜色从灰色一直延伸到红色。使馆建筑材料的选择与贞洞的设计相似，门前的人行道一直延伸到德寿宫的围墙。对古树的保护在建筑设计中尤为重要，建筑设计将对树根的影响减到最小。建筑还采用了以下可持续设计策略：利用现有的城市市场地，将排水导入景观区域，采用本地建材，利用高密度自动化系统将停车场最小化，能效最大化，采用试运行策略等。

Architect: Zeidler Partnership Architects

Completion Date: 2007

Corporate

Photo: Kim Yong Kwan

General view 全景图

Post Tower

The Post Tower, situated in one of the most critical and highly visible sites in the historic central business district of Seoul, is the south gate of Myeongdong. Three major roads and their intersections form a triangle within the city that defines the district.

The 72,000-square-metre building presents a symmetrical face and dramatic silhouette to the primarily axial approach from Namdaemun. The building is developed with a primarily stone wall to convey solidity and permanence to the place and accented with a detailed stainless steel window system to provide shadow, richness and sparkle. A super scaled gentle arch that provides a generous and welcoming entrance for the project defines the lobby and main postal hall. The building is sited to the east of the site to maximise the public space in front of the project and create a memorable urban place that accepts the many pedestrian movements through and across the site.

Programmatically the first three floors contain the public postal spaces and the office lobbies. The next six floors contain postal sorting operations on large floor plates to minimise vertical movements. As the tower breaks into two individual towers, the central postal authorities occupy the south tower and the north tower is speculative office space to generate income for the owner. The tenth floor becomes the community centre for the project and contains dining, meeting and lounge spaces for the users all centred on a large exterior terrace.

邮政大厦

邮政大厦位于首尔最重要的历史文化商业中心——明洞的南大门，十分抢眼。三条主要街道在这里相交叉，形成了一个三角形区域。

大厦总面积为72,000平方米，呈夸张的对称结构，正好坐落在南大门的中轴线上。建筑的石墙体现了它的坚固和持久性，而不锈钢窗口系统则为建筑增添了光彩和丰富了细节。巨大的弧形拱门为大厅和主邮政厅提供了一个温馨而大方的入口。建筑位于空地的东侧，使建筑前方的公共空间得以最大化，营造出一个令人难忘的城市空间，为行人提供了多条通道。

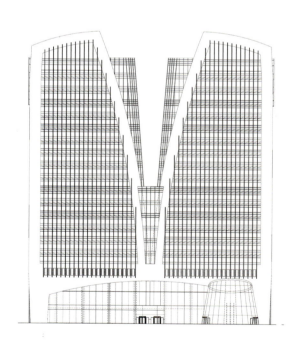

Elevation 立面图

大厦的下面三层是公共邮政空间和办公大厅。往上的六层是邮件分类作业区，楼面布局开阔，适合垂直运动。上方的塔楼一分为二，南侧是中央邮政机构，北侧是用于出租的办公空间。大厦的十楼是社区中心，餐饮、集会和休闲空间都集中在一个开阔的露天平台上。

夜景 Night view

入口 Entrance

门厅 Hall

Photo: Young-Kwan Kim

Architect: DeStefano and Partners, Ltd.

Completion Date: 2009

Commercial

General view 全景图

Kumho Culture Complex – Kring

The designers' intention is to construct an architectural building to create codes for companies and consumers to facilitate communication, then embed an identity in that architecture. "Urban Sculpture for Branding" is the concept applied in Kring compound culture space to realise brand identity for the client. The way to create this building was different from the conventional method. It was about creating an urban sculpture, and then composing a compound space in it.

Uh–Ul–Lim, phonetic pronunciation of Korean word means "harmony", and the brand image of Kumho E&C was transformed as brand identity through the architectural shaping process. Through the architectural interpretation process, the phenomenon of harmony was connected and emphasised to the city and society with undulation, and the notion of undulation was deepened. The designers wanted to gather various elements of nature, life, and city harmoniously and the essence of harmony to rush out to the city creating echo and undulations, and that became the brand scenario of the project. Projecting images to the city and sucking up the energy of the city at the same time, gigantic container for echo was born, which leads to the concept of "Dream" as well. The designers wanted this to be the monument of the city day and night, specifically, a lighting sculpture when it is dark. Moreover, it becomes a pure white space when entered into, and contains all and any types of cultural programme. The pure white space which achieves spatial surrealism is acting as a stage for performance. Just like a stage changes itself to conform to the types and story of a play, this becomes a compound space to fit itself to diverse needs.

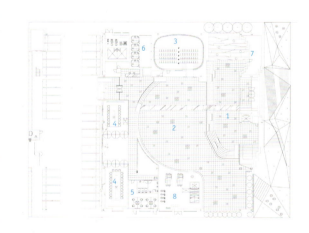

1. atrium	1. 中庭
2. exhibition	2. 展览区
3. cinema	3. 影院
4. meeting room	4. 会议室
5. office	5. 办公室
6. toilet	6. 洗手间
7. storage	7. 贮藏室
8. machinery	8. 机械室

锦湖文化大厦——Kring

设计师试图建造一座能够促进公司和消费者之间交流的建筑。在Kring综合文化空间中，他们采用了"品牌化城市雕塑"这一概念，以求增强客户的品牌识别度。这座建筑的设计方法有别于其他传统方式：首先是创造一个城市雕塑，其次才是在里面设计综合空间。

Uh–Ul–Lim在韩语中意为"和谐"，是锦湖E&C公司的品牌形象。设计师在建筑的构型过程中将它改造成了公司的品牌标识。在建筑诠释的过程中，"和谐"元素与城市和社会的联系通过波纹来显示，波纹的概念被深化了。设计师试图将自然、生命和城市和谐地聚集在一起，让和谐的本质冲出城市，营造出波纹和回音，从而形成项目的品牌特征。向城市发射图像，然后从城市中吸取能量，这就形成了一个巨大的回音收集箱，同时也形成了"梦"的概念。设计师渴望这个建筑能够成为城市不朽的纪念碑，尤其是在夜间的时候。大厦的内部空间是纯白色的，有各种各样的文化设施。纯白色的空间实现了一种空间超现实主义，构成了一个表演的舞台。正如舞台可以改变自己以适应不同的戏剧情节，建筑的内部空间也可以适应不同的需求。

建筑正面 Front view

入口 Entrance

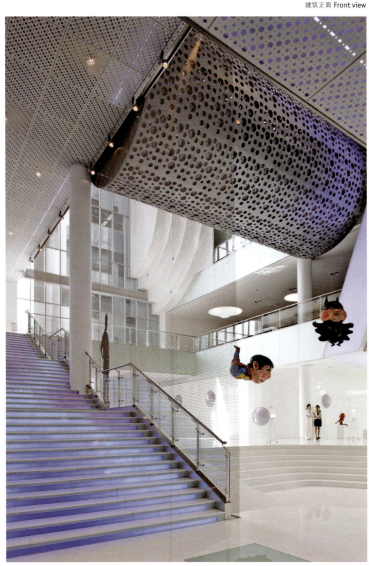

大厅 Hall

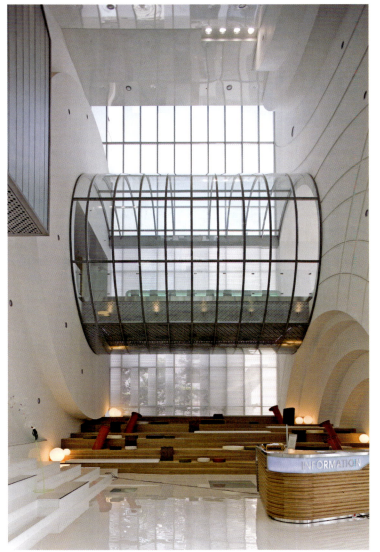

接待处 Reception

Commercial

Completion Date: 2008

Architect: Unsangdong Architects

View from the street 街景

Shin–Marunouchi Towers

Hopkins Architects have designed and built a landmark office, retail and restaurant development in Tokyo's most prestigious location for Mitsubishi Estate Group.

Hopkins Architects worked alongside architects, Mitsubishi Jisho Sekkei, to create a structure divided into three sections – two towers on a podium base. The six–storey podium incorporates retail and restaurants. The building's West façade looks onto one of Tokyo's most exclusive shopping streets. Conran, Prada and Hermes all have shops in the area. The offices are aimed at leading international businesses.

The commission required a mixed–use complex with large shop, luxury boutiques, restaurants, and offices for international business tenants; linked below ground level to the Tokyo Plaza metro station. The urban and intrinsic value of the project is revealed by the lightness of the volumetric relationships, the meticulous care taken over the details, the Miesian classicism of the structure that uses relationships on both the urban and human scales. The steel and glass structure resolves the relationship between architecture and its context.

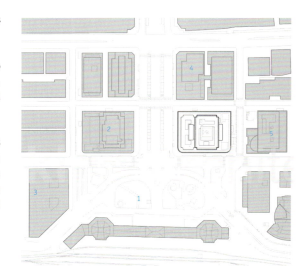

1. Tokyo Station
2. Marunouchi Building
3. Tokyo Central Post Office
4. Tokyo Marine Nichido Building
5. Industry Club of Japan/Mitsubishi Ufj Trust and Banking Building

1. 东京车站
2. 丸之内大厦
3. 东京中央邮局
4. 东京海上日动公司
5. 日本工业协会/三菱信贷银行

新丸之内塔

霍普金斯建筑事务所为三菱地产集团在东京最繁华的区域设计了一座集办公、零售、餐饮于一身的地标式建筑。

霍普金斯建筑事务所与三菱设计共同设计了一个分为三部分的结构——两座塔楼和一个底座。六层高的底座里是商店和餐厅。建筑的西立面正对东京最高端的商业街，康兰、普拉达和爱马仕等品牌都在这里有专卖店。这座办公楼试图达到世界一流的水准。

新丸之内塔地下与东京广场地铁站相连，里面包含大型商场、奢侈品店、餐厅和国际化办公空间。项目的价值体现在其轻快的连接性、精致的细节设计以及具有城市人文尺度的经典密斯式建筑结构里。钢铁和玻璃结构将建筑与周边环境有机地结合在一起。

全景图 General view

外部一角 Façade corner

前廊 Corridor

入口 Entrance

Photo: Ken'ichi Suzuki

Completion Date: 2007

Architect: Hopkins Architects

Towada City Hospital

十和田市立中央医院

Towada City is at the foot of the mountain Hakkoda-san in the Tohoku snowy district. It is distinguished by its fertile nature and the streets which are orderly marked off as the origin of the modern urban planning. In view of designing this hospital, the outward walls are coloured with the earth-colour and the balconies are with the colour of trunk of cherry blossoms to reflect the cherry blossoms at the government office quarter. The outward walls are designed with grid to match with the latticed streets in Towada City. The façade facing the streets uses of many glasses and extends back a long way with reflecting the trees lined in the streets and the streaming through the leaves of trees.

It aimed at making the building long-standing by adopting the PC structure and the quake-absorbing structure in the building frame, discontinuing the building frame burial of piping as much as possible, and adopting the simple partition.

Regarding the hospital rooms, the water-based facilities and water supply are placed along the outside wall. This design ensures that any future structural changes can be flexibly accommodated, that nurses can visually care for patients with greater ease, and that the window-side bed temperature during the winter season can be maintained at an adequate level.

十和田市位于东北雪区内八甲田山脚下，以浓郁的自然景观和规整的街道著称。十和田市立中央医院的设计正是从这一点出发，外墙粉饰以泥土色调，阳台则涂抹着樱树树干的颜色，与周围的樱树交相呼应。此外，外墙更是特意打造成格子形状以便与格子样式的街道相互一致。朝向街道一侧的立面采用玻璃材质打造，映射出路边树木的影子。

为延长建筑的寿命，设计师在框架打造上专门选用了预制混凝土及防震结构，空间之间的分隔结构也尽量简化。病房的设置考虑到了未来的格局调整，灵活性十足。

Externals shining at night　建筑在夜晚格外闪亮

建筑外观与其所在的政府机关区建筑相称 Externals considered in accordance with the streets at the government office quarter

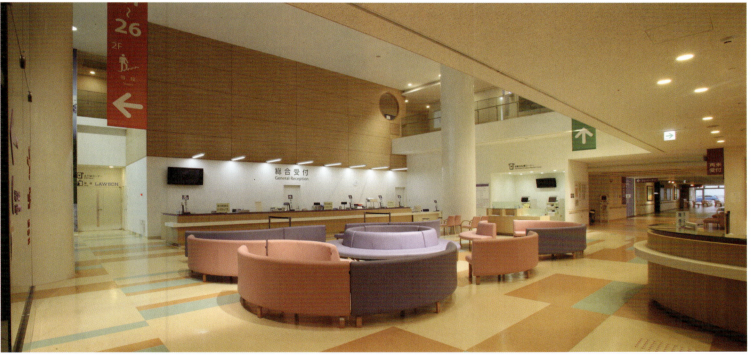

门厅的自然设计 Entrance hall that images nature

入口夜景 Entrance at night

Photo: Mitsuhiro Wada

Completion Date: 2007

Architect: Mikio Chikusa, Norihiro Sawa, Moriki Matsuguma/Showa-Sekkei, Inc

Exterior view 建筑外观

"Steel" House

Using 3.2-millimetre-thick corrugated steel plates, a house of monocoque construction which resembles a freight car was made, without having any beams or columns. The client, Professor Hirose, has been a devoted fan of railroads cars since childhood, and stores a few thousand models of trains in his home. He himself had wanted to live a kind of life in a freight-car environment. Fitting in to the L-shaped site, the house looks like a freight car stopping on a slope, curving into an "L" shape.

The basic idea of the architectural structure is to bend the steel plates to gain strength. By bending them, the detail of the bent parts tells us how soft the material of steel is. If the steel plates were used without being bent and the surface were to be painted, we would not be able to recognise that the material used is steel. There would only be the presence of a white abstract plane, the same as plaster boards or concrete. With such abstract detail, communication does not exist between the substance of steel and people. On the other hand, the detail created by bending the steel establishes communication between steel and us.

钢铁住宅

钢铁住宅采用了3.2毫米厚的波纹钢板，没有采用任何梁柱结构，看起来像一辆厢式货车。屋主广濑教授从小就是有轨电车的爱好者，并且在家里收集了上千辆火车模型。他一直想居住在一个类似厢式货车的环境中。钢铁住宅位于一块L形的场地上，看起来就像是停在坡道上的厢式货车一样。

建筑结构的基本设计理念是通过弯曲钢板来获取力量。在使钢板弯曲的过程中，弯折的细节展示了材料的柔软度。如果未经弯折便使用钢板，并将钢板漆上颜色，建筑外观只会呈现出一块白板，和塑料板或是水泥墙没有区别，人们无法辨识建筑使用了钢铁材料，钢铁和人之间也就无法形成一种对话。相反，弯折的钢板让人与钢铁进行了沟通。

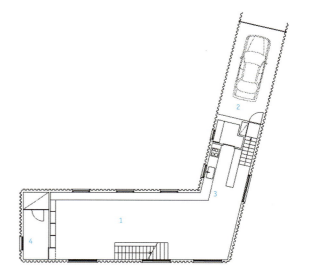

1. living and dining area
2. car parking
3. kitchen
4. wash room

1. 起居和餐饮区
2. 停车场
3. 厨房
4. 洗手间

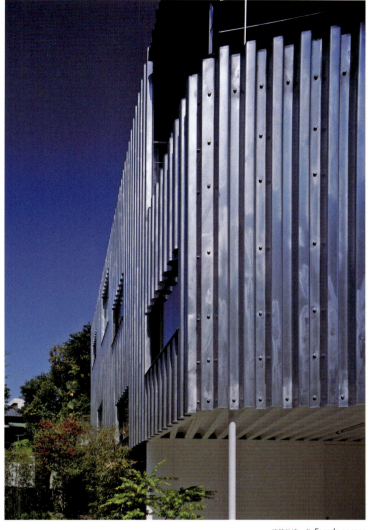

建筑外墙一角 **Façade corner**

底部架高结构 **Stilts**

走廊 **Corridor**

会客区 **Common space**

Photo: Mitsumasa Fujitsuka

Residential

Completion Date: 2007

Architect: Kengo Kuma & Associates

FUJITEC "Big Wing" Headquarters/Product Development Centre

This building was planned for the headquarters to take the hub of the enterprise and for the R & D to make use of the most advanced technology for the products development, in the site along the superhigh speed railway of Tokaido–shinkansen where a rural landscape is stretched around.

In order to strengthen and promote company's growth, the designers aimed to build a space that only showed the company's appeal, but also tried to create an atmosphere in which the employees could work together

The interior of the building is composed of the entrance hall with a working area to the north and a VIP area to the south. The two floors are open and connected, allowing for the working atmosphere to have a sense of continuity and togetherness. The atrium to the east serves multiple purposes besides being a way to move throughout the building, including as a meeting area and a space to take breaks. The atrium's two floors are open, further fostering a sense of workplace cohesion. Additionally, from the bridge that crosses to the lifts of the research tower, one can observe the latest research and development while seeing through the glass the green landscape that the Shinkansen tracks run through.

富士达大翼总部／产品开发中心

该项目旨在为富士达公司设立一个办公总部，并为研发部门提供最先进的产品研发技术支持。项目位于东海道新干线超高速铁路沿线，周边具有浓郁的乡村风情。

为了促进公司的成长，设计师设计的空间不仅满足了公司的日常工作需求，而且营造了一个令员工充满激情和活力的工作环境。

大楼的室内设计以入口大厅为中心，北侧是工作区，南侧是VIP区。两个区域采用开放式布局，并且互相连通，使工作环境具有延续感。东侧的中庭不仅具有通道的作用，而且还包含一个会议区和一个休息区。中庭的两层楼也是开放式的，打造了工作空间的延续性。此外，人们可以在穿过科研楼电梯的天桥上看到最先进的科研成果和新干线与其周边的风景。

General view from west 西侧全景

会议室 Conference room

外墙夜景 Façade, night view

中庭 Atrium

Photo: Yoshiharu Matsumura, Yoshihisa Araki, Fujitec

Architect: Yoshihiro Matsuda, Hideaki Takagi/Showa Sekkei, Inc.

Corporate

Completion Date: 2007

153

Front façade 建筑正面

Front façade 建筑正面

Suntory Museum of Art

A "Japanese-style room" in the city was to be the concept of the museum. A "Japanese-style room" is a comfortable space where people are able to relax on tatamis that are laid on the floor, and among the Japanese traditions, is the most relaxing and relieving environment.

People of the 20th century sought for a museum that was an exaggerated "urban monument", yet people of the 21st century are looking for a peaceful "Japanese-style room". In fact, as the Suntory Museum of Art worked on the theme "Beauty of Life" from an early stage of planning, there is no other museum which is suited to be a "Japanese-style room". This museum should reveal its natural self at the frontier of the global trend.

The "Japanese-style room" building should not be a pretentious bluff. The "Japanese-style room" is constructed by human-friendly materials cherished in our daily lives – for example white porcelain kind to the skin, paulownia which maintains humidity, and white oak used for barrels. A light adjusting device hinted from the design of the traditional Japanese window "Muso-Koshi" was placed on the frontage facing the greenery of the park. This device softens the scenery and light falling into the "Japanese-style room". Japanese people have used these kinds of devices to appreciate the four seasons and the passing time.

三得利美术馆

三得利美术馆的设计理念是打造一个城市中的"和室"。"和室"是日本人休闲放松的空间，里面铺设着榻榻米，是最令人放松的环境。20世纪，人们追求像"城市纪念碑"一样夸张的博物馆；21世纪，人们则更想要一座"和室"一样的博物馆。由于三得利美术馆的主题是"美丽生活"，再没有比它更适合"和室"这一概念的博物馆了。这座博物馆的设计将处于国际潮流前沿。

"和室建筑"不仅仅是建筑师的空想而已。三得利美术馆的设计采用了具有亲和力的材料：白瓷装饰着外墙，桐木保持了室内的湿度，白橡木组成了横梁。美术馆正面的灯光调节装置从日本传统的Muso-Koshi窗中获得灵感，柔化了风景和落在建筑上的光线。

1. 入口
2. 主入口
3. 前厅
4. 美术馆
5. 美术馆（空）
6. 商店
7. 咖啡厅
8. 会员沙龙
9. 大厅
10. 平台
11. 茶室庭院
12. Ryuurei茶室
13. 小茶室
14. 水谷
15. 茶室平台

1. entrance
2. main entrance
3. lobby
4. gallery
5. gallery (void)
6. shop
7. café
8. member's salon
9. hall
10. deck
11. roji
12. ryuurei tearoom
13. small tearoom
14. mizuya
15. tearoom terrace

平台 Terrace

茶室 Tea room

美术馆内部 Gallery

Photo: Mitsumasa Fujitsuka

Cultural

Completion Date: 2007

Architect: Kengo Kuma & Associates + Nikken Sekkei

Façade 建筑外观

Silent Office

The site is located at the side of an elevated highway near Haneda Airport, at the corner defined by the crossing of two roads. The neighbourhood townscape doesn't show any continuity. The architecture is composed of the closed space to catch silence. In a rectangular box is divided by two slits. The interval between those parts leads the air, the wind and the light into the office itself. To ensure an effective internal space, while respecting the legislative outline of the allowed building volume, the ceiling's height of the top floor is higher on the west side than the east one.

Not only to be functional, but aslo to give to the place some complexity, the inside and the outside spaces have been carefully thought out. The parking and the delivery areas are on the west side. The road side provides tree plantations, adding to the surroundings some green space.The ground floor is divided between the warehouse and working space for the personnel. The first floor is devoted to an other part of the warehouse function. The tenant office is on the last and second floor.

静谧办公空间

建筑选址在羽田机场附近的高架公路旁，两条马路的交叉口，周围的城镇景观极其缺乏连贯性。建筑外观为封闭的长方形结构，中央设有开口，便于空气、光线等进入，同时确保室内安静的环境。为满足室内空间功能性，但同时又不违背建筑结构条例规定，设计师特别将顶层屋顶的西侧抬高一些。

另外，为赋予建筑以条理性，室内外空间格局全部经过仔细布置：西侧设置着停车场及配送区；马路一侧栽种植物，增添绿色气息。建筑内部功能分明：一层主要为仓库及员工工作区；二层全部为仓库；三层及四层为租赁办公区。

1. entrance	1. 入口
2. hall	2. 大厅
3. lobby	3. 门厅
4. office	4. 办公区
5. storehouse	5. 贮藏室
6. reception	6. 前台
7. storage	7. 仓库
8. parking	8. 停车场
9. space	9. 空地
10. utility	10. 机械房
11. balcony	11. 阳台
12. terrace	12. 平台

平台 Terrace

窗边空地 Void

办公室 Office room

Photo: Takashi Yamaguchi & Associates

Architect: Takashi Yamaguchi & Associates

Completion Date: 2008

Coporate

External 建筑外观

Uwajima City Hospital

The Uwajima Municipal Hospital had opened in 1910 which has played the core of hospital in the outskirts of the district. It was reconstructed toward the new era of 100 years from now on. The total area of this project is 19,604.79 square metres, with a 11,233.09-square-metre building area. The roof slab occupies 50.38metres. The highest height is 50.98metres. The ground floor is in the underground, the tenth floor on the ground. The materials are PcaPC, steel structure and quake-absorbing structure. There are 435 beds in total.

The outward appearance is designed for representing the blue sky in Uwajima with lightening the size of the structure with a basic tone of a vertical stripe. And the interior appearance is designed rather for the well-ordered and calm hospitals. The colouring of the hospital is composed of being approachable to the regional people.

宇和岛市市立医院

宇和岛市市立医院坐落在市郊，建于1910年，在其成立100年之际，决定加以修复。医院占地面积达19604.79平方米，其中建筑面积为11233.09平方米。建筑屋顶长达50.38米，最高部分为50.98米，一层位于地下，其余九层位于地上。医院共有435个床位。

建筑外观主要突出宇和岛市湛蓝的天空，垂直的条带结构在视觉上营造了轻盈感。室内则以营造秩序性和恬淡氛围为主，在色调选择上多从病人能够接受的角度出发。

1. lecture hall	1. 演讲厅
2. reception	2. 前台
3. office	3. 办公室
4. dialysis	4. 透析室
5. physiological examination	5. 生理检查室
6. specimen lab	6. 样品实验室
7. medical examination	7. 医药检查室
8. central treatment	8. 中心处置室

建筑正面夜景 Night view of the front

建筑正面 Front view

建筑侧面夜景 Side view at night

康复花园 Rehabilitation garden

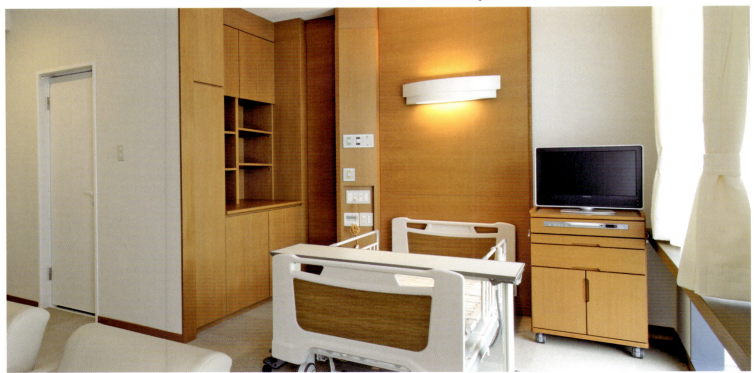
特殊私人病房 Special private room

Bird's eye view 鸟瞰图

Namics Techno Core

Namics is a company that manufactures special paste for semiconductor boards. In the year of its 60th anniversary of founding, this innovative company has decided to rebuild its research centre. Riken Yamamamoto & Field Shop were invited to a design competition, and proposed an iconic, mushroom–shaped structure which appears to float in midair. Namics is a vigorous company which tries to constantly re–establish its corporate image by endeavouring in new projects. As Namics grew rapidly into a multinational corporation, its management sought a building which appropriately symbolises the company's momentum. Another design requirement was to "open up" the building to numerous visiting clients, and the interpretation of this requirement was that the building had to be legible and recognisable as possible.

They proposed an articulated three–layer structure with the ground floor and first floor pulled apart. The ground floor accommodates the research facilities, the middle first floor a floating garden, and the first floor a cafeteria and hoteling office. The research facilities on the first floor are arranged on a grid plan as requested by the client. The middle second floor, or the ground floor roof, is landscaped with a space for exhaust outlets coming from the research rooms and laboratories, which provide a relaxation area for the staff. On top of that there is the second floor mushroom structure. This building is sure to establish Namics' new image.

Namics技术中心

Namics公司主要生产半导体板专用胶。在其成立60周年之际，这个不断创新的公司决定重建其技术中心。随着公司的不断发展，他们一直致力于打造一个全新的建筑，以彰显公司的形象。因此，山本理显建筑事务所受邀为Namics公司设计了一个标志性的蘑菇结构建筑，仿佛悬浮于空中一样。此外，为满足公司面对客户开放的需求，整个建筑必须在视觉上清晰简约。

建筑师设计了一个三个层次的结构，一层和二层相分离。一层用于安放研发设备，二层包括餐厅和酒店式办公区，两层之间是一个悬浮花园。一层设备根据用户的需要呈格子形状排列。两层之间的空间（或者说是一层的屋顶）栽种着景观植物，也设置着实验室的排水口，为员工提供了一个休闲空间。最上面是二层的蘑菇结构。整个建筑体现了Namics的新品牌形象。

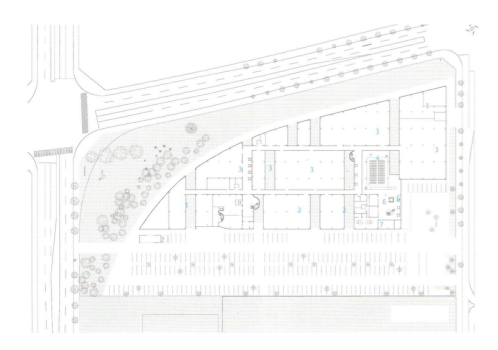

1. 垃圾回收区
2. 机械室
3. 实验室
4. 多功能广场
5. 前台
6. 访客等候室
7. 综合办公室

1. utilities room
2. waste and recycling area
3. laboratory
4. multi-function plaza
5. reception area
6. visitors room
7. general affairs office

建筑外观 External view

建筑外观 External view

综合办公室 General affairs office

Photo: Koichi Satake

Completion Date: 2008

Architect: Riken Yamamoto & Field Shop

Front view 建筑正面

Flower Shop H

H花店

The designers tried to make small rooms without the presence of ceiling, which may make people feel as if they were in exterior space even though they are in interior space. Since it has a limit of height of the building, the designers divided a building into several small volumes. The smaller the footprint of each building is, the higher the impression of the ceiling is.

The context of the site consists of both the city and the park. Since the city and the park are completely different in various points, the designers were in two minds about choosing which context they take. If they design a pavilion, that does not fit in with the city. If they design just a building, that does not fit to the park. They did not want either of them. Thus, the designers decided to accept both of them. The division of the building made the proportion of each building vertical, which made the project resemble to the building around it. The designers love such way of respecting existing landscape. The project does not only copy the man-made context, but also respect the spaces like trees.

设计师试图在这个小空间内不设置天花板，让人们即使在室内也有置身于室外的感觉。由于建筑的高度受限，设计师将建筑分成若干个小模块，每个模块的占地面积越小，天花板所营造的效果就越明显。

项目所在地既是城市又是公园。由于城市和公园是两个完全不同的概念，设计师很难决定建筑的周围环境设计。凉亭与城市不符，单个的建筑又与公园不符。于是，设计师将它们合二为一，设计了多个建筑式凉亭。这些凉亭的结构与周边的建筑相似，完全没有影响周边的环境。项目不仅复制了周边的人造环境，而且尊重了树木等植物的生长空间。

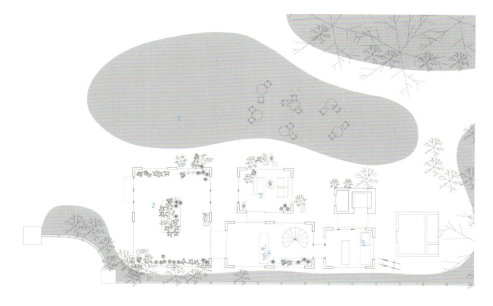

1. 平台
2. 售货区
3. 售货区
4. 收银台
5. 后院

1. terrace
2. retail space
3. retail space
4. cashier
5. backyard

前厅 Lobby

入口 Entrance

入口 Entrance

Photo: Daichi Ano

Commercial

Completion Date: 2005

Architect: Kumiko Inui

Daytime view 日间景色

House in Yugawara

The south part of the slope, where used to be a tangerine farm, is the site. Flexible soil was found after the research and the pile work was necessary. The wooden bungalow was very light and the architects were able to have smaller number of stakes by making the groundwork positioning consolidate and distributed by blocks. With the post on the balanced groundwork, they made the box-shaped walls and the floor and roof.

The box, created as one of the groundwork, presents the softness in the continuous one-room spacing. Besides, the slope presents the multilevel floor, which gives the area some separation by the level. People who would live there can choose their own "space" as they like.

At the TEPCO comfortable Housing Contest, the prize was given especially for the understanding of the construction and consistency of the flatness plan. The project is particularly designed to fit the strange site of the area. The low cost of the construction of the project is also part of the consideration for the winning in this contest.

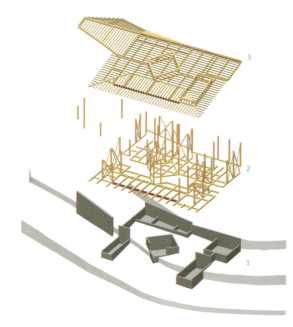

1. base
2. in-between column structure
3. roof

1. 底座
2. 中间柱结构
3. 屋顶

汤河源别墅

别墅坐落在一块坡地上，原址是一个橘子园。由于土质十分柔软，施工之前必须进行打桩工作。建筑师利用分散式木桩加固了地基。墙壁、地板和屋顶都采用了盒结构。

盒形的建筑主体与地基融为一体，体现了一室式空间的延续性和柔和性。坡地地势显示了多层楼面设计，将空间分割开来。别墅的居住者可以根据自己的需求选择适合的空间。

在TEPCO（东京电力）舒适住宅竞赛中，汤河源别墅以其对建筑构造的独特见解和平面布局的相容性获得了优胜。项目的设计完全符合当地奇特的地形。项目的低成本建设也是其获得优胜的主要原因之一。

建筑侧面 Side view

平台 Terrace

楼梯 Stairs

Photo: Hiroshi Ueda

Completion Date: 2006

Architect: Kazuhiko Kishimoto / acaa

Front façade 建筑正面

Night view 夜景

Namba Hips

A client who has developed amusement businesses obtained a plot of a prime location fronting Midosuji, a main road in Osaka. The client planned to build a complex mainly containing amusement facilities and restaurants, and required many architects to submit their ideas in a long term. As the result of the comparison, he finally designated Shin Takamatsu. The idea which made the client decide was so clear. The lucid idea led a consistent architecture from the beginning to the completion. It is a separation from meanings. Takamatsu says without fearing the misunderstanding, this form of the architecture was developed as the theme which thoroughly hates the meaning. The architecture was designed by pursuing the form which ignores architectural urban scale and refered to nothing. The architect was sure that the only existence of unregistered glory can be the symbolic architecture in the city where there is super saturation with registered glory.

南波HIPS大厦

一位娱乐产业开发商获得了一块朝向大阪主要街道——御堂筋的地皮,他计划建造一座集餐饮娱乐于一身的大型娱乐设施,并邀请了多位建筑师参与设计方案的竞标。最终,高松伸获得了该项目的设计权。由于委托人的要求十分简明,建筑的设计过程十分顺利。高松伸所设计的建筑造型完全体现了它的功能,让人一目了然。建筑在城市建设中独树一帜,别出心裁。建筑师认为只有别具一格的造型元素才能在充斥着各种各样建筑的城市中脱颖而出。

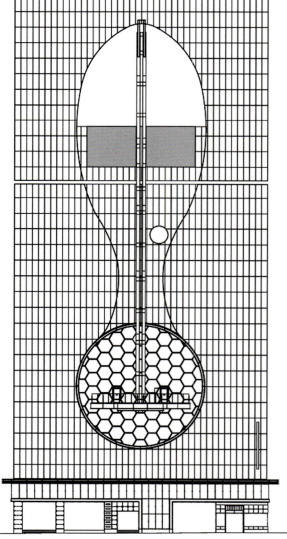

West elevation 西立面

室内 Interior

室内 Interior

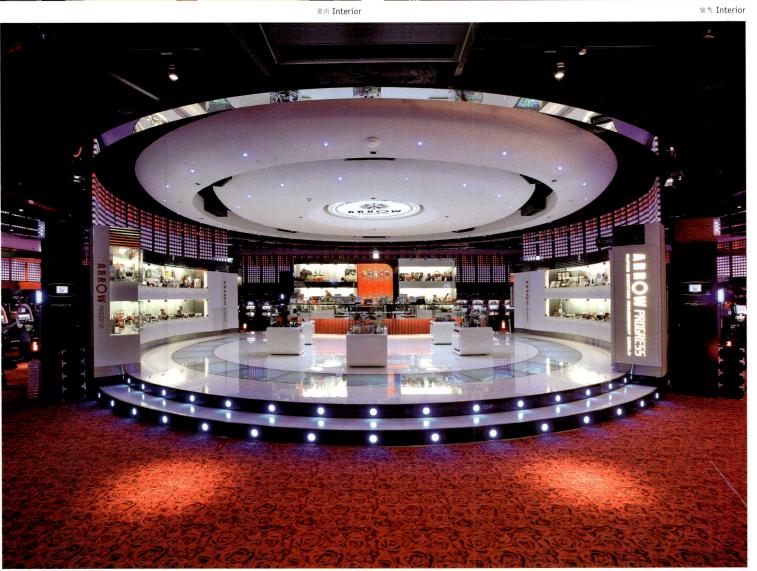

室内 Interior

Architect: Shin Takamatsu +Shin Takamatsu Architect and Associates Co., Ltd.

Photo: No credit

Completion Date: 2007

Recreational

Front view 建筑正面

Miyasaka Residence

The Miyasaka residence bridges the aesthetics of two disparate cultures (Western and Japanese), emblematic of neither, yet summoning echoes of both. Built to accommodate the busy lifestyle of the president of a major commercial building contractor, as well as providing a place of repose for his elderly parents; the house was conceived to appear as a radiant jewel in the midst of an urban garden oasis. From the beginning, the client expressed his wish for a "Western" style home, as well as his parents' desire for Japanese tradition.

The architect brought an approach to design that honours the use of natural materials. The architect's efforts to bring morning sunlight into the master suite and kitchen area, afternoon sunlight into the parent's suite, while preserving the existing landscape ultimately provided a genesis for the plan. Elevating the first storey allowed for the insertion of a continuous band of clerestory glazing between it and the canopy, thereby introducing an incredible amount of natural light to penetrate throughout the interior even during the heavy snow buildup of winter. This resulted in the entire first storey being held proud of the ground, providing the sensation of it appearing to float above its base, like a bird frozen in mid–flight.

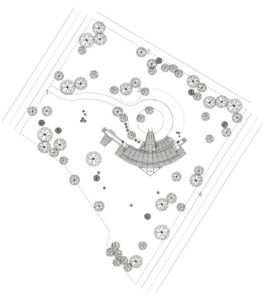

宫坂住宅

宫坂住宅在两种风格迥异的文化——西方文化和日本文化之间搭建了桥梁，引起了二者的共鸣。住宅是为一位工作繁忙的建筑承包商和他年迈的父母所建造的，在城市花园绿洲的中心，宛如一颗璀璨的宝石。委托人想要一座西式住宅，而他的父母却崇尚日本传统。

Site plan	总平面图
1. entre gate	1. 入口大门
2. alley	2. 小巷
3. future carport	3. 未来的停车场
4. street	4. 街道

在设计过程中，建筑师大量使用了自然材料。建筑师努力将早晨的阳光引入了主卧室和厨房，将午后的阳光引入了父母的卧室，同时保留了场地原有的景观。二楼和穹顶之间嵌入了一个连续的侧天窗环形带，即使在冬天也能为室内带来充足的阳光。这让二楼比一楼有更多的优势，使它看起来像悬浮在空中，仿佛一只被冻结在空中的小鸟一样。

Japan

Asia

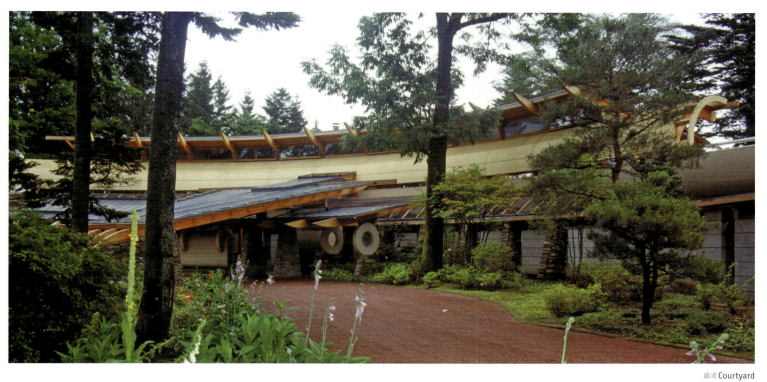

庭先 Courtyard

庭院 Courtyard

卧室 Bedroom

Façade 建筑外观

Akasaka Phoenix

The location of the site is around fifty metres far back from the heavily-trafficked main street, where we get the impression of quiet in a high-density city. The east side of the site faces "city air pocket" where is the open space of the complex facility, and it has a spacious atmosphere. For that reason, there is no concern about eye gaze from neighbours, and the internal space can get enough sunlight and natural ventilation. Since the site level is about six metres up from the road in front, the building has a high level of visibility. The designers have reinforced identity of the building by giving strong design impact on the east side, and tried to differentiate it from surrounding buildings.

It can be said that one of the characteristic design aspects of this building is louvres of the façade. Five louvres are made into a unit and each unit controls outside environment such as direct sunlight, eye gaze from outside and so on. In order to create comfortable indoor condition, the units consist of eight different patterns and each unit is arranged according to the function. For instance, upward horizontal louvre is used abundantly for shielding of direct sunlight.

Moreover, the louvres show various expressions according to the angle and direction of attachment. These variations give motion to the façade and create deep shade which gives dignity to a building. The louvres are coloured in vivid royal blue and a white and their expression change with the positions of eyes. The façade of Akasaka Phoenix is rich in diversity and creates a new city landscape.

赤坂凤凰楼

建筑地点距离交通繁忙的街道50米，远离喧嚣的都市。场地的东侧是一片开阔的"城市气囊"————一片空地。因此，建筑既不会担心隐私被窥视，又可以得到充分的阳光和良好的通风。由于建筑场地离街面有6米高，建筑的视角相当高。设计师在建筑的东面加入了极具个性的元素，使其从周围的建筑中脱颖而出。

建筑设计最显著的特点就是外墙的百叶窗结构。五扇百叶窗形成一个单元，每个单元都能够控制阳光直射、保护内部隐私。为了营造舒适的室内环境，整幢大楼根据不同功能需求采用了八种不同的外墙单元结构。例如，上升式百叶窗被用于屏蔽阳光直射。

不同位置和角度的百叶窗形成了多样化的视觉效果，使外墙看起来丰富多彩，为建筑树立了个性。百叶窗被漆成活泼的宝蓝色和白色，随着视觉角度的变化，它们的颜色也不尽相同。赤坂凤凰楼的外墙以其独具特色的视觉效果为城市添加了新景观。

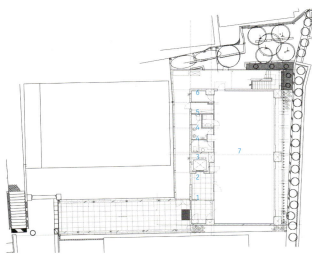

1. 入口
2. 电梯大厅
3. 电梯
4. 洗手间
5. 员工厨房
6. 仓库
7. 办公室

1. entrance
2. lift hall
3. lift
4. WC
5. staff kitchen
6. storage
7. office

建筑细部 Details

不同角度看建筑 DiFerent view

窗户 Windows

建筑内部 Interior

Façade 建筑外观

Rooftecture O-K

The Rooftecture projects, eight buildings designed by Osaka-based architect Shuhei Endo for the Japan-based used car dealership, O-RUSH, are whimsical structures with sweeping cantilevered roofs and continuous suspended steel sheets that minimise interior walls, leaving room to show off the glimmering cars inside.

In a similar use of suspended steel, Endo utilised a one-storey structural steel frame from an existing building and covered it with a new roof for the Kyoto O-RUSH building, Rooftecture O-K. Contact points within the building's frame stabilise the continuous suspended steel sheet roof that begins at ground level on one end, and waves upward to cover the two-storey showroom. The thickness of the steel sheets provides self-weighted stress, opposing gravity. The glazed façade opens the 2,000-plus-square-foot interior to natural daylighting.

The one-storey difference in elevation from the maintenance space to the two-storey showroom of the building is united with a continuous visual context under the sinuous metallic roof.

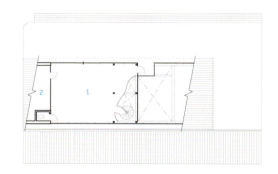

屋顶结构建筑O-K

屋顶结构系列建筑是大阪建筑师远藤修平为O-RUSH二手汽车代理公司所设计的八座建筑。这些建筑以异想天开的大范围悬臂屋顶和连续的悬浮式钢板为特征，里面是明亮的汽车展示空间。

在京都分公司的屋顶结构建筑O-K的设计中，远藤修平利用了与悬浮式钢板相似的，原有建筑遗留下来的单层钢结构作为基座，并在上面添加了一个新屋顶。建筑框架上的节点固定住了连续的悬浮式钢板屋顶。屋顶从地面呈坡形上升，盖住了二层楼高的展厅。钢板的厚度提供了与重力相对的自重应力。玻璃外墙为185平方米的室内空间提供了自然采光。

维护保养空间和展厅之间一层楼高的高度差被蜿蜒而连续的屋顶完美地结合了起来，视觉上十分统一。

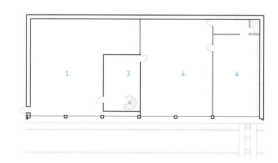

1. 展厅
2. 办公室
3. 会议中心
4. 工作间

1. exhibition space
2. office
3. meeting space
4. workshop

全景图 General view

工作室 Workshop

前台 Reception

Looking towards the elevation facing the scramble junction 从繁忙的交叉路口看建筑

The atrium becomes increasingly transparent at night as it is lit up 夜晚的灯光让建筑看上去更通透

EX

The site of EX faces the front of a four-way crosswalk with heavy pedestrian traffic. The goal of the project was to design a core facility that could change the scenery of the Chiba station area that is suffering from economic decay. The solution to this problem was to design a building with a five-storey glass atrium in the front so the various activities inside would be seen from the outside. Then by having an atrium full of movement, the architects aimed to let the context and the atrium have a stronger connection to each other.

The main circulation for the building is gathered at the front atrium facing the existing shopping district. The atrium then is made with a transparent curtain wall so it visually opens up to the surrounding district. Besides, the escalator facing the atrium is cantilevered to make the space become open. The building's exterior is then clad with metal panels that vary in width to give the building a rhythm and avoids the monotonous surface.

The name of the building is "EX" which is a prefix meaning "out of". This is to emphasise on the designer's design concept: wide range of information and activities are sent out from this building, making the whole community a more attractive place.

EX商场

EX商场朝向一个交通繁忙的十字路口，项目的主要目标是设计一个能够改变千叶市车站区萧条的经济状况的商业设施。商场正面是一座五层高的透明中庭，因此，外面过往的行人都可以看到商场内部的活动，外部环境和中庭被更紧密地联系在了一起。

建筑的主楼梯全部设在面朝商业区的中庭里。中庭透明玻璃幕墙从视觉上让商场向周边的区域开放。此外，朝向中庭的自动扶梯采用了悬臂结构，让空间看起来更开阔。建筑的外部包裹着不同宽度的金属板，形成了一种韵律感，避免了建筑表面过于单一。

建筑被命名为"EX"，在英文中意为"向外"，强调了设计师的设计理念，即源源不断的信息和活动从这座建筑向外涌出，为社区增添了诱人的活力。

外部细节 Close-up of the exterior　外墙上的印花使建筑更具活力 The organic print on the building surface adds playfulness to the architecture

Photo: Kohichi Torimura

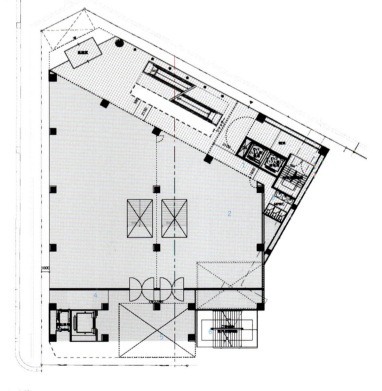

View from the main ground level entrance looking at the cantilevered floors

从一楼看上方的悬臂楼层

1. 电梯
2. 出租商铺
3. 洗手间
4. 中心监控室
5. 垃圾回收区
6. 逃生楼梯

1. lift
2. tenant
3. WC
4. central monitor room
5. garbage place
6. escape staircase

Completion Date: 2008

Architect: SATOH Hirotaka Architects

Façade 建筑外观

Cat Building

This building stands in front of a street called the Cat Street. The Cat Street is a very narrow pedestrian walk, and under it a river lies which is made to a culvert. There are always many people walking around the shops and boutiques along the street. In front of the site, there are tall buildings standing along a major arterial road, and behind the site, there lies a hill with single-family houses at high density. This place is in a neighbourhood where urban city structure changes to a familiar human scale. The site is located at the corner of the Cat Street and can be seen in many angles from the street. This building is located on such an area, and is a commercial building mainly purposed for shops.

In this irregular shaped site, varieties of form regulations determine the outline shape of the building. Outside form of the building is designed as to visualise these regulation lines. It is rather clumsy and violent in shape, but like the innocent form of architecture standing in the dense city of Tokyo. Based on this form of the building, it became a theme to search for a tenants' space, in which co-responds to the dynamism of the dense city and the various aspects of activities inside.

猫大厦

建筑位于一条叫猫街的街道前。猫街是一条非常狭窄的步行街，街下面是涵洞和小河，街道上的商店和精品店吸引了众多的行人。建筑场地前面是沿着主干道的高楼，而后面则是一座布满独栋别墅的小山。这里充满了城市的人文关怀。猫大厦就位于猫街的街角，从各个角度都可以看到。建筑的主要功能就是作为商业建筑。

造型不规则的场地决定了建筑的造型，建筑外部的线条充分反映了这些不规则的线条。建筑的造型十分粗犷，在高密度的东京市区里显得异常纯真而有趣。多变的造型反映了建筑内部租用空间的多样性和城市本身的动态变化。

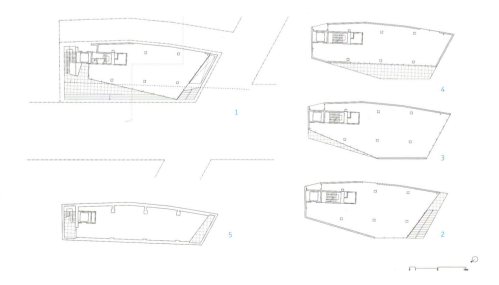

1. 1楼平面图
2. 2楼平面图
3. 3楼平面图
4. 4楼平面图
5. 地下一层平面图

1. 1F plan
2. 2F plan
3. 3F plan
4. 4F plan
5. B1 plan

建築内部 Interior

入口 Entrance

楼梯 Stairs

建築側面 Side view

建築細部 Details

Photo: Koichi Torimura

Completion Date: 2006

Architect: Taketo Shimohigoshi/A.A.E.

Iidabashi Subway Station

The significance of "INDUCTION DESIGN / ALGOrithmicDESIGN" project lies in the search for better solutions to given conditions. There were two types of conditions that "WEB FRAME" programme had to solve.
1. Restrictions on space & Conditions imposed by each component.
2. The intended volume and density of the space.

The first of them was an absolute condition allowing no margin for improvisation. The second condition – spatial requirements – became more flexible parameter. The issues here are different from those of conventional space frames. Simply because the degree of freedom is great, divergences can occur and lead in unpredictable directions.

Freedom can, of course, readily slip over into chaos. An important element of this concept is to give the appearance of chaos while in fact obeying certain regularities.

The coexistence of freedom and harmony! This sounds like a catchphrase put forth at some kind of meeting by people fully aware that such a thing will never come about in reality. This is not an empty slogan. Here we are (just) beginning to see signs that it can be realised.

饭田桥地铁站

纳设计/算法设计"的重点在于在既定条件下寻找更好的解决方法。"网架结构"必须解决以下两个问题：
1. 空间的限制和各成分施加的条件；
2. 预计体量和空间密度。
第一个问题是绝对条件，不容改变；第二个的空间要求则更加灵活。这一问题与传统空间框架不同，因为自主程度更大，会产生更多的分歧和不可预测的发展方向。 自由可能造成混乱。但是设计理念中的重要元素之一就是在混乱中遵循一定的规律。

自由和协调共存！这听起来像会议中的空洞的口号，永远不会实现。但是这绝不是一个空洞的标语，设计师正在项目中逐步将它实现。

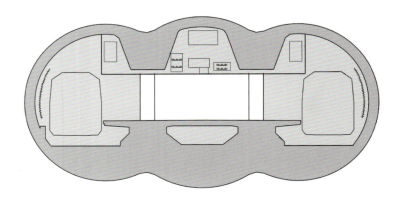

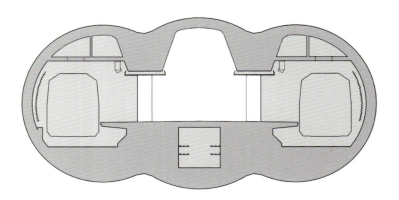

Exterior 地铁外观

Platform 月台

网状结构灯 The light of network structure

通道 Passageway

入口 Entrance

Photo: Makoto Sei Watanabe

Architect: Makoto Sei Watanabe, Architects' Office

View of the clubhouse from the golf course 从高尔夫球场看俱乐部

Unimas Golf Resort Club

The Unimas Golf Resort Club house is part of the unimas master plan in Kota Samarahan, Kuching. The club house and golf course is the centre piece and communal heartland of a university township planned on the successful development of Universiti Malaysia Sarawak.

The facilities of the club house are located on three floors – with the guest arrival and main entrance at upper ground level. The second entrance point is for buggies located at the lower ground level. The upper ground floor houses the banquet hall with a capacity of fifty tables, i.e. 500 persons. Ancillary spaces to the banquet hall are provided in the form of VIP rooms and pre function spaces. The lobby and terrace that wrap around the building serves as both a viewing gallery and dining terrace. The kitchen and its supporting facilities are also located in this floor for easier operation and efficiency of use. The "open–plan" nature of the design means that all the facilities are visually linked with the exception of the services and amenities.

In order to liberate the upper floor plane for the members' viewing and f & b needs, the water courtyard and the amenities are located in the lower ground floor – the changing rooms and toilets are located here for proximity to the swimming pool level and the buggy parks. The second entrance is located at this level, with sixty buggy parks under cover nearby and other twenty-eight buggy parks outdoors. In keeping with the effort to maintain an open plan and maximum view advantage on the main floor, the roomed facilities such as offices, library and conference facilities are located in a mezzanine floor partially housed within the building's roof space.

Unimas高尔夫俱乐部

1. tennis court (future extension)
2. future extension
3. swimming pool
4. interior

1. 网球场（未建成）
2. 未建工程
3. 游泳池
4. 内部

Unimas高尔夫度假俱乐部是大学城总体规划工程的一部分，俱乐部和高尔夫球场作为中心结构。

俱乐部共为三层。主入口位于一层，客人可从这里进入。小型车辆出入口位于地下室。一层设有宴会厅，内共有50张餐桌，可容纳500人同时就餐；大厅及露台围绕在空间四周，便于欣赏风景；厨房同样设在一层，方便使用。开放式的格局设计使得所有区域在视觉上自然连通。为满足客人的视觉及就餐需求，他们便将水岸庭院设置在地下室。此外，这里还包括更衣室、卫生间等，为游泳的客人提供便利。室内小型停车场可容纳60辆车，而室外另有28个车位。为确保空间的开放式格局以及一层的景象不被打扰，办公区、图书室、会议室便设在这一层。

建筑在阳光下产生的光影效果 The new clubhouse speaks of lightness and shade   色调深沉、线条简洁的现代热带风格 Modern tropical style in a sombre palette and tidy lines

游泳池 Resort—swimming pool

Photo: Design Network Architects Sdn Bhd.

Recreational

Completion Date: 2009

Architect: Design Network Architects Sdn Bhd.

Façade 建筑外观

Yeoman's Bungalow

The house is entered from main staircase – an open structure where the timber-lined cantilevered flights of stairs double as a sculptural lookout. At the top of the stairs, one crosses the threshold into the first pavilion that is the conversation area, the kitchen and a dining room that overlooks at lap pool. From here, one crosses a water-court into the second pavilion further up the slope; which is the main living room with a large viewing deck. The spaces in these two pavilions are visually linked as one looks towards the lap pool.

Nestled uppermost in the hillside is the bedroom pavilion which contains the master bedroom suite, guest bedrooms and study. This longish block is flanked by a veranda, which can be turned into an informal living space when the guests slide open the glazed doors to the bedrooms. From this veranda and down a flight of stairs, one comes full-circle and arrives back at the first pavilion.

Ultimately, this is a house to be experienced – by standing in her rooms, by walking through her corridors and looking through her windows, because the architectural language is deliberately uncomplicated and its form is simple, acting merely as a foil to the more commanding presence of Mount Santubong.

耶曼别墅

别墅的门口有一个悬臂结构的开放式楼梯，楼梯也可以作为瞭望台。走上楼梯，踏入门内，便是会客厅、厨房和游泳池上方的餐厅。从这第一个亭馆穿过一个水庭，便到了坡上的第二个亭馆——带有巨大观景台的主起居室。朝向泳池看去，这两个亭馆是连接在一起的。

别墅最高的部分是卧室区，里面包含主卧套房、客房和书房。细长的区域四周环绕着游廊，当客人打开卧室的玻璃拉门，游廊可以作为一个非正式的会客空间。从游廊和楼梯可以重新回到第一个亭馆。

这是一座必须亲身体验的别墅——人们通过站在房间里，穿过走廊，从窗口看出去来慢慢体会别墅的舒适与简约。别墅宛如山都望山上的一件装饰品。

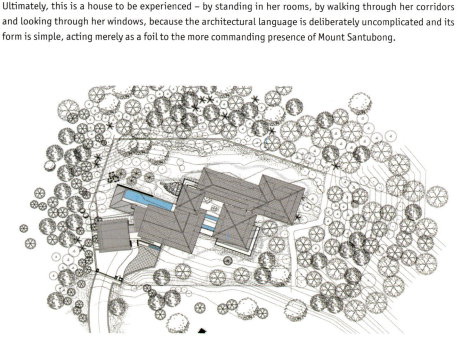

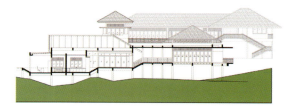

1. 景观
2. 游泳池
3. 建筑

1. landscape
2. swimming pool
3. building

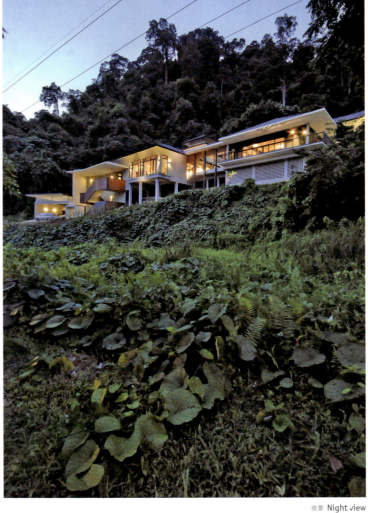

从远处看室内 Savour the vistas

夜景 Night view

楼梯和庭院 Stairs and countyard

室内 Interior

Photo: Design Network Architects Sdn Bhd

Architect: Design Network Architects Sdn Bhd

Residential

Completion Date: 2009

Front view with pool 建筑正面的游泳池

Enclosed Open House

The owners wanted a spacious, contemporary house that would be as open as possible but without compromising security and privacy at the same time. Surrounded by neighbours on four sides, the solution was a fully fenced compound with a spatial programme that internalised spaces such as pools and gardens, which are normally regarded as external to the envelope of the house. By zoning spaces such as the bedrooms and servants' quarters on alternative levels, i.e. upper storey and basement levels, the ground plane was freed from walls that would have been required if public and private programmes were interlaced on the same plane. The see–through volumes allow a continuous, uninterrupted forty–metre view, from the entrance foyer and pool, through the formal living area to the internal garden courtyard and formal dining area in the second volume. All these spaces are perceived to be within the built enclosure of the house.

The environmental transparencies at ground level and between courtyards are important in passively cooling the house. All the courtyards have different material finishes and therefore different heat gain and latency (water, grass, granite). As long as there are temperature differences between courtyards, the living, dining and pool house become conduits for breezes that move in between the courtyards, very much like how land and sea breezes are generated.

闭合的开放式住宅

业主想要一个宽敞的现代住宅，既开放开敞又要能保证足够的隐私和安全。住宅的四周都是建造了各式各样的防护区，如泳池和花园，将住宅和外部的空间隔绝开来。卧室和佣人房被设在了二楼和地下室，一楼没有墙壁，私人和公共空间得以被融合在一起。从入口的门厅和泳池，穿过起居室、中心花园和餐厅，一楼的无障碍景观一直延续了近40米。而这些开放的空间都被包围在闭合的住宅里。

一楼的开敞环境和中间的连接庭院对住宅的被动式通风十分重要。各个庭院所采用的材料、装饰不同，因此它们的热摄入量和影响气候的因素（水、草、岩石）也不尽相同。只要各个庭院之间存在温度差，起居室、餐厅、泳池之间便会产生流动的风，就像陆地与海洋之间风的形成一样。

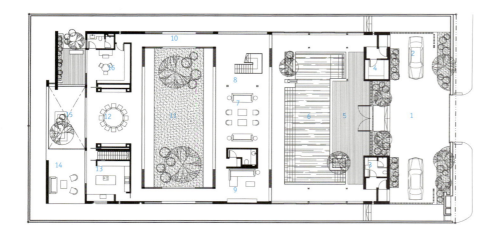

1. driveway
2. charporch
3. changing room
4. store
5. foyer
6. swimming pool
7. living
8. powder room
9. TV area
10. linkway
11. counrtyard
12. dining
13. kitchen
14. outdoor terrace
15. void to basement
16. study room

1. 车道
2. 门廊
3. 更衣室
4. 仓库
5. 门厅
6. 游泳池
7. 客厅
8. 洗手间
9. 电视区
10. 走廊
11. 庭院
12. 餐厅
13. 厨房
14. 户外平台
15. 地下室通道
16. 书房

全景图 General view

入口 Entrance

客厅 Living room

Photo: Albert Lim

Completion Date: 2009

Architect: Wallflower Architecture + Design

Profile 建筑外形

Studio M Hotel

A mere stone's throw away from the Singapore River, Studio M Hotel is in the heart of entertainment districts. Given its high visibility and the accompanying social vibrance of the site, the key idea was to design a trendy and memorable landmark worthy of gracing the historical river next to which it resides. The result is a building façade shaped like a boat sail, with the elevated deck being likened to a cruise deck. These features collectively paint an image of a ship moored by the riverside.

Rooms were designed as live-work-play spaces that cater to the needs of urban travellers. Conceptualised as a "box within a box", each 15 m2 room is compact yet luxurious, with double volume space for added depth and fully functional bathroom "pods" that can be easily reconfigured to create a diversity of room types. A staircase leads up to the furniture deck that either houses a bed or workstation, giving business travellers the added convenience of computer facilities during their stay.

Studio M Hotel's Urban Connector acts as an open space that links the hotel with outdoor areas like the landscape deck and Singapore River. The design concept of the park was to compliment the hotel and its public spaces, with lush greenery acting to buffer the stoic landscape while also reducing heat. Provision for adequate seating also encourages guests to relax in the park while breathing in the beautiful scenery.

M工作室酒店

作为新加坡河河畔的一块界石，M工作室酒店是娱乐区域的中心。由于建筑场地高度的可见性和社会活跃性，设计的重点是建造一个时尚而令人难忘的地标，以使河畔的风景更加优美。建筑的外墙像船帆一样，架高的平台犹如一片甲板。整个建筑看起来像是停泊在河边的一艘船。

酒店的客房为都市旅行者提供了生活、工作和娱乐的空间。客房采用了"盒中盒"的概念，每个15平方米的房间都紧凑而奢华，采用双层空间，里面功能齐全的浴室可以轻易地改变造型，便于营造多样化的房间类型。房间内部的楼梯通往上方的平台。平台上或是床，或是工作台，方便旅行者使用电脑设备。

M工作室酒店公园将酒店和户外区域（如景观平台和新加坡河）连接起来。公园的设计理念是以茂密的绿色植物装点酒店和其周边的公共区域，同时起到降温的作用。充足的座椅让客人可以尽情享受新鲜空气和优美的风景。

1. 入口
2. 前台
3. 门厅
4. 走廊
5. 庭院
6. 休息大厅

1. entrance
2. reception area
3. lobby
4. corridor
5. courtyard
6. lounge

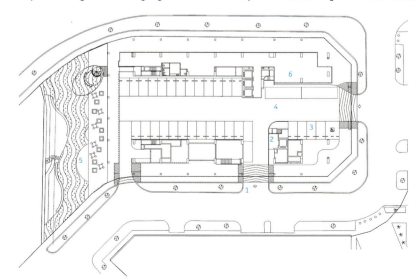

建筑正面 Front

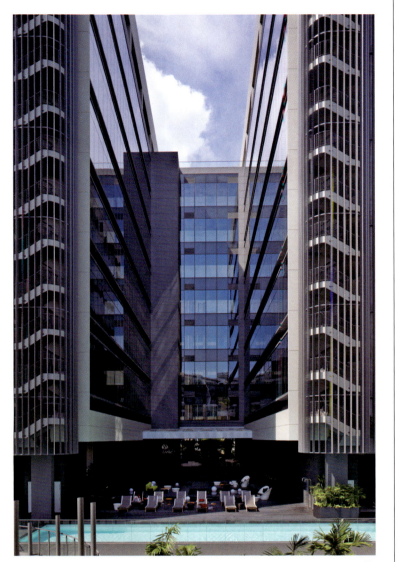

建筑外形 Profile

建筑细部 Detail

卧室 Bedroom

Photo: Derek Swalwell

Completion Date: 2010

Architect: ONG&ONG Pte Ltd.

187

Multipurpose Building For Auroville Papers

This building houses a multipurpose hall and services for 60 workers (kitchen, laundry, lockers, bathrooms etc.) of the existing factory. The services are contained within closed rooms and the hall is left completely open on two sides for maximum cross ventilation and feeling of space at congregations. The hall is also marked by a curvilinear wall on one side and the linear block of services on the other. The roof is lifted from the building to provide a natural draft of air circulation in the room. A small bas–relief design on the ceiling animates the space otherwise left bare. Local natural smooth stones in shades of grey, yellow ochre and green are used in a pattern for the flooring of the main hall. Rough stone is used in bathrooms and the other service areas.

A natural root zone waste water recycling system treats the grey water from the toilets and the water is re–used for irrigation needs of the whole site. Leftover blue metal aggregate of the site with locally found boulders in conjunction with local species of the plants dot the green area around the building. Textured exterior wall paint in shades of orange adds to the tropical context.

奥罗维尔纸业综合楼

这一建筑内包括多功能大厅及员工服务中心区（厨房、洗衣房、储物间、浴室等）。其中，服务中心区全部设置在封闭空间内，而多功能厅则两侧开敞，增添通透性及空间感。此外，大厅内一侧蜿蜒曲折的墙壁以及略微抬高的天花（便于空气流通）特色十足。天花上的浅浮雕装饰带来动态气息，地面采用灰色、黄色以及绿色的天然光滑石子铺设。浴室及其他服务区地面则采用粗糙的石块打造。

生活污水经专门的废水处理系统净化之后用于建筑周围空间的灌溉用水。建筑周围的绿化带采用剩余的施工材质、废弃的石块以及当地植物装饰，橘色的外墙与本地的热带环境相得益彰。

Photo: Pino Marchese

Complex

Completion Date: 2006

Architect: Sherilcastelino, C&M Architects

Entrance Canopy 入口的华盖

Triose Food Court

Angled spaces projected towards different directions encapsulated in an organically folded concrete skin create a two-level building that houses a few retail shops, a food court, two restaurants, a large bar and an entertainment gaming area.

The entire frontage of the site along the main road overlooks large trees and a riverbed and hills. The axis of the building changes constantly from one side to the other allowing each space within to look out towards different views of the surrounding landscapes.

The concrete folded skin that forms most of the building creates large open frames towards the external views and the plans of the building also open out towards these large frames accentuating the beautiful natural surroundings to the inner spaces. A natural slope in the site towards the rear allowed an entire parking level to be created to facilitate the high traffic expected for the building with natural light and ventilation from the rear housing over 100 cars.

The building is created sculpturally from within & externally and is a unique manifestation of abstract volumes that are fluid in the interior and perceived as a dramatic juxtaposition of trapezoidal volumes on the site.

特里塞美食广场

水泥表皮包裹的两层建筑特色十足，角状的结构指向不同的方向，内部包括几家零售店铺、一个美食广场、两间餐厅、一间酒吧及一个游戏区。

沿着主公路一侧可以俯瞰高大的树木、美丽的河岸和葱翠的小山。整幢建筑的轴线变化十足，巧妙的设计使得内部的每一个空间都可以欣赏到不同的风景。

水泥表皮蜿蜒折叠，形成了开敞的结构，将外面的美丽景致引入到建筑内部。建筑后部自然的坡度地形为停车场的建造带来了便利条件。建成的停车场光线充足，通风良好，可容纳100多辆车，在很大程度上缓解了大厦内部停车问题。

此外，建筑内外强调流畅统一，使得抽象的建筑形式得以独特而清晰的诠释。

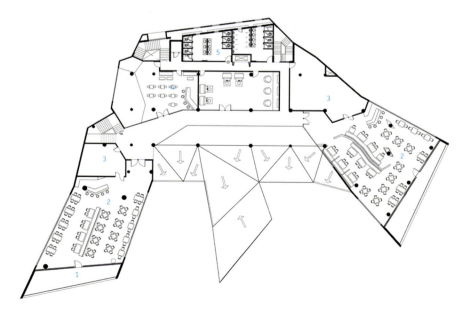

1. 入口
2. 餐厅
3. 厨房
4. 酒吧
5. 洗手间

1. entrance
2. restaurant
3. kitchen
4. pub
5. WC

建筑景观 Landscape with the building

水池 Water

美食广场 Food Court

Photo: Vinesh Gandhi

Completion Date: 2008

Architect: Sanjay Puri, Mamata Shelar

77/32, Gurgaon

古尔冈77／32号

Located in Gurgaon, the office hub in the suburbs of New Delhi, the building moves away from the typical office typology, and provides an alternative with interweaving open social spaces and closed workspaces.

The design is envisaged with two types of informal spaces – one at the public level and the other at the individual office level. The ground floor is designed to be a recreational, informal meeting space which defines the entrance as well. A passive cooling strategy is adopted to create a modified environment which is non-air-conditioned. This is done through the creation of water bodies, and allows for built mass only on two sides and the remaining two sides are left open to allow for wind movement. A café is designed as a part of the recreation zone. Each individual office has been provided with a terrace garden which becomes its private, informal breakout zone.

To address the environmental issues that concern the contemporary office, orientation is optimised in the creation of built volumes. The East and West sun are blocked off with the help of solid stonewalls that act as a thermal buffer. The two long sides, north and south, are provided with glazing and punctures respectively. Each floor plate is designed to be fifteen metres wide to allow for daylight penetration. The use of post-tensioned beams allows for the creation of column-free spaces, which permits maximum flexibility within the office space.

这座建筑位于新德里的卫星城古尔冈。与传统的办公楼不同，项目以一种全新的形态将开放式社交空间和封闭式办公空间结合起来。

项目设计有两种非正式的空间，一个在公共区域，另一个在独立办公室。建筑的一楼是一个富有休闲感的非正式会面场所，同时也是建筑的入口。被动式制冷系统的运用营造了一个无空调的室内空间。建筑只朝向两面，另两面空间空出来保证了流畅的通风。这一设计和水体设计一起保证了建筑的制冷。休闲区还设有一个小咖啡厅。每个独立办公室都配备了一个平台花园，形成了私人休闲区域。

为了体现现代办公的环保特征，设计优化了建筑的朝向。高高的石墙作为一个热缓冲器隔绝了东西方向的日光。南北方向的墙面上采用了玻璃幕墙和穿孔结构。每层楼面都有15米宽，以将自然采光的区域最大化。后张力横梁替代了柱子，让室内办公空间开敞而具有灵活性。

Exterior view at night 建筑外观夜景

景观区域 Landscaped area

外部空间 Informal breakout spaces

入口区域 Entrance area

1. entrance
2. exit
3. food court
4. landscape garden
5. landscape court
6. reception
7. lift lobby

1. 入口
2. 出口
3. 美食广场
4. 景观花园
5. 景观庭院
6. 前台
7. 电梯大厅

Photo: André J Fanthome

Corporate

Completion Date: 2009

Architect: Morphogenesis

Exterior view 外景图

Centra Mall

Chandigarh is a city with a high degree of urbanisation and most of the commercial areas in Chandigarh are concentrated in specifically-designated sectors. Hence, there arises a need for developing other commercial centres within the city. The site for Centra Mall is located in the Industrial Sector on the most visible corner plot on the main sector road and there are excellent linkages from the airport, railway station and the residential and institutional areas of the city, thus making the site an extremely vital component of the city fabric.

The design intent for the mall has two predominant aspects viz. the socio-cultural aspect and the environmental strategy. Traditionally, malls have been approached with a Box-type morphology that excludes people and is not democratic in nature. Hence, the approach has been to reverse this morphology and open up the Box instead towards the site context and surroundings. The built form enclosure is straight-off the road and the design approach is directed towards providing a High Street nature to the mall which opens up. An atrium along the premium road access helps to attain the public disposition, whilst maintaining the mall typology. The retail shops and the entertainment block face the atrium and being wholly transparent, it allows visibility of all the shops from the road. State-of-the-art escalators and lifts allow for the vertical movement and build in pedestrian linkages inside the mall. The layout is extremely efficient with very low super area loading. There are two levels of basement with adequate parking and centralised building services.

中心购物广场

昌迪加尔的城市化程度相当高，主要商业区域都集中在特定的区域。因此，城市需要再开发其他的商业中心。中心购物广场位于昌迪加尔工业区一个最显眼的转角处，交通方便，与机场、火车站、住宅区和行政区都有紧密的联系，是一处极佳的新商业中心选址。

购物广场的设计主要出于两方面的考虑：社会人文和环境策略。传统的商场总是被设计成盒形结构，将人排除在外，缺乏大众亲和力。中心购物广场的设计恰好与之相反，设计师将盒子打开，面向周边的环境敞开。建筑直面街道，充分与繁华的大街融为一体。沿街的中庭吸引了公众的目光，并具备了典型的商场特征。零售店和娱乐空间朝向中庭展开，透明的橱窗让街道上的行人可以清楚地看到店内的情形。自动扶梯和电梯的设置方便了顾客上下楼，而室内人行道则连接了商场内部。商场的内部布局十分具有效率，丝毫没有浪费空间。商场地下的两层是停车场和配套的大楼服务设施。

1. 商场入口
2. 自动扶梯
3. 电梯大厅
4. 公共洗手间
5. 中庭
6. 影院售票处
7. 楼梯
8. 消防楼梯

1. mall entrance
2. escalators
3. lift lobby
4. public toilets
5. atrium
6. cinema ticket counter
7. staircase
8. fire escape staircase

自动扶梯 Escalator

外景图 Exterior view

走道 Passageway

Photo: André J Fanthome

Completion Date: 2008

Architect: Morphogenesis

India Glycols Limited

The office design for the corporate office for India Glycols embodies the issues concerning the workplace today, and explores the paradigm of the office space as a social activity. Sited in a non-contextual suburban area of Delhi, the setting led to the development of an introverted scheme that would address environmental and socio-economic issues from first principles.

As is the nature of most bespoke corporate developments, the building had to exemplify the brand identity and corporate ideology of equity and transparency in the workplace as an integral part of the architectural vocabulary. Conceived as a solid perimeter scheme with a more fluid interior, the morphology blurs the interface between the inside and outside. The built form optimises the natural day lighting and helps to define the programmatic requirements of the office. A stacking system is used to generate a variety of open spaces, such as courtyards, verandahs and terraces, which help to structure the office spaces. A central spine traversing the built volume serves as the common activity zone, with other departments branching out. The design's conceptual strength comes from the spatial organisation which creates overlaps between the exterior and the interior and between the various programmatic requirements, hence creating a vibrant and creative work environment.

印度乙二醇公司办公楼

印度乙二醇公司办公设计体现了当代办公特色，也在办公空间中融入了社交因素。公司坐落在德里市郊，整体设计以环保和社会经济元素为首要考虑。

作为公司发展之根本，办公楼的设计必须要体现公司的品牌形象和公正、透明的经营理念。建筑呈环形结构，室内设计采用不固定形态，整体造型模糊了室内和室外的界限。建筑的造型优化了自然采光，满足了办公所需的光线要求。堆叠系统产生了一系列开放式空间，庭院、游廊、平台等设施也共同塑造了办公空间。横跨建筑的中央通道可以作为公共活动空间，各个部门都围绕着它而设置。设计的重点在于室内外重叠而产生的空间感和多样化的办公环境，这是一个别具活力与创意的办公空间。

Night view of the inner courtyard　内部空间夜景

External spaces　外部空间

外墙上的裂缝形窗户 Shaded outer façade with small slit windows

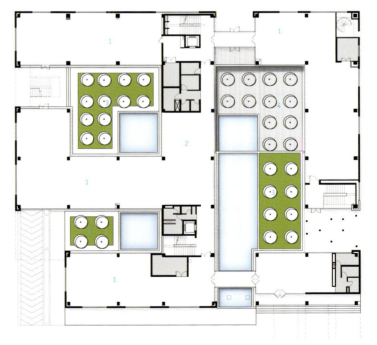

建筑外观 Exterior shot

接待处 Reception area

1. workstation
2. common spine
3. landscape greens
4. deck below

1. 工作台
2. 走道
3. 景观绿化
4. 下方平台

General View 全景图

Avana Apartment

阿瓦纳公寓

This sixteen–storey apartment project is located in Kemang area in South Jakarta, an established neighbourhood famous for its collection of mature trees and vast greeneries. The initial concept of an eight-storey apartment with balconies was abandoned for half a year in 2006 before the client decided to appoint Aboday Architects (Indonesia) to revive the project. As its first project in Indonesia, Aboday decided to retain only the project's internal modular unit plan, while changing the rest of external design, including rising it to sixteen storeys as a result of intense negotiation with local building regulator.

Living in an apartment unit way above ground is some kind of new experience for local people who used to the idea of staying in landed property, hence the introduction of extensive balcony surrounding each unit, mimicking a "yard" in "normal" house. Comprising sixty-four units ranging between 180 square metres to 460 square metres (for the penthouse), with each spacious private space and service area, this apartment is an epitome of landed house that stacked on top of each other creating a home in the sky. The idea of a fluid internal–external space is also explored in every space creation on this building. A six-metre-high lobby, which opens directly to its public pool and garden, starts this experience on the ground floor when people enter the building, leading to a glass–enclosed private lobby. This concept of internal–external space is further explored in each unit above. There is almost no visible boundary between internal unit, balcony (and even the sky...), as a result of extensive use of tempered clear glass panel for door, window and balcony's railing.

这幢16层的公寓位于雅加达南部一个绿树成荫的街区，其主要建筑特色为凸出的阳台以及四周的私人游泳池。其中阳台环绕着公寓四周，间距约为3米，为欣赏雅加达的街景提供了完美的场所。游泳池规格为25X11米，四周采用透明玻璃围合，给住户带来舒适的游泳体验。小花园和池塘犹如在单独住宅和整体背景中建立了一条纽带，将两者融合在一起。设计师摒弃了雅加达地区公寓的典型样式，营造纹理清晰、质感十足的外观，突出温馨柔和的气息，与这一街区的绿树青草交相呼应。此外，住户们还可以在自家的阳台上栽种植物，打造空中花园。郁郁葱葱的植物与混凝土玻璃材质的建筑相互映衬，使得整幢公寓看似犹如巨大的绿色织锦，格外引人注目。

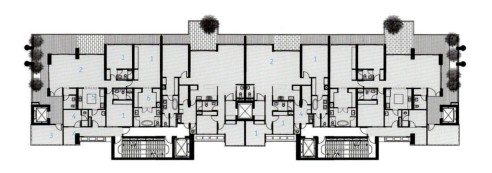

1. 卧室
2. 客厅和餐厅
3. 书房
4. 服务区
5. 厨房
6. 主卫

1. bedroom
2. living & dining room
3. study room
4. service area
5. kitchen
6. master bathroom

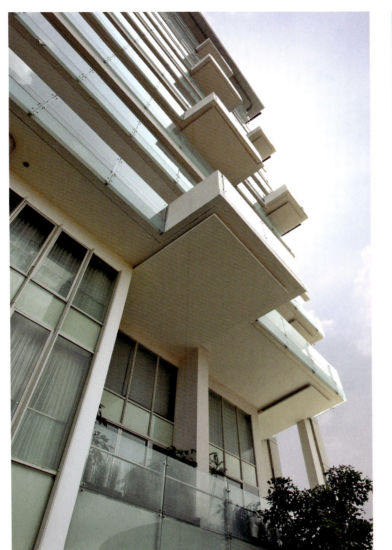

建筑细节 Details

外立面 Façade

内部空间 Interior

Photo: Happy Lim Photography

Completion Date: 2009

Architect: ABODAY (www.aboday.com)

Villa Paya-Paya, Seminyak, Bali

This villa is located in Seminyak, a bustling residential area in the heart of Bali, Indonesia. The site bordered on the north by a six-metre public road, and by a pangkung (dried, old river in Balinese) on the south. The client requested to have a holiday home for the small family of four, with a simple programme: large living and dining, large servant quarter, a master bedroom with huge bathroom and two smaller bedrooms.

The sloping site gives an advantage to the design. Aboday, as an architect, doesn't want to have an imposing building. The villa needs to respect human scale and main road as main way to the temple. This road is sometimes crowded during the Hindu celebration as a shortcut to the nearby temple, and anything taller than coconut trees will be an intrusion to their ritual. The two-level villa appears as a friendly single-storey building from the road, sunk in the rest of the room programme on its ground level. Instead of evoking the surrounding typical Balinese building of slooping coconut leaf roof, Aboday choose a simple concrete white box as the façade of the building. The traditional sloping roof is still used in the master bedroom pavilion with its wood structure, hidden behind the white box façade, as an element of surprise among the domination of white forms.

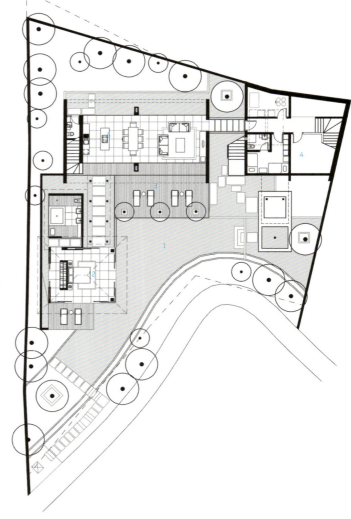

帕雅别墅

帕雅别墅位于巴厘岛上一片繁华的居民区——水明漾，北面是6米宽的公路，南面是一条干涸的河流。委托人一家共有四口人，需要一座简单的度假别墅：开阔的客厅和餐厅、宽敞的佣人房、一间带有浴室的主卧室和两间略小的卧室。

具有坡度的建筑场地为设计提供了便利。建筑师并不想设计一座过于壮观的建筑，别墅要具有人文尺度，也要尊重通往寺庙的道路。这条路在印度教庆典时会变得异常拥挤，而过高的建筑（高于椰子树）会为仪式带来困扰。由于斜坡的原因，这座两层高的别墅看起来就像只有一层一样，与周边的环境十分相称。别墅并没有采用巴厘岛经典的椰子叶屋顶造型，而是采用了简约的白色混凝土盒结构。传统的斜面屋顶将被运用在主卧室区，隐藏在白色盒结构里面，作为一个惊喜元素。

1. swimming pool	1. 游泳池
2. master bedroom	2. 主卧室
3. pool deck	3. 泳池平台
4. store room	4. 贮藏室
5. dining room	5. 餐厅

Photo: Happy Lim Photography

Architect: ABODAY (www.aboday.com)

Completion Date: 2008

Residential

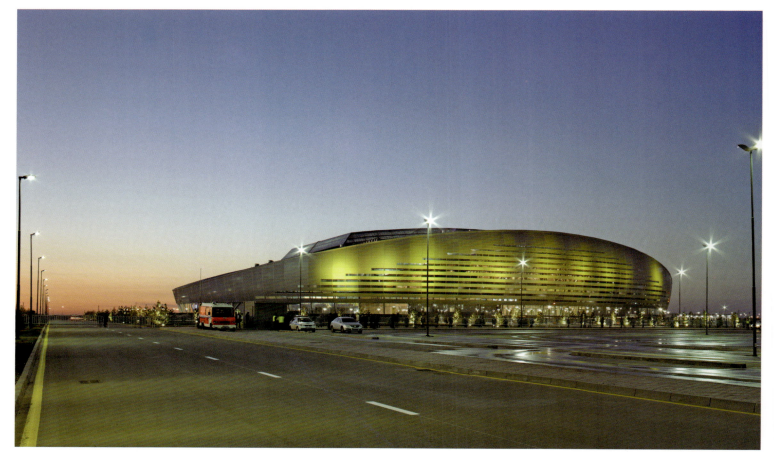

A new scene in Astana tallying with the synergy of sports, games, hospitality, nature and contemporary architecture 体育场是城市的新景观，这个现代建筑融合了比赛、运动、服务业、自然

Astana Arena

阿斯塔纳体育场

Apart from its functional features, the Astana Arena is designed specifically for Astana City to be a symbolic building that reflects the modern and contemporary aspects of the new capital of Kazakhstan. The design introduces innovative solutions adopting high technology principles for operational management, interaction with the environment and especially with harsh climatic conditions of the geography. An operable roof system that functions independently from the fixed roof is programmed in order to protect the green area and provide eligible conditions for the spectators and players.

The Arena can be used for different sports and gathering purposes but it is mainly programmed as a soccer field, which will be covered with high-quality artificial grass that fits the FIFA and UEFA criteria.

Aligning with the elliptical outer form, the circulation line creates dynamic and peaceful areas. Entering and leaving the Arena, people will be able to walk in or out via a safe and secure path and spectators will be welcomed and discharged via twenty-four separate portals around the Arena. 30,000 people can enter and exit at the same time. Six separate zones, behind the different levels of the stand areas, are spared as concourse that open to food kiosks, restaurants and sufficient number of restrooms. Car parks and service roads for 1,411 cars are planned on 71,650 square metres, outside of these secure spaces where the ticket offices are located. A 9,000-square-metre VIP and media car park is secured behind the west wing.

除了实用功能之外，阿斯塔纳体育场还被设计成为阿斯塔纳市独特的标志性建筑，反映了哈萨克斯坦新首都现代时尚的一面。体育场的设计采用了创新方案和高技术含量的经营管理系统，使建筑能够适应环境和当地恶劣的气候。可控式屋顶系统可以独立于固定屋顶运作，保护绿化区域，为观众和运动员提供适合的环境条件。

体育场内可以进行不同的体育比赛，但它的主要作用还是作为足球场。体育场内的高品质人造草坪完全符合国际足联和欧足联的标准。

与椭圆形造型相一致的流通路径营造了充满活力而和平的区域。观众可以通过安全通道从环绕体育场的24个独立大门进出。安全通道可以保证30,000人可以同时进出。看台后面的六个独立区域里设置着小吃亭、餐厅和数量充足的洗手间。售票处外面的停车场和辅道占地71,650平方米，共有1,411个停车位。体育场的西翼还有一个占地9,000平方米的VIP和媒体停车场。

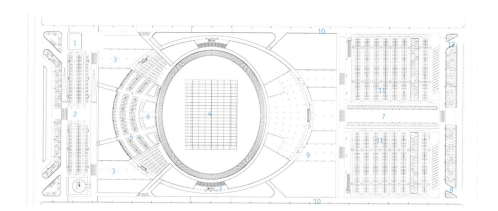

1. heating Centre	1. 供暖中心
2. public car park , 340 cars	2. 公共停车场，340个车位
3. home team entrance	3. 主队入口
4. media car park , 65 cars	4. 媒体停车场，65个车位
5. VIP car park, 136 cars	5. VIP停车场，136个车位
6. mixed zone	6. 混合区
7. pedestrian walkway & landscape	7. 行人走道和景观
8. public exit	8. 公共出口
9. emergency entrance & exit	9. 紧急出入口
10. emergency ring road	10. 紧急环形通道
11. public car park, 532 cars	11. 公共停车场，532个车位
12. public entrance	12. 公共入口

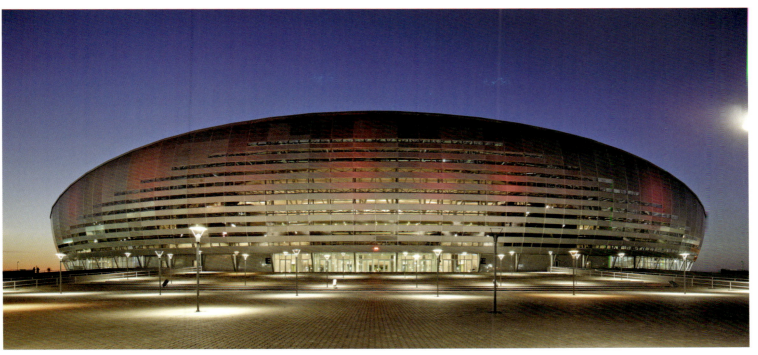

建筑反映了哈萨克斯坦新首都的现代化特征 A symbolic building that reflects the contemporary aspects of the new capital of Kazakhstan

椭圆造型上方的可控式屋顶 Operable roof system over the elliptical outer form

入口之一，全部入口可供30,000观众进出 One of the spacious entrances for 30,000 spectators to enter and exit

Photo: Cemal Emden

Cultural

Completion Date: 2009

Architect: Tabanlıoalu Architects, Melkan Gürsel & Murat Tabanlıoalu

Business Centre Tbilisi

第比利斯商业中心

A successful businessman in Georgia planned to build a "business centre" as the base of his business in a hill overlooking Tobilisi city, and held an international architectural design competition. As the result, Shin Takamatsu was designated. The business centre is not a mere office building. It consists of guest house, multipurpose hall, large and small conference room, fitness gym, pool, dance hall, indoor tennis court, relaxation space, and the owner's enormous residence. Besides, the client required the best security system and the best existence of architecture. It is a kind of a fortress. The architect is strong at creation of architecture as fortress. Though sizes are different, the architect replied to the requirements with all cultivated methods. The method was said "to be the Strengthening method of Forms". The effect is proved by the metallic dignified appearance which suddenly appeared on the hill.

一位出色的格鲁吉亚商人计划在俯瞰第比利斯城的山上建一座商业中心，于是举办了一次国际建筑设计竞赛。最终，高松伸获得了项目的设计权。这座商业中心不仅仅是办公楼，它由宾馆、多功能厅、大大小小的会议室、健身中心、游泳池、舞厅、休闲空间和业主的豪宅组成，拥有最好的安保系统和最佳的建筑形式。它类似一个堡垒，建筑师也努力营造一种堡垒式的建筑。建筑师以各种先进的手段对这个商业中心进行设计，强调了建筑的造型。最终，一座金属建筑以一种华丽的姿态出现在山顶。

1. 入口
2. 卫生间
3. 大厅

1. entrance
2. WC
3. lobby

Photo: Shin Takamatsu + Shin Takamatsu Architect and Associates Co., Ltd.

Architect: Shin Takamatsu + Shin Takamatsu Architect and Associates Co., Ltd.

Complex

Completion Date: 2008

建筑外墙 Façade

夜景 Night view

0–14

Located along the extension of Dubai Creek, occupying a prominent location on the waterfront esplanade, 0–14, a 22–storey–tall commercial tower perched on a two–storey podium, broke ground in February 2007, and comprises over 300,000 square feet of office space for the Dubai Business Bay. The concrete shell of 0–14 provides an efficient structural exoskeleton that frees the core from the burden of lateral forces and creates highly efficient, column–free open spaces in the building's interior. By moving the lateral bracing for the building to the perimeter, the core, which is traditionally enlarged to receive lateral loading in most curtain wall office towers, can be minimised for only vertical loading, utilities and transportation. Additionally, the typical curtain–wall tower configuration results in floor plates that must be thickened to carry lateral loads to the core, yet in 0–14 these can be minimised to only respond to span and vibration. Consequently, the future tenants can arrange the flexible floor space according to their individual needs.

The main shell is organised as a diagrid, the efficiency of which is wed to a system of continuous variation of openings, always maintaining a minimum structural member, adding material locally where necessary and taking away where possible. This efficiency and modulation enable the shell to create a wide range of atmospheric and visual effects in the structure without changing the basic structural form, allowing for systematic analysis and construction.

0–14办公楼

0–14办公楼选址在迪拜河沿岸，22层的商业建筑"栖息"在两层的基座上，总面积达27871平方米，于2007年二月动工建设。水泥外壳构成了主体框架，使得中央结构免受侧面压力，同时营造了无廊柱的室内空间。侧面支撑结构被移到四周，这样进一步减轻了中央结构的压力。此外，幕墙结构的使用助于租户根据需要改变空间格局。

建筑正面幕墙为斜肋构架，与连续变换的开口系统相互呼应。十足的功能性以及可调制特色使其在不改变基础结构的同时，便可营造出不同的氛围及视觉效果。

1. rooftop garden
2. mechanical
3. prayer room
4. offices
5. landscape
6. offices
7. interior bridges
8. drop–off
9. parking entrance

1. 屋顶花园
2. 机械结构
3. 祈祷室
4. 办公室
5. 景观
6. 办公室
7. 室内廊桥
8. 下降区
9. 停车场入口

夜间鸟瞰图 Bird's eye view by night

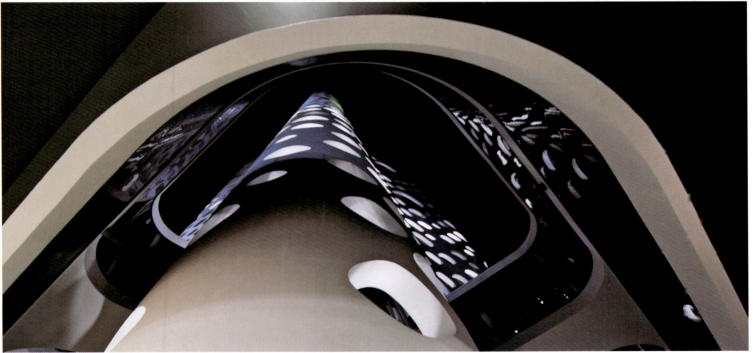

外墙细部 Façade details

外墙细部 Façade details

Photo: Reiser + Umemoto (1st photo), and Sebastian Opitz (photos 2–5, night photos)

Architect: Principals, Jesse Reiser + Nanako Umemoto; Design Team, Mitsuhisa Matsunaga, Kutan Ayata, Jason Scroggin, Cooper Mack, Michael Overby, Roland Snooks, Michael Young; Assistants and Interns, Tina Tung, Raha Talebi, Yan Wai Chu

Completion Date: 2010

Corporate

Abu Dhabi Investment Authority Headquarters (ADIA)

阿布扎比投资局总部

ADIA's future requirements demand a large single floor plate with a large central common zone on each floor for meeting and social interaction. This central zone has become the vertical atrium garden, the heart of the scheme unifies the whole building and represents the organisation and its openness. The design concept further refines the idea of creating an indigenous form, a form inspired by the special character of this waterfront site and the buildings importance as an international headquarters. The key to the design is the acknowledgement of the profound importance of the sea in the development of the site and of the urban plan as a garden city. The growth of the city begun at the old fort besides the headquarters.

The gardens of the fort and those adjacent to them create a wide green zone with further reinforces the seaward connection. The site is incorporated into this garden zone. The scheme has two great arms or fingers reaching into the sea; the extensive ground floor planting reinforces the original urban landscape strategy, tying the site to the urban plan. The success of this urban design strategy for landscaping has earned Abu Dhabi the title "Garden of the Gulf".

根据阿布扎比投资局未来的发展计划，建筑师设计了一个开放式楼面，每层楼中央都有一个开阔的会议互动空间。这个中心区域就是垂直的中庭花园，它统一了整座大楼的设计，展现了大楼的组织结构和开放性。建筑拥有一个本土化造型，这一造型的灵感来源于其滨水的地理位置和其国际化总部的地位。设计的重点承认了海洋在城市发展中起到的重要作用。阿布扎比正是起源于阿布扎比投资局总部旁边的老堡垒。

堡垒的花园和其临近区域共同营造了一片开阔的绿地，建筑和这个花园被结合在了一起，加强了建筑与海洋的联系。这片绿地规划有两条直通海洋的道路，仿佛手臂或手指一样。建筑周围的绿化区和城市景观规划相辅相成，为阿布扎比赢得了"高尔夫花园"的美誉。

1. 楼梯
2. 办公室
3. 会议室

1. stairs
2. office
3. meeting room

General view 全景图

建筑正面 Front view

走廊和中庭 Corridor and courtyard

楼梯和中庭 Stairs and courtyard

楼梯 Stairs

Photo: H.G. Esch

Corporate

Completion Date: 2007

Architect: Kohn Pedersen Fox Associates (International) PA

General view 全景图

Night view 夜景

Bahrain World Trade Centre

The Bahrain World Trade Centre forms the focal point of a master plan to rejuvenate an existing hotel and shopping mall on a prestigious site overlooking the Arabian Gulf in the downtown central business district of Manama, Bahrain. The concept design of the Bahrain World Trade Centre towers was inspired by the traditional Arabian "Wind Towers" in that the very shape of the buildings harness the unobstructed prevailing onshore breeze from the Gulf, providing a renewable source of energy for the project.

The two fifty-storey sail-shaped office towers taper to a height of 240 metres and support three twenty-nine-metre diameter horizontal-axis wind turbines. The towers are harmoniously integrated on top of a three-storey sculpted podium and basement which accommodate a new shopping centre, restaurants, business centres and car parking.

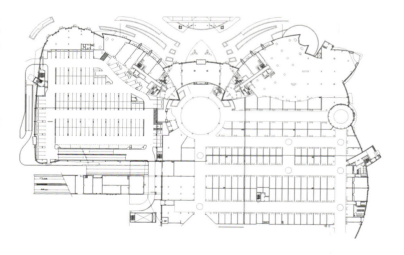

The elliptical plan forms and sail-like profiles act as aerofoils, funnelling the onshore breeze between them as well as creating a negative pressure behind, thus accelerating the wind velocity between the two towers. Vertically, the sculpting of the towers is also a function of airflow dynamics. As they taper upwards, their aerofoil sections reduce. This effect, when combined with the increasing velocity of the onshore breeze at increasing heights, creates a near equal regime of wind velocity on each of the three turbines. Understanding and utilising this phenomenon has been one of the key factors that has allowed the practical integration of wind turbine generators in a commercial building design.

巴林世贸中心

巴林世贸中心是巴林首都麦纳麦的一处市区商业中心重建工程的焦点，可以俯瞰阿拉伯湾的景色。世贸中心双子楼的设计灵感来自阿拉伯传统的"风塔"设计，建筑的造型利用海湾的海风为世贸中心提供了可再生能源。

这两座50层的船帆造型办公楼，高240米，底部装有直径29米的横轴风轮。两座双子楼共享下面的3层底座购物中心、餐厅、商业中心、停车场和地下室。

椭圆形造型和船帆式剖面形成了一个翼面，过滤了二者之间的海风，形成了一个向后的负压力，进而加快了两座楼之间的风速。在垂直方向，双子楼的造型同样形成了气流动力。随着向上逐渐变窄，它们的翼型截面也逐渐缩小。这一效果与随着高度而增加的风速为三个风轮提供了相等的风速。这一原理的理解与运用是在商业建筑设计上使用风力发电机的关键。

连接处 Connection

入口 Entrance

大厅 Hall

Photo: Atkins

Completion Date: 2008

Architect: Atkins

Suliman S. Olayan School of Business, American University of Beirut

贝鲁特美国大学Suliman S. Olayan商学院

The building includes, first, a large green oval carefully located on the axis of existing steps that will become a major access to the sea, connecting students from the Faculty of Engineering and Architecture, and beyond to the Corniche's elevated edge. Second, the design creates an L-shaped four-storey building with a traversable ground plane consisting of four enclosed pieces. These are grouped around the school's central space, a triangular open courtyard. Porous and transparent, the ground floor promotes collegiality, containing the school's lobby, auditorium, café and terrace, as well as student facilities, mailboxes and related social programmes.

To clarify wayfinding and the building's legibility, the undergraduate education facilities are located on the first floor, graduate education, the MBA programme on the second, and the Executive Education programme on the third floor, which also contains the Dean's Office in its corner. The triangular courtyard joins these three levels, and each overlooks the space, enriching it with their different lives.

The image of the building is one of vernacular precedent and contemporary vision. The "hanging" façade, made of pre-cast blocks replicates the warmth of the local Forni limestone present in the campus, while the openings of the screen-like skin recall the wooden mashrabiya that are characteristic of the region.

建筑的中心是一大片椭圆形的绿地。绿地所在轴线上的台阶是通往大海的主要通道，它连接了工程建筑学院和海滨大道的高架路。建筑呈L形，围绕着中心的三角形庭院，共分四个封闭的区域。一楼的外墙是透明而多孔的，里面设置着学院的休息大厅、礼堂、咖啡厅、露台、信箱以及相关的社交项目。

建筑的结构导向系统划分十分清晰：二楼是大学本科教育设施，三楼是研究生教育和MBA教学设施，四楼是高阶主管教育项目和院长办公室。三角形的庭院将三层楼连接在一起，每层楼都能够俯瞰到庭院的景色。

建筑的现代化造型在黎巴嫩具有先锋意义。"高悬"的外墙由预制混凝土制成，与校园里温馨的弗尼石灰岩十分相似。外墙上的镂空质感让人想起了极具当地特色的mashrabiya木格纹。

Corniche view from northwest 从西北侧的滨海路看建筑

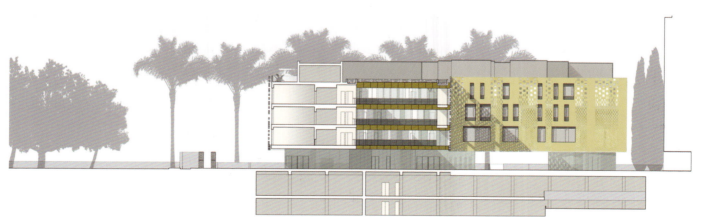

Section through atrium 中庭剖面图

建筑西入口，下方是海滨路 Building entrance from west with cornice below

礼堂，外面是庭院 View of auditorium with courtyard beyond

一楼的玻璃空间，后面是内部庭院 View of ground-floor glass volumes with interior courtyard beyond

Photo: Fares Jammal, Clement Tannouri, Machado and Silvetti Associates

Educational

Completion Date: 2009

Design Architect: Machado and Silvetti Associates, Boston, Massachusetts
Executive Architect: Builders Design Consultants, Beirut, Lebanon

Night view of west exterior 建筑西侧外观

Deichmann Centre for Social Interaction & Spitzer–Salant School of Social Work at Ben–Gurion University of the Negev, Beer–Sheva, Israel

The Spitzer–Salant School of Social Work and the Deichmann Centre for Social interaction were designed and built as an integrated project located on the campus of Ben–Gurion University in Beer–Sheva, the main city and gateway to the Negev desert in south Israel. The goal of the Centre for Social Interaction is to open the university to the city, to tie academic life with the history of Beer–Sheva, and to encourage all forms of art and education, while providing a sustainable environment.

Responding to its urban context, the sculpted façade establishes a distinctive identity to the campus's new entrance through a piazza that links the town of Beer–Sheva with the campus. By sinking the piazza and the complex below street level, the adjacent highway noise is buffered. Rainwater catchments were situated throughout the site, where the collected runoff is used to supplement landscape irrigation.

The complex has a bold and sculptural spirit, with a tilted concrete wall sitting in water supporting a floating zinc clad structure. A freestanding concrete ramp originating at the piazza level leads through the buildings to the main campus. Since concrete and zinc are environmentally friendly building materials, they were a natural choice for the sustainable construction of this project. The materials perform well in desert climate, as they are UV–resistant, maintenance–free and guarantee longevity and timeless beauty. The first layer of concrete walls on the west elevation defines a coherent formal edge to the piazza. Visitors to the building experience an alternating rhythm between solid concrete walls and several visual voids recessed in cast shadows. The physical presence of these concrete walls creates a structure that feels grounded yet shielded from the desert sun, while meeting budget constraints through the use of common and inexpensive local building techniques.

Beyond the first layer, walls become organic free forms and transparent, with a curved metal wall penetrating through large sheets of glass into the buildings, establishing an undisturbed visual connection between indoor and outdoor. Natural light is introduced into the interiors, through the shifted layers of walls accentuated by the change of materials between concrete glass and zinc.

Ground floor plan	一楼平面图
1. courtyard	1. 庭院
2. entrance	2. 入口
3. exhibition area	3. 展览区
4. theatre/auditorium	4. 剧院/礼堂
5. computer lab	5. 计算机实验室
6. seminar	6. 小组讨论室
7. reading room	7. 阅览室
8. computer room	8. 计算机室
9. art class	9. 艺术教室
10. staff office	10. 办公室
11. storage	11. 仓库
12. mechanical	12. 机械室
13. dressing	13. 更衣室
14. sitting area	14. 休息区
15. working open space	15. 开放办公空间
16. water	16. 水景

萨伦特斯学校戴希曼中心

萨伦特斯学校社会服务中心及社会活动戴希曼中心是一项整体工程，位于别是巴（以色列南部城市）本•古里安大学校园内。戴希曼中心的设计目标即为使大学面向城市开放，将学术氛围同城市历史结合，同时促进艺术及教育的发展。校园与城市之间通过一个广场相连，而戴希曼中心恰好在入口处形成了一个独特的标识，其雕刻般的外观格外引人注目。设计师特意将广场及中心低于街面，这样就免去了临近公路上噪音的影响。集水盆地结构遍布于整个场址，用于浇灌周围的景观植物。

整幢建筑在造型上凸显大胆的理念——倾斜的水泥墙矗立于水中，支撑着上面恰似浮动的镀锌结构。独立的水泥坡道从广场处伸展出来，穿过建筑，一直通往校园的中心。之所以选择水泥和钢材，其一是从环保理念出发。另外，这些材料因其独特的性能，如耐紫外线、易维护、长寿、永恒美等而极为适合沙漠气候。西立面上最外层的水泥结构与相邻的广场形成统一感。来访者走进建筑内，会充分感受到墙壁的"层次节奏感"。如此的设计使其免收了沙漠强光的照射。

在外层水泥结构，墙壁在形状设计上极为自由而且变成透明状。其中一道弯曲的金属墙壁"穿越"层层玻璃，一直伸向建筑内部，在室内外之间营造了一条视觉纽带。

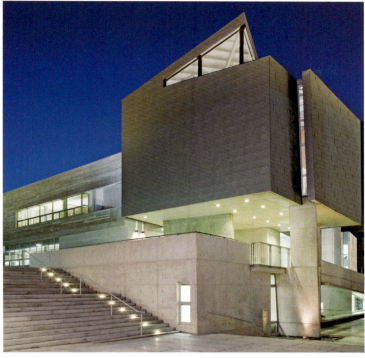

夜景 Night shot of the cute

外部环境 Exterior context

建筑东侧场景 East exterior view

剧院 Theatre

Photo: Amit Geron
Post Production: Paul Chamberlain

Educational

Architect: Vert Architects / Raquel Vert, Principal Architect
In Collaboration With Axelrod–Grobman Architects / Irit Axelrod, Yasha Grobman (Israel)
Consultants: Piazza–Chyutin Architects, Yaron Ari Landscape Architect

Completion Date: 2008

Buildings melt with nature 建筑融于自然之中

University of Cyprus – School of Economics and Public Administration and School of Classical Studies

塞浦路斯大学——经济行政学院和古典文学学院

The Schools for Public Administration and Economics are designed as two long and narrow buildings, parallel to each other and perpendicular to the central pedestrian spine defined by the master plan, while the School for Classical Studies and the teaching halls are placed in a third building forming an angle with the other two, in order to create open space between them and allow for views towards the pedestrian way at the north-west site boundary. They form part of the new University campus master plan of the city of Nicosia.

The basic design concept is the creation of a diagonal axis (bridge) which interconnects the schools, while at the same time intersecting them in such a manner that independent units can be positioned on either side of the axis. The buildings follow the natural slope of the ground and form two-storey volumes as seen from the pedestrian spine to the north-west and four-storey volumes as seen from the south-east limit of the site. The composition includes parking facilities with planting as a prolongation of the Schools of Public Administration and Economics, and a lake for the creation of a pleasant microclimate and the reduction of temperatures during the hot summer season. Agreeable comfort conditions within the buildings are provided by the careful design of natural lighting, ventilation and shading.

经济行政学院是两座狭长的大楼，两座大楼相互平行，垂直于校园的中心人行道。古典文学学院和教学大厅和这两座大楼垂直，在中间空出了一片开阔的空间。它们都是塞浦路斯大学新校园规划的一部分。

设计的基本理念是用一个对角轴（廊桥）将两个学院连接起来，独立的部门可以被设置在轴线的两侧。建筑依场地的自然坡度而建，从西北的人行道看去是两层，从东南侧看去则是四层。学院的规划包括经济行政学院旁一个植有绿植的停车场和一个人造湖。人造湖为校园营造了一个令人愉悦的微气候，在炎热的夏季为校园带来了一丝清凉。学院楼的内部设计也经过了深思熟虑，自然采光、通风和遮阳让室内环境极为舒适。

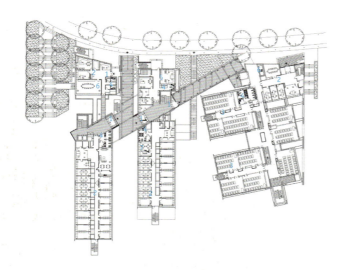

1. 入口
2. 办公室
3. 教室
4. 洗手间/厨房
5. 实验室
6. 会议室
7. 图书馆

1. entrance
2. offices
3. training rooms
4. toilets/kitchen
5. laboratories
6. meeting room
7. library

连接桥 Connecting bridges

楼梯 Stairs

阶梯教室和礼堂 Lecture theatre and auditorium

Photo: Alexandros N. Tombazis and Associates Architects Ltd, Athens, Greece

Architect: Alexandros N. Tombazis and Associates Architects Ltd, Athens, Greece + Alekos Gabrielides and Associates Architects, Nicosia, Cyprus

Completion Date: 2010

Educational

Façade 建筑外观

Dalaman International Airport Terminal

达拉曼国际机场航站楼

The project which strived to deal with the boredom and feeling of emptiness created by the standardness of terminal buildings, aimed to problematise the international airport conventions in the project by making use of the region's rich landscape, climatic characteristics and the specificity of its tourism activities.

The plan was developed by differing from the customary massive orders of terminal buildings, which are conditioned by the disproportionate sizes of the narrow and long piers and the relatively shorter and wider halls; the design formed man-made valleys of the gaps between the interior spaces and the fragmented exterior masses. These gaps enabled the continuity of the region's landscape using its natural form outside and its abstraction inside. Other significant inputs were the fact that the terminal, which has a capacity of five million passengers, would almost only be used during the summer, and that the circulation of arriving and departing passengers were envisaged to be on different floors, the visual fluidity between the interior spaces and different levels, ensuring that the commercial units be attractive.

这一项目旨在打破航站楼空旷、乏味的标准化设计，利用该地区丰富的景观资源、独特的气候以及旅游文化为达拉曼国际机场设计增添全新的活力。

传统航站楼大多具有不成比例的细长支柱和相对短而宽的大厅；这一项目则不同，建筑师在室内空间和零碎的室外结构之间设计了人造山谷。这些缺口以其外部自然形态和内部抽象形态延续了区域的自然景观。航站楼能够接待500万客流量，并且几乎只在夏天使用。到达和离开的旅客将被安排在不同的楼层，旅客可以看到室内空间不同的景色，同时也保证了各个楼层的商业空间都能吸引更多的注意力。

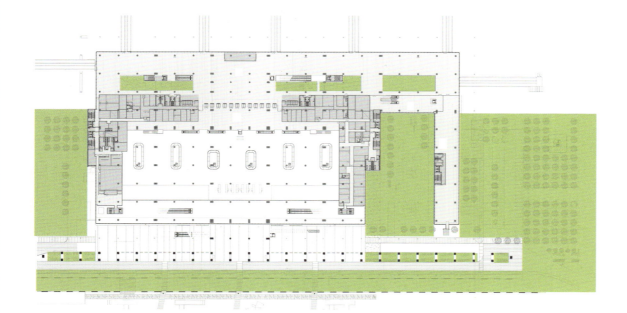

廊桥 Bridge

大厅 Hall

Photo: Ali Bekman, Thomas Mayer

Transportation

建筑背立面 Back view

Architect: EAA—Emre Arolat Architects

Completion Date: 2006

External view 建筑外观

Ipekyol Textile Factory

Ipekyol纺织品工厂

The service entrance leads into the building from the road to Kırklareli and pedestrian access from the busy E-5 Highway on the other side. The sales unit, situated close to the road, again in the direction of the E-5, was connected to the main building by an overhang which covered and thus defined the walkway.

As are usually used in similar buildings, here too the main components were the vertical reinforced concrete load-bearing systems, a lightweight steel structure cover placed on top of them and the coffered system on the façades; the exterior surface took shape through a grammar established by the clear distinction between areas open or closed to the exterior.

The administrative section, which commonly is visually detached from the production building through the use of different surface languages, due to the conventional approach to such facilities, was more directly associated with production in this project and thus, instead of different buildings a large mass took shape. This mass, which reaches the outer borders of the lot, because of the constraints of the land, was implicitly loosened thanks to linear gardens located between sections. The main purpose of these gardens was that they be used by the staff during breaks and that natural light and air enter work places; it was intended that the gardens separate areas and that thanks to their transparent frames visual fluidity would be achieved. And due to the limitations of local production possibilities, innovative experimentations in building materials and production methods were especially avoided.

沿通往克尔克雷利大路一侧的入口或另一侧E-5公路人行道进入便可到达建筑内部。销售区紧邻大路一侧,而临近E-5公路一侧通过悬臂结构与主建筑相连。

与同类建筑一样,大楼主要部分为混凝土承重结构,上面覆以轻型钢,外观以格子图案装饰。通常说来,设计中往往会通过运用不同表面装饰将行政楼与生产车间在视觉上分离开来,但这一项目却恰恰相反——设计师打造了一个统一的"大结构"(出于地形本身的限制)。花园的设置为员工提供了休憩之所,透明的框架结构实现了视觉流畅性。另外,由于当地条件的限制,设计师在材质选择上尽量选择传统的方式。

Building detail 建筑细部

外观局部 Exterio detail

1. production area
2. inner garden
3. entrance hall
4. cafeteria
5. offices
6. staff entrance
7. changing room
8. service entrance

1. 生产区
2. 内部花园
3. 入口大厅
4. 餐厅
5. 办公室
6. 员工入口
7. 更衣室
8. 服务入口

室内 Interior

俯瞰整个室内空间 Looking down the whole interior

Santral Istanbul Museum of Contemporary Arts

Istanbul Bilgi University seeked for the renovation of Silahtaraga Power Plant, a typical modern industrial setting to be transformed into a museum, recreational and educational centre. Among the various buildings that were dealt with in this context, the two large boiler houses were handled with an interpretation that implied to their new function, in a way of abstraction in the design. The two buildings, that were detached but stood very close to each other to complete the surrounding building mass, were planned in a way proper to the volumetric existence of their older functions, but with a kind of "timeless" approach on surface qualities.

Just like the old buildings, new structures are composed of a dense and heavy inner core and a light, semi-transparent exterior sheathing that covers the core without touching to the possible extend. Instead of the punced state created by the walls and windows on the surfaces of the old buildings, a metal mesh that this time homogenises the sense of the whole building is simply placed on the concrete base. In this sense, it was considered that the buildings should evoke a kind of insignaficance by intervening into the aura of the environment at daytime, but should turn out to be a simple lighthouse with the interior lighting of the museum that makes the metal mesh invisible at night time.

伊斯坦布尔Santral现代艺术博物馆

伊斯坦布尔比尔及大学试图翻新Silahtaraga发电厂，将其改建成一个集博物馆、休闲、教育于一身的中心设施。在发电厂里各种各样的建筑中，两座大型锅炉房的翻新设计十分具有抽象效果，从外形上展示了自身的功能性。这两座建筑虽然各自独立，但是相距很近，它们的设计与其自身的体量特征相符，在外观上体现了一种"永恒"的特点。

与旧建筑改造相同，新建的结构同样拥有厚重感的室内结构和轻盈的半透明外墙。与旧建筑的外墙和窗户所营造的块状结构相反，新建筑的外墙在混凝土底座上添加了一层金属网结构。这样一来，建筑在白天看起来平淡无奇，与周边的建筑无异；但是到了夜晚，博物馆的室内照明模糊金属网，建筑将变身成为一座简洁的灯塔。

Photo: EAA–Emre Arolat Architects, nevzat sayın (nsmh)

Cultural

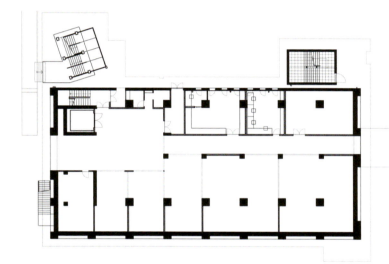
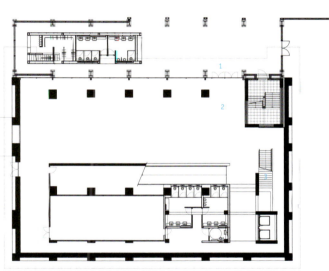

1. entrance　2. hall　3. stairs
1. 入口　2.大厅　3. 楼梯

Completion Date: 2007

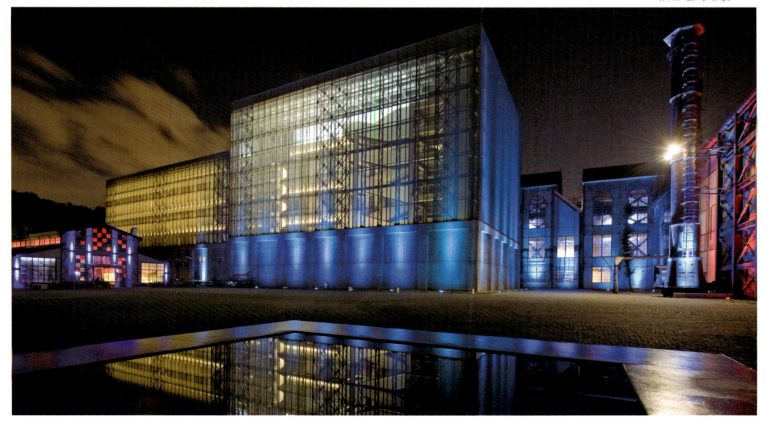

Architect: EAA–Emre Arolat Architects, nevzat sayın (nsmh)

General view 全景图

Bursa Wholesale Greengrocers and Fishmongers Market

The municipality of Bursa required a new, modern facility for the wholesale trade of fruits and vegetables, as well as separate facility for fish and other seafoods. The building would consolidate these commercial activities, providing the city with a centralised control point from which to monitor the Bursa's food supply. The design of Bursa's wholesale greengrocer's and fishmonger's market maintains the idiom of the high, vaulted bazaar, connecting the new buildings symbolically and functionally with long-standing Central Asian architectural and cultural traditions. The complex patterns of vehicle, material, and pedestrian traffic are carefully coordinated within fluid, elliptical shapes, which in turn are bordered by brokers' offices. The ra-tional form of the 350-metre-long greengrocer's market is designed to facilitate easy orientation, efficient exchange and optimal routing of foodstuffs from suppliers to retailers and restaurateurs – all of then keep down transaction costs. But it is also a good place to work: an animated space and architecture that is representative of the energy and productivity of the labourers, as well as of the city of Bursa.

At the same time, the configuration of the naturally-ventilated spaces allows the municipality to ensure the efficient, safe distribution of food products to its citizens. By consolidating the wholesale trade of produce and fish for the city of Bursa in a single location, the municipality is able to monitor the goods for quality and also to ensure that health regulations are followed.

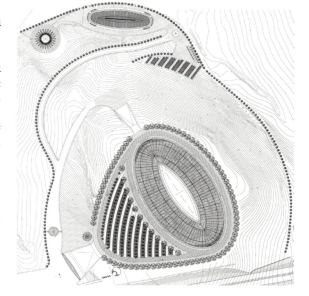

布尔萨果蔬鱼类批发市场

布尔萨市需要一个新的现代化果蔬、海产品批发市场。市场将促进商业活动发展，同时为城市提供一个可监控食品供应的中心控制点。布尔萨蔬菜鱼类批发市场的设计采用了高穹顶圆形集市的造型，极具中亚建筑文化特色。车辆、材料和行人通道等被精心地设计在流畅的椭圆形空间里，市场的外围是中间商的办公室。350米长的果蔬市场设计简洁明快，方便了交通运输，降低了食品从供应商到零售商和餐厅经营者的交易成本。市场是一个极佳的工作空间：充满生气的空间和建筑结构象征着能量和生产力，也象征着生机勃勃的布尔萨市。

这个采用自然通风的交易空间为市民提供了便捷、安全的食品。通过将农产品和海产品交易集中起来，布尔萨市当局可以进一步监控食品质量和安全，保证食品健康相关条例的执行力度。

鸟瞰图 Bird's eye view

建筑外部 External view

建筑内部 Interior view

Photo: Tuncer Çakmakli, Aziz Safi, Gürkan Akay, Fehmi Ferit, Seyfettin Bal

Complex

Completion Date: 2007

Architect: Tuncer Çakmakli

General view 全景图

7800 Cesme Residences & Hotel

Main mass of 7800 Cesme Residences & Hotel has been made close to the border of the road, so that the frontal large beach and the natural environment has left as it exists as possible. The linear block of five storeys, has been transformed double sided by an internal street on which both vertical and horizontal circulation are organised.

Instead of being self-centralised and prominent by visual structural form, attractive and awaiting to gain its power by this kind of attention, what is aimed in the project is a kind of structure that tends to hide behind the landscape layer which covers it and in this way chooses to get rid of all the burden of concepts that might be defined as style, taste and genre of architecture.

Two different blind systems were designed in order to prevent the north and south façades that constitute the units' point of view, from sunlight and wind. Both of these systems, which are made as simplified as possible, became the most important elements of the exterior perception.

切什梅7800酒店公寓

切什梅7800酒店公寓的主体建筑离公路边缘很近，因此，从酒店公寓可以尽情欣赏海滩的自然景色。建筑呈长条形，共有五层，中间的走道将建筑一分为二，走道上设置着水平和垂直流通路径。

与那些以自我为中心、单纯依靠外部造型来吸引眼球的建筑不同，这一项目将自己隐藏在层层景观之后，抛弃了所谓的时尚、品位、风格等概念的包袱。

南北两侧不同的遮阳板结构既保证了公寓窗口所获取的景观，又将刺眼的阳光和狂风阻挡在外。这些极简的遮阳板成为了建筑外观最显要的结构元素。

1. dock
2. landscape
3. hotel

1. 码头
2. 景观
3. 酒店

廊桥 Bridge

黄昏景色 View of sunset

建筑外观 Façade

General view 全景图

Forum Çamlik

Forum Çamlik, in the city of Denizli (Çamlık), has a 3,420-square-metre supermarket and 32,000-square-metre lease able areas. It has a total construction area of 73,000 square metres. In the mall exist 7,700-square-metre Anchor, 600-square-metre MSU, 9,700-square-metre shop, 2,355-square-metre restaurants and food court, 800-square-metre leisure area and 2,140-square-metre cinemas.

Mall has four main entrances, one on the ground floor from "Democracy Square", two on the first floor and the last one on the second floor. The entrances and exits for the closed car park will be made on the side of "Democracy Square"; open car park entrances and exits will be positioned on the south of the building, where the food court is located. The Face Veneer is a colour combination of the materials of glass, metal, bricks and stones. The building has exciting movements within vertical and horizontal changes in placement. Especially the cinema side view has an artistic sculpture with its turquoise colour.

Camlik购物中心

Camlik购物中心位于土耳其代尼兹利市，包含3,420 平方米的超市和32,000平方米的出租店铺，总建筑面积为73,000平方米。购物中心共有7,700平方米的专卖店、600平方米的多用户共用单元、9,700平方米的商店、2,355平方米的餐厅和饮食广场、800平方米的休闲娱乐区和2,140平方米的电影院。

购物中心有四个主要入口：一楼的入口可以从"民主广场"进入，另外还有两个二楼的入口和一个三楼的入口。室内停车场的出入口也在"民主广场"一侧，室外停车场的出入口则设在大楼的南侧，靠近美食广场。建筑的外表面由色彩斑斓的玻璃、金属、砖和石块堆砌而成。建筑在水平和垂直方向的布局都充满了动感和活力。电影院的侧面被漆成了蓝绿色，宛如一块美丽的绿松石。

外观细部 External detail

室内 Interior view

Photo: Chapman Taylor España

Completion Date: 2008

Architect: Chapman Taylor España

Commercial

Exterior view   建筑外观

Enter

Enter is an IT college in the town of Sipoo, close to Helsinki. The L-shaped pavilion-like building forms the last corner of an existing school campus.

The urban setting is dominated by two curved yards: a larger one towards the campus garden and a smaller courtyard connected to the main road. The two curved glass façades open up the school towards the community. The students and teachers see the community and are seen. The campus garden serves as learning area, featuring a wireless network and green islands. Fruit trees with white flowers celebrate the graduation in the end of May.

The street-side façade is broken up in smaller volumes to relate to the scale of the surrounding villas. The warm tone of the wooden surfaces is reminiscent of early-20th-century school buildings.

A-cast-on site concrete stair forms the centre of action in the building. It is topped by a large conical top light. A small mediatheaque and a café in the southern wing open up towars the central lobby. The class rooms are reduced in their material palette into industrial IT workshops.

恩特信息技术学校

恩特信息技术学校位于芬兰赫尔辛基附近的Sipoo镇。设计者K2S事务所取得这一项目的设计权。建筑顶部是很大的圆锥形照明设备。木质表面的暖色恰似20世纪早期的学校建筑，混凝土楼梯构成了建筑内部活动的中心。

建筑靠近街道一面的立面分割成小块，与周边的别墅相对应。两座弯曲的庭院起到主导作用——较大的一座向学校花园开放，较小的一座则与主马路连接。花园也是学习区。设置了无线网络和"绿岛"（Green Islands）。五月，果树上白色花朵竞相开放，预示着一个学期的结束。

建筑南侧结构上设置着咖啡厅，朝向中央大厅。教室设计在材质选择上尽量简单化，如同IT工作室一般。

1. IT classroom	1. IT教室
2. library	2. 图书馆
3. café	3. 咖啡厅
4. lobby	4. 大厅
5. hall / languages	5. 语言大厅
6. theory classroom	6. 理论教室
7. janitor	7. 警卫室
8. IT laboratory	8. IT实验室
9. servers	9. 服务器
10. mechanics	10. 机械室
11. welding	11. 焊接室
12. electrical laboratory	12. 电子实验室
13. storage	13. 贮藏室
14. facilities	14. 设备室
15. dressing rooms	15. 更衣室
16. technical facilities	16. 技术设备
17. teachers' room	17. 教师办公室

建筑外观 Exterior view

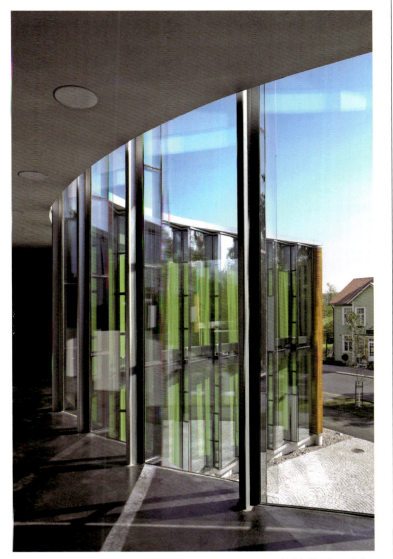

窗方景色 View from window

建筑外观 Exterior view

楼梯 Staircase

Viikki Church

The church is located at the termination point of a narrowing landscape space, along the edge of a new square. The architecture evokes impressions of the Finnish forest. The eaves following the roof shapes reflect the forms of the treetops surrounding them. The approach to the church from the square takes a route past the bell tower and arrangements of vines. In the halls of the building the rising lines of the timber structures, resembling foliage, meet the systems of beams that define the space with light filtering through the structural members. The spaces, made of a single material, are hollowed out within the building like clearings in a forest.

The church design is based on the winning entrance to an architectural competition for the Latokartano Centre in Viikki, organised in 2000. The competition aimed to find suitable townscape and functional concepts for the civic and public service buildings of the area. The design brief also included proposals for the organisation of a public square, a park and commercial buildings on the competition site. In the chosen entrance the public buildings form rectangular shapes that delineate the square and the park, their light-coloured brick surfaces differentiating them from the wooden church rising in their midst.

维克教堂

该教堂位于一处狭长的景观区域的末端，沿着一个新广场的边缘而建。这座建筑使人想起芬兰森林。屋檐顺着屋顶的形状，呈现出周围的树冠的造型。从广场通向教堂的小路要经过钟楼和种植整齐的藤蔓织物。建筑内的大厅里，木质结构的线条模仿树叶的形状，跟限定空间的横梁相一致，阳光透过横梁照射到室内。建筑内部的各个空间都采用统一的材料，有一种挖空出来的效果，就像一块块林中空地。

教堂的设计以2000年为设计维克地区拉托卡塔诺中心而举办的建筑竞赛的获胜方案为蓝本。设计要求也包括规划一个公共广场、一个公园以及该地的几座商业建筑。入选的方案中，公共建筑呈矩形，界定出广场和公园的边界，浅色的砖石表面跟中间拔地而起的教堂的木质结构区分开来。

Exterior 建筑外观

1. church hall	1. 教堂大厅
2. parish hall	2. 教区大厅
3. entrance hall	3. 门厅
4. sacristy	4. 圣器收藏室
5. hall porter	5. 门房
6. office	6. 办公室
7. meeting room	7. 会议室
8. club room	8. 俱乐部
9. kitchen	9. 厨房
10. waiting room	10. 等候室
11. storage	11. 仓库
12. bell tower	12. 钟楼
13. technical facilities	13. 技术设施

外墙 Wall

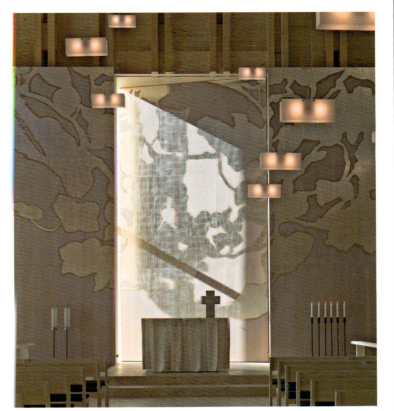

讲坛和艺术作品 Altar and art work

屋内看景观 View to landscape

俯视教堂大厅 Axial view of church hall

Photo: Arno de la Chapelle photos, Kimmo Räisänen photos

Cultural

Completion Date: 2005

Architect: JKMM Architects

Outside view 外景图

Knut Hamsun Centre

Dedicated to Norway's most inventive 20th-century writer and recipient of the Nobel Prize in Literature, the 2,700-square-metre Knut Hamsun Centre is located above the Arctic Circle by the village of Presteid of Hamaroy, near the farm where Hamsun grew up. The building includes exhibition areas, a library and reading room, a café, and a 230-seat auditorium for museum and community use.

The concept for the museum, "Building as a Body: Battleground of Invisible Forces," is realised from both inside and out. The wood exterior is punctuated by hidden impulses piercing through the surface. The concrete structure with stained white interiors is illuminated by diagonal rays of sunlight calculated to ricochet through the section on certain days of the year. Strange, surprising and phenomenal experiences in space perspective and light will provide an inspiring frame for the exhibitions.

The tarred black wood exterior skin alludes to Norwegian Medieval wooden stave churches, and in the roof garden, long chutes of bamboo refer to traditional Norwegian sod roofs. The spine of the building body, constructed from perforated brass, is the central lift, providing handicapped and service access to all parts of the building. The building includes a community auditorium which is connected to the main building via a passageway accessed through the lower lobby, which takes advantage of the topography, allowing for natural light along the circulation route.

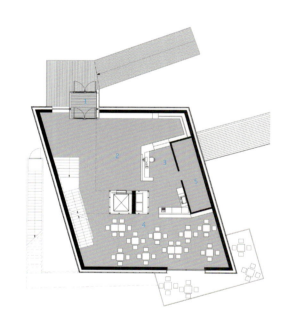

克努特·汉姆生中心

总面积为2,700平方米的克努特·汉姆生中心位于北极圈北部的Hamaroy，临近汉姆生的出生地，是为了纪念20世纪挪威最具创新性作家、诺贝尔文学奖获得者克努特·汉姆生而建造。中心包括展览区域、图书馆、阅览室、咖啡厅和拥有230个坐席的礼堂。

博物馆的设计主题是"身体一般的建筑：隐形军队的战场"。这一主题在室内外设计中被贯彻到底。零星的突起刺穿了木质外墙，阳光透过玻璃突起在混凝土结构内部粗糙的白色墙面上打下斜纹光影。奇异的空间体验和光影效果外展览提供了一个令人兴奋的框架背景。

焦黑色木质外墙与中世纪挪威的木板条结构教堂相似；屋顶花园上长长的竹竿模仿了传统的挪威屋顶。由穿孔黄铜所构造的中心电梯结构让整幢建筑实现了无障碍交通，建筑的礼堂与主建筑通过一个走廊相连接。底层通往走廊的大厅充分利用了地理优势，为博物馆的流通区域提供了自然采光。

1. entrance
2. lobby
3. reception
4. café
5. kitchen

1. 入口
2. 门厅
3. 前台
4. 咖啡厅
5. 厨房

外景图 Outside view

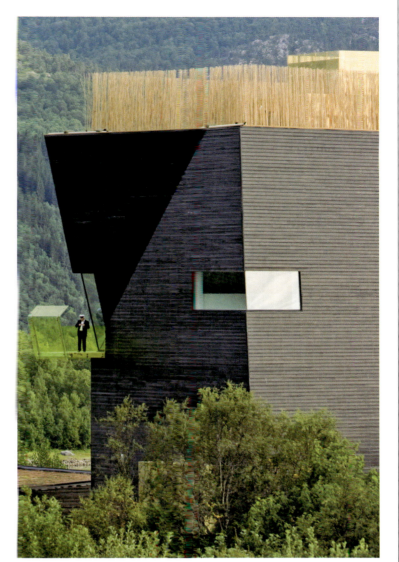

外景图 Outside view

外景图 Outside view

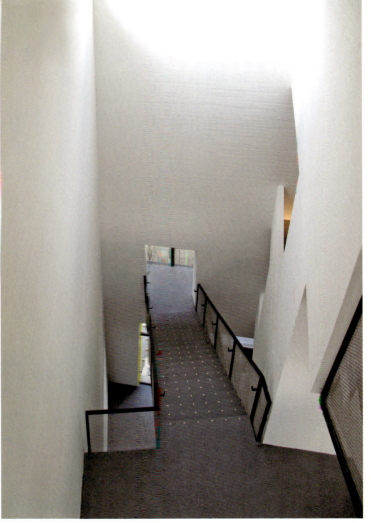

室内 Interior

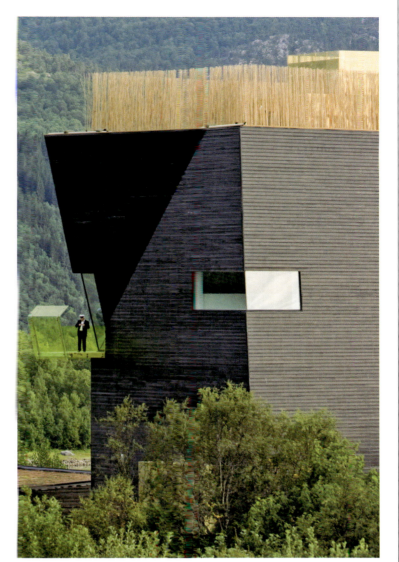

Photo: Iwan Baan

Cultural

Completion Date: 2009

Architect: Steven Holl Architects

General view 全景图

The Energy Hotel

The power plant and adjacent dining hall, designed by the renowned architect Geir Grung in the 1960s is listed among the twelve most significant modern buildings in Norway. New owners took over the dining hall to transform it into an exclusive hotel in 2006. Its generally good condition led to a philosophy of minimal intervention. Only non-original or elements added later have been replaced or modified. New elements have been designed by re-using materials from the power plant, e.g. cables, wires and ceramic insulators. Helen & Hard invited three artists to develop parts of the interior with the intention to add more layers of exquisite value; the light weight, moveable reception by Marli Mul, a flexible textile roomdivider of Yngve Holen and furniture in the conference room of Randy Naylor. In this process Helen & Hard had the role of a curator and negotiated between the artists and the clients interests.

能量酒店

20世纪60年代，著名建筑师吉尔·格朗设计了这个发电厂和其旁边的餐厅。这个项目是挪威12座著名现代建筑之一。2006年，餐厅的新业主决定将它改造成一个高档酒店。项目本身的优秀条件决定了翻修工程只需进行微调，设计师只替换、改良了一些非原始元素。项目新添加的元素重新使用了发电厂遗留下来的材料，例如：电缆、电线和陶瓷绝缘体等。Helen & Hard建筑事务所邀请了三位艺术家来进行室内设计，以使项目看起来更精致。马力·摩尔设计了轻质可移动前台；扬威·霍伦设计了灵活的布艺隔墙；兰迪·内勒设计了会议室中的家具。在这一过程中，Helen & Hard建筑事务所充当了监理人角色，并且在艺术家和业主之间进行沟通协商。

Part view 建筑一部分

可移动前台 **Moveable reception**

大厅 **Lobby**

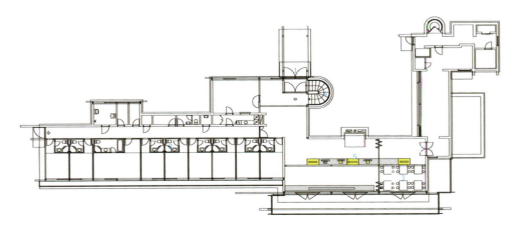

Photo: Helen & Hard

Completion Date: 2007

Architect: Helen & Hard

1. 入口
2. 洗手间
3. 餐厅
4. 楼梯
5. 接待区

1. entrance
2. washing room
3. dining
4. stairs
5. reception

View from the south 建筑南侧外观

Advice-House

Advice-House is the first completed building in the Lysholt Park, a new business-park north of Vejle, and is, with its proximity to the motorway, designated to act as landmark and eye-catcher for the entire development. The Advice-House interior is 5,000 square metres of open and flexible office layout, where various tenants share the same large space, which offers dramatic perspectives and angles. The building is shaped around two angled office wings, separated by an equally angled atrium, resulting in a plan resembling a hexagon with one angle pushed inwards. The two wings are connected by walkways across the atrium, and the floors' continuous window-bands give a high degree of freedom in the space-planning. A large, north-facing glazed gable gives passers by a glimpse into the dynamic void, day or night, and the open and transparent interior is also naturally ventilated.

The building's unusual geometry makes for a dramatic and changing appearance when cars are passing by on the motorway, and this mutability in form and shadows is further heightened by the colouring and texturing of the façades, designed to catch the light. The cladding-strips are composed of a "random" sequence of a total of thirteen differently-proportioned cladding panels, some of which are folded diagonally to create a triangulated pattern. The panels are mounted horizontally at staggered intervals, creating a glittering array of colours, light and shadows.

建议大厦

建议大厦是维积利北部的新商业园区——利肖尔特商业园第一座建成的建筑。由于其位置靠近高速公路，建议大厦也是整个园区建设工程的地标。建议大厦内部拥有5,000平方米的灵活开放办公空间，为各式各样的租户呈现了多样化的景观和角度。建筑由两个呈角度办公翼楼组成，二者中间隔着一个中庭，整体呈现出一个凹六边形结构。建筑的两翼由中庭之间的走道相连接，不间断的玻璃窗带为空间规划提供了高度的自由。朝北的三角形结构在白天和黑夜都能给人以动态的空间感。开放而通透的室内具有自然通风系统。

当车辆从高速公路上驶过时，建筑会呈现出变幻多端的造型，而能够捕捉光线的外墙则进一步为建筑增添了光影效果。建筑外墙上装饰条共由13中不同比例的墙板组成，其排列次序完全是随机的，其中一些墙板还被弯折成了三角形。这些墙板交错排列，光影交错、熠熠生辉。

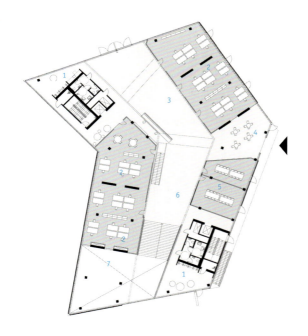

1. lounge	1. 大厅
2. office	2. 办公室
3. reception	3. 前台
4. entrance	4. 入口
5. meeting room	5. 会议室
6. atrium	6. 中庭
7. canteen	7. 餐厅

从高速公路上看建筑 Seen from the passing motorway

两侧翼楼由走道连接 The two wings connected by walkways

建筑室内空间开阔而灵活 The interior is an open and flexible office layout

Green Lighthouse

Green Lighthouse is Denmark's first CO_2-neutral public building is home to the Faculty of Science at the University of Copenhagen. The building's circular shape and the adjustable louvres of the façade mirror the course of the sun. The sun being the predominant source of energy is the overriding design concept behind the new building. Green Lighthouse is based on a whole new experiment with an energy concept, consisting of a supply combination of district heating, photovoltaics, solarheating and cooling and seasonal storage.

To achieve carbon neutrality, many green design features were incorporated to reduce energy use and provide a holistic and healthy indoor environment for students and faculty. The building itself was oriented to maximise its solar resources, while windows and doors are recessed and covered with automatic solar shades to minimise direct solar heat gain inside the building. Plentiful daylight and natural ventilation are provided by means of the carefully-placed VELUX skylights, Velfac windows and the generous atrium. Finally, sensibly-integrated state-of-the-art technology has been applied: heat recovery systems, photovoltaic panels, solar heating, LED lighting, phase change materials, geothermal heat are just some of the technologies that are seamlessly integrated into the building. Seventy percent of the reduction of the energy consumption is the direct consequence of architectural design.

绿色灯塔楼

绿色灯塔楼是丹麦首座达到二氧化碳平衡的公共建筑，是哥本哈根大学理学院的教学楼。建筑的圆形造型和可调节的百叶窗式外墙可以反射阳光。绿色灯塔楼启用了全新的试验性能源理念，结合了集中供热、太阳能光电板、太阳能采暖和制冷、季节性存储设施。

为了达到碳平衡，建筑采取了多种绿色设计策略来减少能源消耗，为师生提供了一个全盘的健康室内环境。建筑的朝向保证了太阳能资源的最大化，凹进的门窗顶上装置着遮阳板，减少了太阳直射。VELUX天窗、Velfac窗和宽大的中庭为室内带来了充足的光线和自然通风。此外，建筑还采用了一系列的最新科技：热回收系统、太阳能光电板、太阳能供暖、LED照明、相变材料、地热系统等。建筑所节约能耗的70%都来自于这些有效的节能设计。

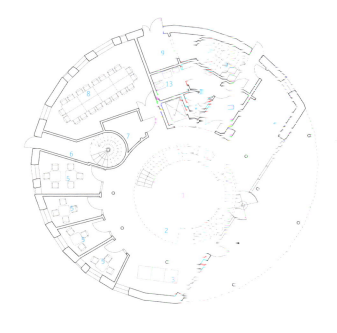

1. main stairs
2. students' lounge/working area
3. students' lounge/quiet area
4. reception
5. meeting rooms
6. fire escapes
7. printing room
8. auditorium
9. plant room
10. garderobe
11. lavatory
12. WC
13. disabled WC

1. 主楼梯
2. 学生活动工作区
3. 学生休过需静区
4. 接待台
5. 会议二
6. 安全口二
7. 打印室
8. 礼堂
9. 机房
10. 衣帽
11. 盥洗室
12. 洗手间
13. 无障碍卫手间

Photo: Adam Moerk

Completion Date: 2009

Architect: Christensen & Co.

Night view 夜景

Birkerod Sports and Leisure Centre

The Birkerod Sports Centre in Rudersdal Municipality close to Copenhagen is a modern sports and culture complex that sets new standards in terms of both practicality and architecture. Fitness, yoga, team handball, concerts and other cultural events are all in a setting of one modern sculptural entity designed by SHL architects.

The façade's long sweeping lines and striking sculptural roof contours evoke a sense of movement and activity, creating a direct link between the building's design and its core function. Upon entering, the space appears bright, airy and open. The interiors are filled with natural light, and transparency creates a sense of permeability and activity.

The sports and activity centre is a multifunctional structure. The new building includes a large multipurpose hall with enough space to accommodate two handball courts with accompanying mobile spectator stands, as well as a VIP lounge. The centre also houses two smaller halls. This means the complex can not only accommodate major sporting events, concerts and other cultural events, but also be adapted for school sporting events and local sports initiatives requiring smaller, more intimate settings.

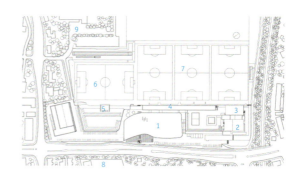

1. Sports Centre	1. 体育中心
2. existing public swimming pool	2. 原有的公共游泳池
3. tranining and hot water pool	3. 训练用热水池
4. existing stand	4. 原有的看台
5. grand stand	5. 大看台
6. show pitch	6. 表演场地
7. training pitch	7. 训练场地
8. public school	8. 公立中小学区
9. high school	9. 高中区

比克勒运动休闲中心

比克勒运动休闲中心位于哥本哈根附近的Rudersdal自治市，是一座实用的现代体育文化建筑。中心由SHL建筑事务所设计，里面可以进行健身、瑜伽、手球、音乐会和其他文娱活动。

建筑细长的线条和极具雕塑感的屋顶轮廓具有运动的流线感，充分体现了建筑的设计和功能。建筑内部空间轻快、明亮而宽敞，充满了自然光线，极具通透性。

运动休闲中心是一个多功能结构，巨大的多功能厅容纳了两个手球场、移动式看台和VIP休息室。中心还有两个较小的大厅，这意味着它不仅能够举行大型体育、演艺活动，而且适合举行中小学运动会和本地小型运动会。

General view 全景图

入口 Entrance

游泳池 Swimming pool

接待处 Eeception

Photo: schmidt hammer lassen architects

Completion Date: 2008

Architect: schmidt hammer lassen architects

Stepped ramp and canopy on south side 南侧的台阶坡道和华盖

Vogaskóli School

沃加斯高利中学

The new extension is founded on the principles of open schooling. The building is centred on a double–height hall surrounded on the lower floor by library, music, kitchen and administration areas and teaching zones for the youngest and middle age groups on the upper. Divisions between areas are minimised and if necessary are of glass or movable partitions. A grand stair connects the hall to the upper level and this can be used either as an audience platform or a stage.

As a consequence of the deep plan, the periphery is predominately glazed with full height windows interspersed with attenuated grilles. The exception is the more massive, north fair–faced–concrete façade due to the proximity of the noisy Skeidavogs road. The heart of the building receives additional daylight through the clearstorey windows in the hall.

The main entrance is located at the junction of the existing building and the new extension. Students may be dropped–off securely in the entrance court and the same route serves the basement staff car park. The entrance is also connected to the school grounds on the south side of the buildings where the land has been lowered to create an external space for teaching and play, sheltered from the inclement weather. On fine days teaching on the upper levels can be extended outside on the east–facing balcony that also doubles as a secure fire escape route.

沃加斯高利中学扩建工程以打造开放式学校为目标。建筑的中央是一个双层楼高的大厅，周围环绕着图书馆、音乐教室、厨房、行政区和中低年级教学区。各个区域之间尽量极少使用隔断，多数都被玻璃门或是可移动门所替代。宽大的楼梯将大厅和上面的楼层连接起来，楼梯也可以作为观众席或是舞台。

建筑外墙的大部分是点缀着细长栅格的落地窗，唯有朝向喧闹的Skeidavogs大街的北侧墙面是由美观的混凝土构成。大厅的天窗为室内提供了足够的自然采光。

建筑的主入口和原有的教学楼相连，学生可以由入口中庭或是教职工地下停车场进入。入口同样地与建筑南侧的操场相连。操场的地势较低，带屋顶的保护区可作为户外教学或运动之用。在晴朗的天气，师生们可以在二楼东侧的阳台上活动，阳台也可作为防火安全通道。

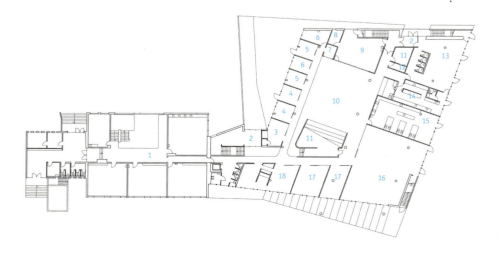

1. existing building	1. 原有建筑
2. entrance lobby	2. 门厅
3. reception	3. 前台
4. headmaster	4. 校长室
5. office	5. 办公室
6. meeting room	6. 会议室
7. DJ booth	7. DJ室
8. music practice room	8. 音乐练习室
9. music	9. 音乐教室
10. hall	10. 大厅
11. store	11. 商店
12. cleaners	12. 洗衣房
13. staff room	13. 更衣室
14. kitchen	14. 厨房
15. teaching kitchen	15. 教学厨房
16. library	16. 图书馆
17. seminar	17. 小组讨论室
18. special teaching	18. 特殊教学教室

南侧的台阶坡道和华盖 Stepped ramp and canopy on south side

入口 Entrance

大厅 Hall

In the daylight 日光下的建筑

The Long Barn Studio

The building is a simple glazed rectangular unit with frameless 3.2-metre-high glazed panels along its main elevations, which hold a very slight green tint to reflect the seasons of the surrounding glorious landscape. Internally, they provide all those who are lucky enough to enter with a panoramic view that is breathtaking.

Like bookends, the building is capped at both ends with full height larch cladding, the course of the building intersects a series of larch timber "pods" which house meeting rooms, a library, a printing area and a WC. Harping back to the adjacent barn building's history, the new studio utilises cor-ten detailing, further enhancing the sense of place, and reflecting the old and discarded agricultural machinery and steelwork.

Floating Wenge storage pods sit dividing reception, to kitchenette to work spaces, Knoll floating wide and deep desking with bespoke cabinets integrating designers files, tools and wate bin, married with wireless screen built-in computers and wireless keyboard ensure all is at hand in this exclusive user friendly environment.

The studio utilises rainwater harvesting, its own wind turbine, whole building air heat recovery circulation system, central vacuum and centralised lighting control. Utilising low energy lighting, organic paints and non-toxic chemical sealers further reinstates the philosophy of Nicolas Tye Architects. Their passion for this is clearly evident in their own building, an achievement that many companies long for.

长谷仓工作室

建筑是一个简单的长方体结构，正面外观使用3.2米高的玻璃板装饰，映射出周围环境四季的美景。同时这些宽大的玻璃结构更开阔了室内工作人员的视野，使他们可以尽情欣赏四周令人惊叹的美景。

工作室恰似一个书夹，屋顶和地板都铺了落叶松木板，室内布置着几间会议室，一个图书馆，打印区和一个公用洗手间。

宽大的办公桌，旁边定制的柜子加上设计室的文件夹、工具和废纸篓，连同无线电脑系统为员工创造了一个友好的独立的工作环境。

工作室配有雨水收集设备、风力发动机、空气热能恢复循环系统、中央清洁系统和中心照明控制系统。低能照明灯，有机涂料和无毒的化学密封剂的使用符合了 Nicolas Tye Architects 的原则。他们的理念在自己的办公建筑中清晰地体现出来，这也是许多公司所希望获得的成就。

1. 池塘
2. 菜园
3. 芦苇地
4. 风轮机
5. 长谷仓工作室
6. 户外就餐区
7. 机动车通道
8. 停车场
9. 长条形车库

1. pond
2. vegetable garden
3. reed bed
4. wind turbine
5. Long Barn Studio
6. external dining area
7. vehicular access from road
8. studio parking
9. the long barn

夕阳下的建筑 In the sunset

办公区 Office

书架 Book shelves

Photo: Philip Bier / Bier Photography London

Completion Date: 2007

Architect: Nicolas Tye Architects

Corporate

Exterior view 建筑外观

University of Oxford New Biochemistry Building

The distinctive 12,000-square-metre facility with its glass façades and coloured glass fins brings together 300 lecturers, researchers and students previously based in a number of separate buildings. Inside, a 400-square-metre atrium with breakout spaces and specially commissioned artworks encourages collaboration between the researchers.

The Biochemistry Department at Oxford University is the largest in the UK and is internationally renowned for its research in the understanding of DNA, cell growth and immunity. Previously the department's scientists have had to conduct research in outmoded buildings spread across the Science Area in the centre of Oxford.

The brief for the new building was to achieve a new ethos of "interdisciplinary working" where the exchange of ideas is promoted in a large collaborative environment. At the same time space was required to enable the research groups to focus on their cutting-edge work in state-of-the-art laboratories.

牛津大学新生物化学系教学楼

新建的生物化学系教学楼总面积为12,000平方米，外墙上装饰着彩色玻璃片，将300名曾分散在不同教学楼里的师生聚集在了一起。楼内400平方米的中庭里设有休息区和独特的艺术品，促进了研究人员之间的相互交流。

牛津大学生物化学系是英国最大的生物化学研究基地，在DNA解码、细胞生长和免疫研究方面享有盛名。在新的教学楼建成之前，科学家们不得不在牛津大学科学区分散的建筑里进行研究。

新建筑的设计实现了"跨学科工作"，在这个巨大的合作式环境中，各种理念得以有效地交换。高科技的实验室保证了研究小组所进行的尖端研究工作。

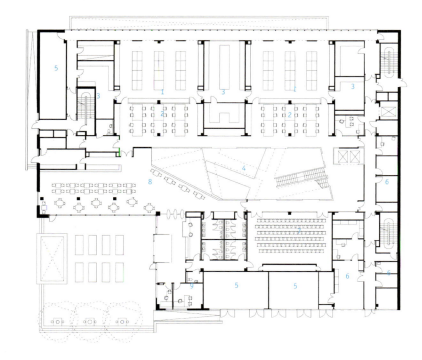

1. 主实验室
2. 写作室
3. 辅助/特殊实验室
4. 中庭
5. 机房
6. 附属空间
7. 小组讨论室/会议室
8. 咖啡厅
9. 办公室

1. main laboratory
2. write up spaces
3. support/specialist laboratory
4. atrium
5. plant
6. ancilliary space
7. seminar/meeting room
8. café
9. offices

Coloured glass fins fixed vertically within the mullions wrap the full perimeter of the building

垂直于窗框安装的彩色玻璃片装饰着整建筑的外围

The exterior glass façade around the entrance courtyard, with Nicky Hirst's artwork The Glass Menagerie

前庭的玻璃外墙，装饰着尼基·赫斯特的作品《玻璃动物园》

Nicky Hirst's artwork Portal features on the glazed wall of the atrium

中庭的玻璃墙上装饰着尼基·赫斯特的作品《大门》

The atrium with the installation, 10,000,000 by Annie Catrell

中庭

Photo: Tim Crocker (opposite, top-left), Keith Collie (top right, bottom left, bottom right)

Educational

Completion Date: 2009

Architect: Hawkins\Brown

Façade 外观

BDP Manchester Studio

The building provides large open-plan studio space and ancillary accommodation including a hub space at ground floor level. This interactive area including café, staff restaurant and extended reception space, overlooks the canal at raised ground level.

A striking feature of the building is the punctuated stainless-steel south façade that rises above the Ducie Street colonnade to contain the open-plan studio areas before sweeping over to form the roof of the building. The reflective external finish, heavily insulated build-up and narrow vertical apertures all serve to minimise solar heat gain, and to maximise privacy with the residential buildings opposite.

By contrast, the northern façade of the building is transparent. The floor-to-soffit glazing takes maximum advantage of north light to illuminate the full extent of studio spaces and reveals wonderful views of the city centre. A fully-glazed circulation staircase cantilevered over the canal provides the circulation for all floors.

Sustainability has been a key driver in all aspects of the design and delivery of the new studio which is an expressive response to context and microclimate. Rainwater is harvested from the roof and used to flush toilets throughout the building. It is the first naturally ventilated and night time cooled office building in Manchester to achieve an Excellent BREEAM rating.

BDP工作室

建筑内部为宽敞的开放式格局，一层为中心服务区，包括咖啡厅、员工餐厅、接待台。建筑最为显著的特色即为南侧外立面的不锈钢结构，似乎从迪西大街处的石廊柱处拔地而起，一直伸展到建筑顶端，形成屋顶结构。光滑的外表、厚重的隔热层以及垂直排列的细孔旨在反射强烈的光线，同时增强私密性。

北侧立面恰恰与此相反，特意设计成透明样式。光线透过宽敞的玻璃结构照射进来，使得室内空间更加明亮，同时市中心的美丽景致亦随之涌入。全玻璃的楼梯将不同楼层连通起来。

尤为重要的一点是，可持续发展是贯穿这一设计的主要元素。雨水经过屋顶上的特殊装置贮藏起来，用于冲洗整幢大厦的卫生间。

Working area 工作区

接待区 Reception

正立面 Front view

Royal Alexandra Children's Hospital

皇家亚力山德拉儿童医院

The site was a very tight handkerchief of land wedged between two roads and a number of existing buildings. Planning constraints dictated a maximum height. The very steep hill overlooking the Channel was the location's great saving grace, affording a sunny panoramic sweep over the sea and historic Brighton. The introduction of an atrium at the heart of the hospital initially seemed unfeasible as it sacrificed a significant amount of area. Addressing this shortfall the designers generated the idea of bowing the building out as it grew upwards, which chimed nicely with the desirability of shallower plan spaces on the lower levels.

As the design began to take shape, it soon became known amongst the team as the "Children's Ark", an image that crystallised a number of important themes: from the idea of a "sustainable" community centred around the family to the nautical spirit of Brighton and the boat-like form of the building. With its soft, rounded corners and sheltering roof, it projected a reassuring and optimistic image, around which the designers could integrate all the elements of the brief in a coherent and appealing way that would resonate with children, families and staff. At the centre of the hospital, the atrium binds all levels together, ensures good day-lighting throughout the building and creates a strong sense of the whole, which is immediately perceived by anyone entering the Alex or moving up through it in the public lifts.

建筑所处地块呈现楔形，"夹"在两条公路与一系列建筑之间。特殊的条件限制了高度的延伸，而俯瞰海峡的优越位置则满足了视觉的享受。中庭的引入最初看起来似乎没有任何意义，又占据了大量的空间，基于这一点，设计师便构思了如下理念——随着高度的上升，建筑结构随之向外"扩张"。

当建筑逐步成形之后，便被誉为"儿童方舟"。这一设计蕴含了多重理念——民居中的"环保社区"，布赖顿的"地域特色"以及独特的"船形结构"。柔和的圆形墙角以及屋顶给人安全而又积极向上的感觉，所有的元素全部以流畅的方式融合，满足儿童、家人及工作人员的需求。中心处，心房将不同楼层连接起来，确保整幢建筑的光照，同时突出统一感。

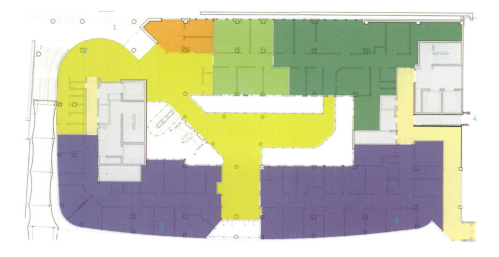

1. 主入口
2. 咖啡馆
3. 楼梯
4. 相邻的医院剧院楼
5. 门诊部

1. main entrance
2. café
3. stairs
4. adjoining main Hospital Theatre block
5. outpatients department

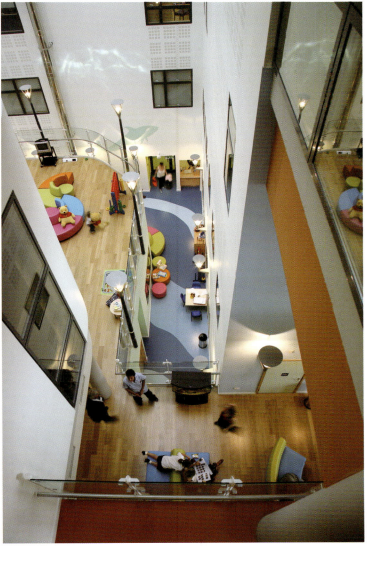

Photo: David Barbour

Completion Date: 2007

Architect: BDP

General view 全景图

Institute of Cancer Studies, UCL

伦敦大学学院癌症研究所

Conceptually, the building design was influenced by its role as a cancer institute and the relationship between science and the study of cancer. In particular there was an interest in the images that have been generated by the processes used in modern medical research and which have now become part of the culture. Images of cells, wave patterns and the chromosome permeate the forms of the building. For example, the terracotta louvre-bank suspended across the main façade has a rhythm that can not only be read as a vertical "bar code" configuration or genetic sequence image, but also reflect the waveform that is so significant to modern science. Likewise, concrete soffits are left exposed, retaining the details of their own construction and revealing points of reinforcement reflective of the mechanisms and cellular structures of biology: they are literally scooped out where the material serves no structural purpose.

The main entrance is located at the juncture between the reinstated wall flank of the Grade II listed building and the new Institute. This gap is glazed with a single glass sheet spanning from the new structure to the old. It reveals a highly engineered staircase that provides the main circulation core and an architectural focal point. The transparency of this circulation area articulates it as an open, shared space that is visually accessible from the street. This emphasises the Institute as a live and active building. The staircase itself consists of cast stainless steel treads cantilevered from a structural spine of pre-cast concrete to create a dramatic feature.

这一建筑设计受到其自身用途——癌症研究所及科学与癌症研究之间的关系的影响。特别指出的是，设计师们似乎对现代医学研究中产生的图像十分热衷，将其大量运用到建筑中——细胞图像、波动图示及染色体形状"充斥"其中。举个例子，建筑正面悬浮的一排排赤色百叶窗板节奏感十足，好比是"条形码"或"序列图像"或是现代科学研究领域至关重要的"波动图示"。

建筑主入口设在一幢二级文物建筑及新建研究所之间，单体玻璃结构横跨新旧建筑。设计精良的楼梯位于其中，构成了核心结构与主要通道。通道区开阔而通透，可以欣赏街边的景致，进一步突出了建筑本身的活泼与动感。楼梯台阶由不锈钢材质打造，悬浮在预制水泥"脊柱"上，独特的设计令人不禁讶异。

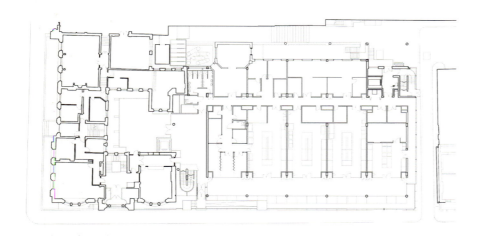

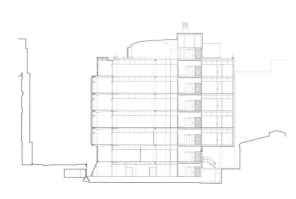

Ground floor plan 一楼平面图

建筑正面 Front view

走廊 Corridor

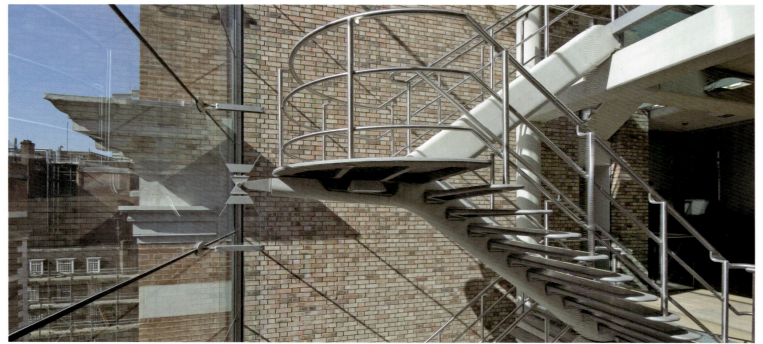

楼梯 Stairway

View form Sandford 从桑福德路上看建筑

Multi–Purpose Hall, Sandford Park School

桑福德公园学校多功能厅

The Multi–Purpose Hall creates a new identity for the school from Sandford Road, advantageously reinforcing the school within its suburban context of Ranelagh. Located at the middle ground between the gate lodge and school house, it allowed for the insertion of a contemporary architectural volume to mediate between the different generations of school appendages.

The brief was to provide a multi–purpose hall for use by various school departments including sport; music; recitals, theatrical performance, examinations, assembly, and ancillary presentations. The space was for an equivalent Department of Education and Science 406-square-metre sports hall allowing for three badminton / single basketball court, with ancillary storage and changing facilities. Due to the context and functions, the quality of both internal and external acoustics was critical. Noise penetration onto neighbouring properties required an increase in the mass of internal walls and roof, and reverberation times internally were reduced through the softening of internal finishes.

Externally the expression of materiality has been reduced to enhance the juxtaposition between the hall and the immediate context. The use of a fibre cement cladding meant the entire form could be homogenously wrapped with only the entrance screen and Profilit glazing of the hall being inserted into the skin. A visual connection is created between the footpath and the hall through the suppression of the internal floor level. It is through this separation of ground that an enhanced spatial arrangement is made between the external school environment and the new hall.

从桑福德大路望去，新建的多功能厅构成了学校的全新标识，使得学校在周围的背景环境中更加突出。多功能介于校舍与大门之间，现代风格的结构犹如纽带一般连接着学校内不同的建筑。设计的理念即为建造一个供学校不同学科部门共同使用的多功能大厅，包括羽毛球馆、篮球场、储存柜及更衣橱等。由于其特殊的用途，室内外音响设备显得尤为重要。

外观材质的运用凸显简约，借以强调建筑与背景环境的跳跃感。清一色的水泥覆层使得入口犹如镶嵌在表面一般。设计中尽量降低减少内部楼层，在视觉上将整个建筑与四周的小路连通，营造统一感。

Front elevation's night view 建筑正面夜景

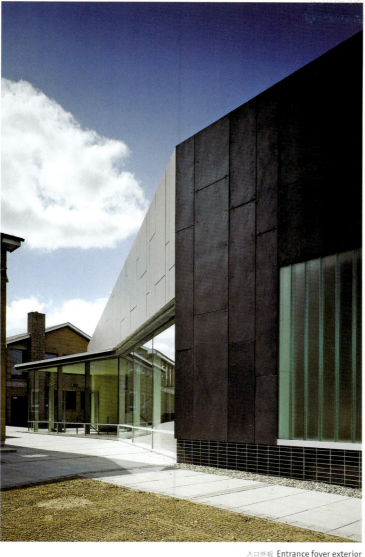

入口外观 Entrance foyer exterior

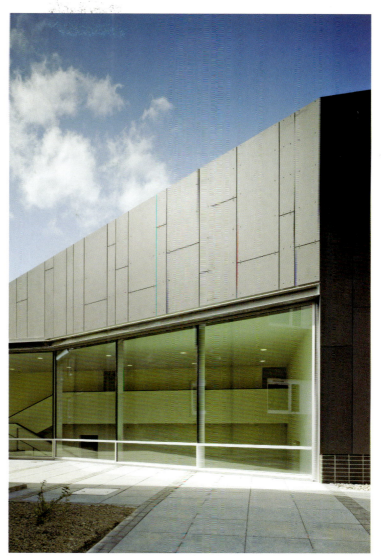

内外连接处 Exterior + interior connection

大厅，集会空间 Foyer, assembly space

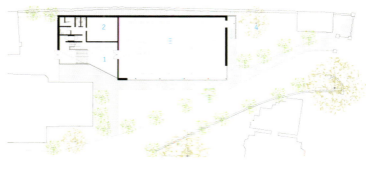

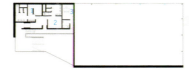

Ground floor plan
1. plant
2. store
3. multi-purpose hall
4. foyer

一楼平面图
1. 机房
2. 仓库
3. 多功能厅
4. 大厅

Upper floor plan
1. changing 1
2. changing 2
3. sports office

二楼平面图
1. 更衣室1
2. 更衣室2
3. 体育厅

Photo: Rós Kavanagh

Completion Date: 2009

Architect: DTA Architects
Team: Colin Mackay, Derek Tynan, Dermot Reynolds, David Graham

View across the street 从街对面看建筑

Salvation Army Chelmsford

The project pioneers modern methods of construction in its use of cross-laminated timber panel system KLH. The system is akin to jumbo plywood and offers all the advantages of reinforced concrete construction without the environmental cost. All walls and floor plates arrived on site as prefabricated panels with cut-outs for doors and windows, ready for quick assembly, allowing the building's frame to be erected in just twenty-four days.

The two most conspicuous elements of the scheme which maximise the cross-laminated technology are the entrance canopy on the north elevation, which provides lateral stability to the front elevation, and the building's signature undulating butterfly roof, which rises to accommodate six generously sized dormer windows, measuring 4.2 metres wide. The butterfly roof is further dramatised by a zinc cladding, which cloaks the building and sweeps down and anchors it on its north elevation on Baddow Road and south elevation on Parkway. The zinc cloak forms a striking enclosure, which gives the building a very robust toughened and urbanistic character and distinguishes it from the surrounding brick buildings.

The building's toughened zinc shell breaks at the Baddow Road entrance lintel where floor-to-ceiling glazing creates a dialogue between the foyer café and the street, projecting an image of openness crucial to the work of the organisation.

救世军教堂

救世军教堂的设计引领了现代建筑方式的潮流——运用交叉层压木板结构。这一结构类似于大体积的胶合板，拥有混凝土具备的一切优点，同时又极为环保。教堂完全采用此类预制板材打造，因此整个框架仅用24天就全部建完。

这一设计中最为显眼的两个结构即为北立面的出入口顶篷（增添整体结构的稳定性）以及蜿蜒起伏的蝶形屋顶（用于容纳6扇大型屋顶窗）。此外，屋顶更采用锌片材质覆层，一直延展到南北两侧里面，并加以固定。锌片覆层使得整个建筑犹如披上一件"坚硬"的外衣，将其与周围的砖结构区分出来。入口处，大幅的玻璃结构增强了室内外的联系，更突出了建筑的通透感。教堂东西两侧立面均采用颜色亮丽的岩板饰面，朝向Goldley 大路一侧的立面呈现CNC控制平面图案，格外引人注目。

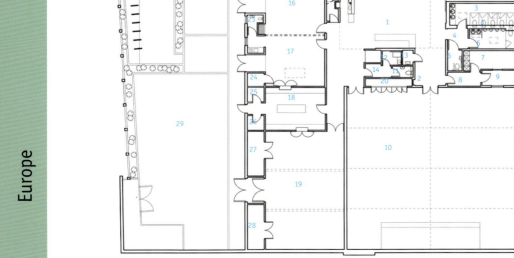

1. foyer	1. 门厅
2. multi-purpose room	2. 多功能室
3. female WC	3. 女洗手间
4. toilet lobby	4. 盥洗大厅
5. male WC	5. 男洗手间
6. disable WC	6. 残疾人洗手间
7. centre manager	7. 中心控制室
8. lobby	8. 大厅
9. drop in space/cry room	9. 告解室
10. worship hall	10. 礼拜堂
11. WC	11. 洗手间
12. baby change	12. 婴儿更衣室
13. cleaner	13. 清洁房
14. lobby	14. 大厅
15. 20.22. 24.27.28.store	15. 20.22.24.27.28.贮藏室
16. lounge	16. 休息室
17. 19.hall	17. 19.大厅
18. kitchen	18. 厨房
21. bin store	21. 回收处
23. baby WC	23. 婴儿洗手间
25. cool store	25. 冷贮藏室
26. dry store	26. 干贮藏室
29. courtyard	29. 庭院

鸟瞰图 Aerial view

入口 Entrance

礼拜堂 Worship Hall

Complex

Photo: Keith Collie

Completion Date: 2009

Architect: Hudson Architects

Folkestone Academy

福克斯顿学院

The new Folkestone Academy will offer a range of curriculum, teaching and learning facilities that combine the best features of the independent education sector with the Pastoral System of intra-school support and encouragement. A core curriculum – based on the National Curriculum – will be supplemented by specialisations in the Creative Arts and European Culture and Language. As with other schools in the academy programme, Folkestone Academy will also contribute to the regeneration of the local area by offering a range of services to the wider community.

A total of 1,480 pupils – aged between eleven and nineteen – will be accommodated in eight Houses. The sophisticated pastoral system lies at the heart of the Academy's philosophy, with each student benefiting from a level of individual support that comes from a small readily-identifiable group which offers them guidance. Pupils will be encouraged to identify with their House, returning there for breaks and meals – when healthy diets for all pupils is a priority. The architectural design supports this system by providing each House with separate dedicated facilities, including outside space for social activities, within the overall building form.

新建的福克斯顿学院将提供一系列教学设施，融合了独立教育系统的最佳特征，具有多种辅助设施。学院将以英国国家课程为基础，以创意艺术和欧洲语言与文化为专业课程。与其他学院相同，福克斯顿学院也将为社区提供一系列服务设施。

1,480名11到19岁之间的学生将被分配在八个部门之内。紧密的牧师学院系统是学院的设计核心，每个学生都会得到一个小型团体的指导。学院鼓励学生在自己的部门里享用早、午餐。建筑的设计也保证了这个系统的进行，每个部门里都有独立的设施，甚至包括楼内的露天社交活动区。

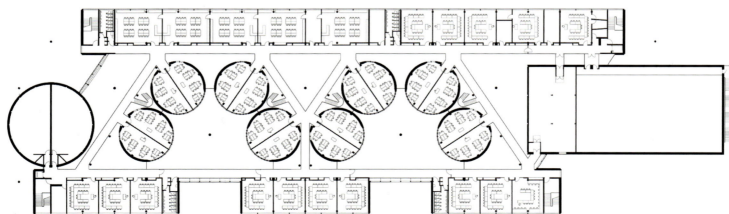

Aerial view of school 鸟瞰图

Hazelwood School

黑泽伍德学校

Hazelwood School is an education facility for up to sixty students with multiple disabilities. The students are aged between three and nineteen and the school provides education from nursery through to secondary stages. The school itself and the Life Skills House (an independent facility used for life learning and respite) have a combined area of 2,665 square metres and are set within a landscaped green adjacent to Bellahouston Park.

The design of the building has focused on creating a safe, stimulating environment for its pupils and staff. The focus and ambition from both client and architect at the outset was to develop a building that would eliminate as much as possible the institutional feel that a project of this nature inherently possesses. The designers worked to avoid conventional/standard details, creating a solution bespoke to the project requirements and developing a building that entirely embodied the users' needs.

The existing site was surrounded by mature lime trees and had a large lime tree and three beach trees in the centre. The building snakes through the site, curving around the existing trees. Its form creates a series of small garden spaces suitable to the small class sizes and maximises the potential for more intimate external teaching environments. Internally the curved form of the building reduces the visual scale of the main circulation spaces and helps remove the institutional feel that one long corridor would create – in addition, this also significantly reduces visual confusion by limiting the extents of the space.

黑泽伍德学校是为60余名残障学生所设计的教学基地。学校为这些3岁到19岁不等的学生提供从幼儿园到中学阶段的教育。学校和校园内的生活技能学习基地总面积为2,665平方米，位于绿树成荫的贝拉霍斯顿公园。

建筑的设计旨在为学生和教职工创造一个安全又能激人奋进的环境。项目委托人和建筑师试图建造一个有别于其他教学设施、充分与自然环境相融合的学校。建筑师在设计过程中有意地避开了传统校园的细节，营造了一座符合项目要求、满足使用者愿望的建筑。

建筑场地四周环绕着高大的菩提树，场地中央也有一棵菩提树和三棵棕榈树。建筑蜿蜒于场地之中，没有破坏任何一棵树木。建筑的造型自然地勾勒出一系列与教室规模相匹配的小花园，营造了更多的户外教学空间。在室内，建筑蜿蜒的造型缩减了主流通区域的视野，减少了又长又直的走廊所营造的教学气息。此外，弯曲的走廊也限制了空间范围，减少了空旷空间所造成的视觉混乱。

1. hydrotherapy pool	1. 水疗池
2. shower	2. 淋浴
3. gym	3. 健身房
4. physiotherapist room	4. 理疗室
5. disabled WC	5. 无障碍洗手间
6. doctors room	6. 诊疗室
7. kitchen	7. 厨房
8. depute head teacher	8. 副校长办公室
9. head teacher	9. 校长办公室
10. staff entrance	10. 员工入口
11. dining/multi-purpose space/assembly	11. 多功能餐饮集会空间
12. staff room	12. 办公室

北立面 North elevation

鸟瞰图 Aerial view of school

移动隔断墙 Sensory wall designed to aid mobility training

Photo: Andrew Lee

Completion Date: 2007

Architect: gm+ad architects

Educational

Langley Academy

The Langley Academy is an exemple of sustainable design, a theme which is showcased by the building itself. The Academy's curriculum highlights rowing, cricket and science and is the first academy to specialise in museum learning. As well as running its own museum, ancient artefacts and objects are brought into the classroom to spark questions, debate, analysis and provide connections across the curriculum. The scheme also provides unparallelled access to significant cultural institutions across the country, involving hundreds of students.

With an enclosed full-height atrium at the heart of the three-storey building, the social life of the school revolves around this assembly space for 1,100 students. A recurrent element in several other Foster + Partners' academy buildings, the atrium is defined by a sense of transparency and openness – like a gallery of learning – which in this case also resonates with the museum theme. Inside the atrium there are three yellow drums raised above the floor on circular columns. These two-storey pods house the Academy's ten science laboratories, reinforcing the importance of science teaching. A dedicated sports and culture block contains specialist facilities for music and drama including a fully-equipped theatre, a TV and sound recording studio, soundproof practice rooms and a rehearsal space, sports hall and lecture theatre. The academy's two light and airy covered streets extend from the atrium and are lined with thirty-eight classrooms.

兰勒学院

兰勒学院是可持续设计的典范。学院课程以赛艇、板球和理科为主，是英国第一所专修博物馆学的学院。

三层高的建筑中心是一个中庭，中庭可以为1,100名学生提供社交集会空间。与Foster + Partners所设计的其他教学建筑相同，学院的中庭通透而开阔，仿佛一个教学美术馆，与学院的博物馆学研究达成了共鸣。中庭内有三个黄色的鼓形圆柱结构，里面共有10个理科实验室，强化了学院的理科教学。专门的体育文化楼里有为音乐和戏剧学习所专设的专门设施，包括剧院、电视录音工作室、隔音练习室、排练室、体育馆、阶梯教室等。学院中庭走廊的两侧是38间教室。

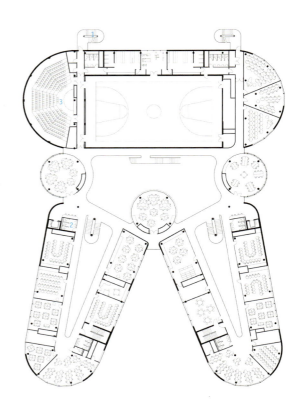

一楼平面图
1. 楼梯
2. 洗手间
3. 多功能厅

Ground floor plan
1. stairs
2. WC
3. multi-purpose hall

Architect: Foster + Partners

Completion Date: 2008

Façade 建筑外观

David Mellor Design Museum

大卫·梅勒设计博物馆

The new David Mellor Design Museum is at Hathersage in the Peak District National Park. The new building has been designed to show the whole historic collection, privately owned by the Mellor family, of David Mellor's designs form the 1950s onwards and includes more recent work of his son, Corin Mellor. More broadly the building will help to reinvigorate local Sheffield traditions of cutlery making, forming a visible link between modern design history and present-day production. The building will have the effect of safeguarding existing and creating new jobs.

The main structure of the building is composed of a steel frame supporting a linear lead-pitched roof. Onto the steel frame reclaimed pitch pine, which the client already owned, was bolted to form a composite flitched structure for both columns and roof. Beneath an overhanging steel gutter, which also serves to shade the building, full-height glazed doors allow the linear building to be opened up. At the bottom of these doors, both inside and out, a pitch pine bench is supported from the steel frame to provide a sheltered space along the south front of the building. Toilets, shops and the exhibition displays are accommodated in the rear, north-facing section of the building and the interior design and fabrication of the building was undertaken by the client.

新建的大卫·梅勒设计博物馆皮克区国家公园内，展示了梅勒家族所收藏的自1950年以来大卫·梅勒以及其子柯林·梅勒的设计作品。博物馆将帮助谢菲尔德复兴其传统的餐具制造业，将现代设计史与当前的产品联系起来。博物馆将具有保护现有产业、创造新的就业机会的效果。

建筑的主体结构是钢框架支撑着一排斜屋顶。建筑师用委托人原有的回收松木板条装饰了钢框架，在屋顶内部和梁柱上形成了复合板条结构。外悬的钢板之下是落地式玻璃门，这一设计使线性建筑看起来更加开阔。靠近玻璃门的内外两侧都设置着由钢架支撑的松木长椅，在建筑的南面为游客提供了一个阴凉的休息空间。洗手间、商店和展览厅都集中在建筑内侧朝北的一面。室内设计和内部装饰则由委托人亲自操刀。

1. 原有的窗户
2. 墙壁
3. 原有的商店
4. 展览区
5. 灵活展览区
6. 休息区
7. 户外休息区
8. 新游客中心
9. 墙壁
10. 原有的建筑

1. existing window
2. wall
3. existing shop building
4. exhibition
5. flexible exhibition space
6. seating
7. outside seating
8. new visitor centre
9. wall
10. existing retort building

外部的树木 Exterior with trees

室内 Interior

夜景 Night view of exterior

Southwest daytime view 西南日间建筑

Fawood Children's Centre

法伍德儿童中心

The Fawood Children's Centre is an integral part of the overall masterplan for the Stonebridge estate, in Harlesden North London. Planned demolition of the neighbouring housing blocks took place in 2007 and a new park was developed. The Fawood Children's Centre is sited within the new parkland where acts as a focal point in the landscape. It was proposed that the new park will include nature trails and adventure play areas, and that the play areas within the park be located adjacent to the Children's Centre for shared use.

The Children's Centre is located adjacent to the site for a proposed new Health and Community building, also designed by Alsop, which is intended to be an important community focus within the overall regeneration of Stonebridge.

The brief from Stonebridge Housing Action Trust (HAT) was for a nursery school that would replace and expand existing nursery and community facilities on the Estate. The new Fawood Children's Centre, was to provide, under one roof a nursery for three–five year olds: nursery facilities for autistic and special needs children, and a Children's Centre with adult learning services – a base for community education workers and consultation services. This brief was in line with the Government's Sure Start proposals, which advocate combining, within one facility, play and educational experiences for nursery age children, with supporting amenities for the local community, parents and childcare workers.

法伍德儿童中心是伦敦北部夏里斯登石桥开发项目的重要组成部分。石桥房产开发基金要求建筑师设计一个幼儿园来替代原有的幼儿园，扩建社区服务设施。法伍德儿童中心将为3到5岁的自闭症儿童和特殊需要儿童提供幼教设施。儿童中心内还将为社区教育工作者和咨询服务业提供成人教育服务。

法伍德儿童中心半透明、半封闭的屋顶覆盖着整片场地，与网眼墙一起划分出开阔的游戏场和被分割成独立模块的室内幼教设施。

预制式室内环境的组合促进了项目的建造过程，以低成本为儿童打造了一个灵活而令人难忘的日常活动空间。

1. 入口平台
2. 接待大厅
3. 行政副校长办公室
4. 校长室
5. 会议室/办公区
6. 厨房
7. 更衣室/儿童洗手间
8. 无障碍洗手间
9. 幼儿园住宿处
10. 原有建筑
11. 贮藏室
12. 幼儿园住宿处: 帐篷

1. entrance deck
2. reception lobby
3. administration/deputy head teacher office
4. head teacher office
5. meeting / office
6. kitchen
7. cloarkroom /children WC
8. unisex disabled WC
9. nurserry accommodetion
10. existing retort building
11. bin store
12. nursery accommodation: yurt

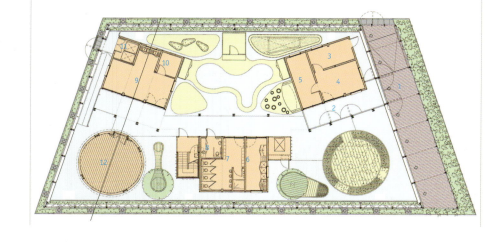

建筑南面 South elevation

舞台 Stage

Photo: Alan Lai of Alsop Design Ltd, Roderick Coyne of Alsop Design Ltd.

Educational

Completion Date: 2005

Architect: Alsop

帐一目 Yurt 269

Main façade 主建筑外墙

Bryanston School

The New Science School is located within the picturesque grounds of Bryanston School. It houses the three main science disciplines, part of the maths department, and a shared lecture theatre.

The semi-circular plan for the building grew out of the aspiration to conclude the formal axis of the Norman Shaw main building and its later additions, whilst addressing the informal series of individual buildings along the route of the main drive. The three-storey building makes use of the naturally sloping landscape by having the main entrance on the middle floor so that only two levels are visible from the main school courtyard. The building focuses on a central south-facing science garden that features a pond and geodesic greenhouse.

The building has a simple construction of load-bearing brick and block walls, concrete floor slabs and a zinc-clad pitched roof. The main façade to the inner courtyard is solid load-bearing brickwork in English Bond using a high lime-content mortar. The concrete floors bear via pads onto brickwork piers, and flat arches span above window openings.

布莱恩斯顿学校

新建的理学院楼位于风景如画的布莱恩斯顿学校校园内，里面设置着三个主要理科学科教室、部分数学系教室和一个公用的阶梯教室。

这座半圆形建筑的设计灵感来自于将诺曼·肖主楼与其附加工程的中轴线，将主要车道上一系列零散的建筑连接了起来。建筑利用了自然坡度，将主入口设在了二楼，从学校的主庭院看去，三层高的建筑就像只有两层一样。建筑的中心是一个朝南的科学花园，里面的水塘和温室独具特色。

建筑采用了简单的砖砌承重墙、混凝土楼板和镀锌板斜屋顶。内庭的主墙采用了英式砖砌法，并运用了大量的石灰泥。混凝土楼板通过衬垫支撑着砖石结构，窗口则采用了平拱结构。

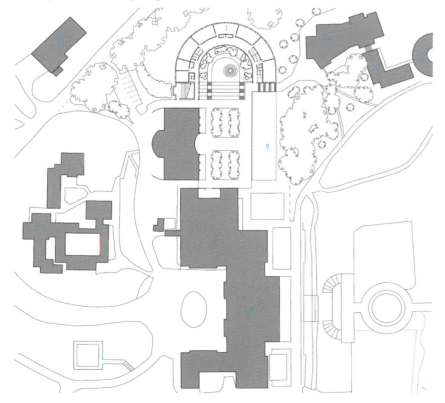

1. 新理学院
2. 主楼
3. 民主科技中心
4. Coade大厅
5. 艺术和戏剧教室
6. 艺术学院
7. 音乐学院
8. 其他地产
9. 未来的音乐学院

1. new science school
2. main building
3. CDT
4. Coade hall
5. arts and drama
6. art school
7. music school
8. estates
9. future music school

Photo: Anthony Weller

Completion Date: 2007

Architect: Hopkins Architects

Falmouth School Design & Technology Building

Designed by Urban Salon, the project came out of the Sorrell Foundation's 'joined up design for school programme where pupils are given control and responsibility as clients. The 95-square-metre extension is a solid prefabricated cross–laminated timber construction with the timber exposed to the interior and standing seam pre–weathered zinc to the exterior. The timber structure was chosen for its environmental performance and was designed with half lap joints and has no visible fixings in the building. The ceiling within the extension rises from 3.2 metres to 5.2 metres, allowing even north light to flood the space, and creates the distinctive saw–tooth form of the building. The only steel structural elements are T–sections that support the ridge of the roof panels from the timber beams below. Specialist contractor KLH manufactured the timber structure off site. This modular construction method meant that the structure was assembled on site within two weeks.

The existing Design & Technology studio was upgraded as part of the scheme. The refurbishment rather than wholescale removal of the existing workshops had the added benefit of less waste and lower embodied energy in building materials required. Coupled with the use of the prefabricated elements, this led to less construction time and less disruption to students and their education. The existing block now has a new roof, wood–fibre insulation, double glazing to prevent heat loss in winter and solar shading to the south of the building to prevent summer overheating.

法尔茅斯学院设计和技术楼

95平方米的扩建结构采用了预制交叉胶合板结构，内部是木板，外部是立缝抗风化锌板。木结构具有优秀的环保性能，半叠拼接技术让建筑内部看不到一个固定螺栓。扩建结构的天花板高3.2米至5.2米，不符合自然规律，为建筑创造了独特的锯齿形结构。唯一的钢铁结构元素是支撑屋顶的T字断面。承包商KLH在工厂里制造了这些木结构。这种模块化施工让整体现场组装可以在两个星期之内完成。

原有的设计和技术工作室同样进行了升级。对原有建筑的翻新比重建减少了废物的产生和能源的消耗。与预制式元素相结合，减少了大量的施工时间，避免了对学生和教学活动的影响。设计师为原有建筑添加了一个新屋顶。木纤维隔热板和双层隔热玻璃有效避免了冬季的热流失；南面的遮阳板隔绝了夏季多余的热量。

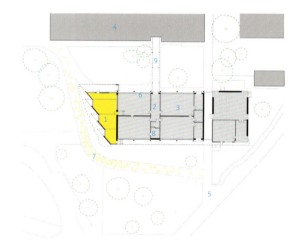

1. design studio extension
2. original design technology block a
3. original design technology block b
4. main school buildings
5. pedestrian entrance
6. polycarbonate canopy above
7. new pathway with markings
8. new disabled toilet
9. existing covered walkway

1. 设计工作室扩建结构
2. 原有技术楼a
3. 原有技术楼b
4. 学校主楼
5. 行人入口
6. 树脂弯顶
7. 新建的带标记的路
8. 新建的无障碍洗手间
9. 原有的走道

Educational

Completion Date: 2008

Architect: Urbansalon Architects

Exterior 建筑外观

John Lewis Department Store and Cineplex and Pedestrian Bridges

约翰·路易斯百货公司和电影院

Department stores are conventionally designed as blank enclosures to allow retailers the flexibility to rearrange their interior layouts. However, the physical experience of shops is an increasingly important consideration to compliment the convenience of online shopping. The concept for the John Lewis Department store is a net curtain, providing privacy to the interior without blocking natural light.

The design of the store provides the retail flexibility required without removing the urban experience from shopping. The store cladding is designed as a double-glazed façade with a pattern introduced, making it like a net curtain. This allows for a controlled transparency between the store interiors and the city, allowing views of the exterior and natural light to penetrate the retail floors whilst future-proofing the store towards changes in layout. Thus, the store is able to reconfigure its interiors without compromising on its exterior appearance.

In order to establish a consistent identity between the cinema and department store, the curtain concept is extended to the cinema. This curtain both associates the cinema and department store and resonates with the theatre curtains which were a traditional interior feature of cinemas.

依照惯例，百货公司都是一个空白的外围设施，以便于零售商能够灵活地进行室内布局。但是，面对便捷的网上购物的挑战，商场给人带来的实际体验变得日益重要起来。约翰·路易斯百货公司的设计理念是设计一个印花幕墙结构，既能保证内部的隐私，又不阻挡自然光。

百货公司的设计既保证了零售区域的灵活性，又保持了城市购物体验的优势。商场的外墙采用了带花纹的双层玻璃结构，看起来像幕帐一样。这一设计控制了百货公司室内和城市直接的透明度，室外的风景和自然光可以通过玻璃进入零售区域。百货公司可以随意改变室内的造型和布局，而又无需改变外观。

为了在电影院和百货公司之间建立统一的联系，电影院也使用了幕墙设计。这一幕墙既联系了百货公司和电影院，又与影院的幕布有异曲同工之妙。

1. 服务庭院
2. 零售区
3. 服务区
4. 百货公司
5. 零售出租区
6. 电影院
7. 沃恩路

1. service yard
2. retail
3. services
4. department store
5. retail units
6. cinema
7. Vaughan Way

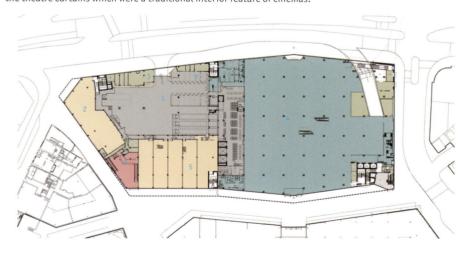

从街道上看左气 View from the street

立标外号 Façade with bridge

建筑细部 Details

Photo: Satoru Mishima, Peter Jeffree, Helene Binet and Lube Saveski

Commercial

Completion Date: 2008

Architect: Foreign Office Architects

Edinburgh Academy – New Nursery and After School Facility

The nursery is arranged entirely on the ground floor and the main teaching accommodation is in three linked rooms, one of which, for the two to three-year-old group, is capable of sub-division. Each of these rooms gives onto landscaped play terraces and also has external covered space and internal bay window spaces. In the centre of the plan are all the necessary services and storage and on the entrance elevation is a large cloakroom arrival point for distributing the children to the different classrooms. The entrance lobby is supervised by the staff room for security. A small dining room and kitchen is also shown for the use of the two to three year olds; the remainder of the children use the junior school's refectory facilities.

The After School and Holiday Care facilities for older children are also to be relocated on the first floor above the new nursery school and these consist of a series of linked rooms for group and individual activities. A floating roof covers the whole building with a line of roof light illuminating the circulation in the middle of the plan.

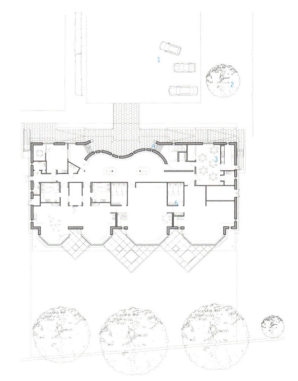

1. park
2. landscape
3. dining
4. entrance
5. WC

1. 公园
2. 景观
3. 餐厅
4. 入口
5. 洗手间

爱丁堡学院幼儿园和课后活动设施

幼儿园设在一楼，主要教学空间是三间连在一起的大教室，其中一间为2-3岁的孩子设计的教室可以进行再划分。这些教室都可以直接通往景观游戏平台，教室外面有一个屋顶下的空间，里面则有一个巨大的凸窗。入口处巨大的衣帽间是设计的核心，孩子们将衣物存储在这里，然后前往不同的教室。出于安全的考虑，从员工工作室可以监察入口大厅的情况。幼儿园还为2-3岁的孩子设置了一个小餐厅和厨房，其他的孩子则可以到小学食堂就餐。

为年龄较大的孩子设计的课后活动设施设在幼儿园的二楼，有一系列不同的集体/个人活动室。这座建筑采用漂浮式屋顶，一排屋顶天窗为活动区域提供了自然采光。

Photo: Richard Murphy Architects Ltd.

Educational

Architect: Richard Murphy, James Mason, Dominik George

Completion Date: 2009

Stratheden Eighteen-Bed Dementia and Mental Health Unit

Stratheden医院痴呆与精神健康病房

The proposal is to provide a low security residential dementia unit in the grounds of Stratheden Hospital. The accommodation consists of eighteen single bedrooms with en-suite shower rooms and associated accommodation for both patients and staff. The building is single storey with a U-shaped plan. Bedrooms are split into the two wings and the communal facilities located centrally. The whole building focuses on a south-facing secure garden for patients with the southern edge of the garden walled and framing the view south towards Walton Hill and White Hill. The design allows for patients to wander freely around the building and into the secure sensory garden. The circulation around the garden has small alcoves with built-in seating for patients to stop and sit and look out into the garden.

Each bedroom has a bay window with views out into the grounds of Stratheden. Secondary light comes into each individual room via a roof light located below the pitched roof ridge line. Each bedroom is identifiable by individual pitched roofs giving the patients a sense of their own identity within the building. All the other accommodation and circulation areas have flat roofs at different heights.

项目为Stratheden医院的寄宿制痴呆病房，共有18间为病患和工作人员所提供的配备浴室的套房。建筑呈U形结构，病房设在两翼，而公共设施设在中央。整个建筑朝向南面的带围墙安全花园，可以看到沃尔顿山和怀特山上的风景。患者们可以在建筑周围或是花园里随意走动。花园四周环绕着小凉亭，患者可以坐在里面休息和欣赏花园的风景。

每个套房都有一个凸窗，可以看见Stratheden医院里的景象。斜屋顶的屋脊上有一排天窗为室内提供自然采光。每个套房的斜屋顶都不尽相同，让患者产生了一种独特的归属感。其他的设施和公共区域则采用了高低错落的平屋顶。

1. 卧室
2. 病房
3. 工作区/咨询区
4. 服务区

1. bedroom
2. patient areas
3. staff/consulting
4. services

Exterior 建筑外观

庭院 Courtyard

花园 Garden

接待处 Reception

General view 建筑全景

Kirkintilloch Adult Learning Centre

Kirkintilloch成人教育中心

The aim of the project was to provide an open learning environment which draws in potential students from all sectors of society and to provide a new home for the classes already located within Kirkintilloch by Strathkelvin Further Education Centre.

The building is laid out as a linear plan with a two-storey block to the south side housing classrooms and offices. Against this sits a lean-to structure which houses the open learning facilities and projects out onto the canal bank.

The entrance to Phase 1 (previously to one end of the linear plan) is now positioned in the centre of the completed building. The reception orientates the visitor and first-time student within the main double-height space, allowing a clear reading of the building to be made with the majority of the classrooms and computer resource space to the left and the café and other administration offices to the right. A void behind the reception, through which the stair descends past a seated area, which protrudes out over the canal, makes connection to the lowest level which houses the open resource areas.

项目为社会各界的学生提供了一个开放式学习空间，也为Strathkelvin继续教育中心新课程提供了教学场地。

教学楼呈线形结构，南侧的两层里面是教室和办公室，对面的单坡屋顶建筑是开放式教学设施。

第一阶段工程的入口位于教学楼的中央，从前台大厅可以清晰地了解到建筑的内部结构。建筑室内左侧的大部分空间都是教室和计算机机房，右侧则是行政办公室和咖啡厅。前台后面的空地连接着通往下面悬在运河上的休息处，空地也连接着底层的开放式资源区。

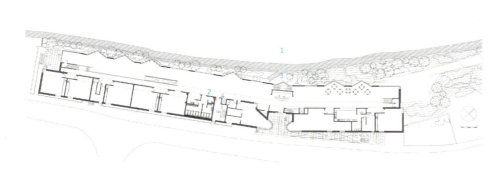

1. 入口
2. 卫生间
3. 服务区
4. 楼梯

1. entrance
2. WC
3. services
4. stair

建筑外观细部 **Exterior detail**

入口 **Entrance**

成人教育中心 **Adult learning space**

Photo: Richard Murphy Architects Ltd.

Educational

Completion Date: 2009

Architect: Richard Murphy Architects Ltd.

General view 建筑全景

Wakering Road Foyer

Wakering Road Foyer provides 116 bedsits for disadvantaged young people. The foyer tackles social exclusion among vulnerable sixteen to twenty-five-year-old through housing and training programmes that transform their lives. It accommodates young people with a range of support needs, including young parents, ex-drug users, wheelchair users and those with minor mental-health difficulties. Residents spend up to two years there to gain independence before moving on to a home of their own.

The massing allows sunlight into all of the dwellings, distances the dwellings and offices from the noise of Northern Relief Road, and reinforces the street edges. The L-shaped building, made up of a two- and a nine-storey block, creates a sheltered garden at ground level and a sunny roof garden. The interaction of the two- and nine-storey blocks is expressed by two distinct materials. Fibre-reinforced-concrete rainscreen cladding dresses the taller wing, while curtain walling defines the low-lying wing.

The landscaped garden is for residents and staff, with a secure play area reserved for the adjacent crèche. Trees and an existing embankment screen the garden from Northern Relief Road. The second-floor roof garden, for the young parents and their babies, includes a lawn and "living roof" to encourage biodiversity, as well as a pavilion for all-weather use. The landscaping promotes sustainable urban drainage, reducing surface-water run-off with planting and permeable paving.

维克灵路救助中心

维克灵路救助中心为年轻的弱势群体提供了116个套间，旨在通过提供住宿和培训计划改变他们的生活，解决16–25岁的青少年的社会排挤问题。救助中心收容各类需要帮助的年轻人，包括年轻的父母、前吸毒者、残障人士和轻微精神疾病患者。这些年轻人通过在救助中心里生活一到两年来获得自立的能力，以建立自己的家庭。

建筑的布局和朝向让内部所有的住宅都能获得充足的阳光，使居住和办公空间远离喧闹的北侧辅路，也装点了街道。L形的建筑由二层和九层的结构组成，形成了一个地面遮阳花园和一个屋顶日光花园。二层和九层的结构分别采用了截然不同的材料。较高的一侧使用了纤维加固混凝土雨幕外墙，而较低的一侧则使用了幕墙结构。

景观花园除了供住户和职员休闲娱乐之外，还有一块供隔壁的幼儿园使用的小型游乐场。树木和原有的路堤将花园和北侧辅路隔离开来。二楼上面的屋顶花园供年轻的父母和他们的小孩使用，花园有一个种满了各种不同植物的草坪和一个可以遮风避雨的凉亭。项目的景观设计优化了城市排水系统，通过植被和可渗透铺装减少了地表径流。

1. reception	1. 前台
2. refuse room	2. 垃圾房
3. office for partner organisations	3. 合伙组织办公室
4. public WC	4. 公共洗手间
5. crèche	5. 托儿所

从街道看建筑 View from street

主入口 Main entrance

主入口外观 Main entrance from outside

Photo: Nikhilesh Haval

Cultural

Completion Date: 2009

Architect: Jestico + Whiles

North view 建筑北立面

Kielder Observatory

The design brief called for an inexpensive building, not only suitable to house two telescopes and a warm room, primarily intended for amateurs and outreach work, but also suitable for scientific research. The design had to achieve a positive relation to the exposed setting on top of Black Fell overlooking Kielder Water and had to include both the facilities needed in this remote site and a "social space" for interaction and presentations, while being accessible both literally and culturally.

Timber was chosen as the material for the observatory early in the design process. Besides being a low carbon material and the obvious relation to its forest setting, the architects wanted a low-tech engineering aesthetic for the observatory, the opposite of the NASA-inspired world of high tech, high-expense and exclusive science. Instead, the architects wanted to evoke the curious, ad hoc structures that have served as observatories down the ages, and to the timber structures of the rural/industrial landscape at Kielder, the pit props of small coal mines and the timber trestle bridges of the railway that served them. The architects felt that a beautifully handcrafted timber building with "Victorian" engineering would be more inspiring in this setting than seamless, glossy domes.

基尔德天文台

项目要求天文台能够摆放两架天文望远镜并设置一个温暖的房间，它既要能满足业余爱好者的需求，又要适合科学研究。天文台的设计要与其所在的布莱克伐木场建立积极的联系，在这个偏远的地区提供基础设施和互动社交空间，使人们从各个层面上走进天文台。

天文台采用了木材作为主要材料。木材是一种低碳材料，又与森林有着显著的联系。此外，建筑师想以一种低科技的工程美学来建造天文台，正好与美国国家航天航空局所运用的高科技、高费用技术形成鲜明对比。建筑师沿用了自古以来天文台的奇特而随意的结构，这些木材与基尔德地区的木结构建筑景观、小煤矿的支柱以及铁路的木栈桥交相映衬。在建筑师的眼里，一座美丽的手工建造的木结构建筑和维多利亚时期的工程设计远比无缝、光滑的穹顶更具吸引力。

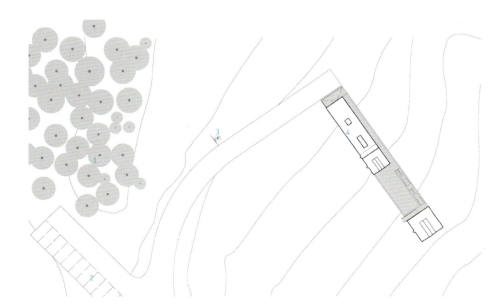

1. 森林
2. 停车场
3. 风力涡轮
4. 天文台

1. forestry plantation
2. parking
3. wind turbine
4. observatory

望远镜 Telescope

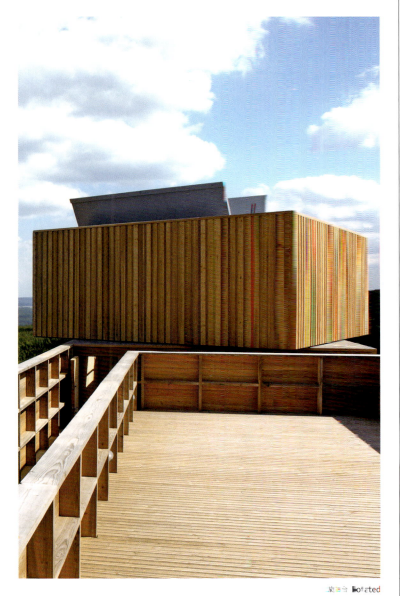

旋转台 Rotated

南南立面 South West view

New Flower Market – Mercabarna–Flor

While the roof is the big integrating element of this market, in the interior three conceptually different markets are located, each of them with its own specific characteristics and logistic and technical conditions, according to the product on sale. One part is meant for the Cut Flower Market, with modern industrial cooling systems, where the temperatures can be maintained between 2 ˚C and 15 ˚C, since the product has a fast turnover with a selling time of only three days.

On the other end of the complex the Plant Market is located, designed with heating systems with a radiant industrial floor, one of the biggest in Europe with 4,000 square metres. It has passive cooling systems that introduce humidity, which guarantee that the temperatures will never be below 15 ˚C, or above 26 ˚C, especially designed for the needs of this product that requires more selling time, about fifteen days. This means that besides being a vending zone, this sector is also a storage zone or greenhouse during this period of time.Finally in the middle of these two opposite sectors the Accessory Market is located, an especially delicate sector, because of its elevated fire risk. Due to the fact that they work with dried flowers and that the sale requests a considerable storage area, this subsector has especially been designed to detect and extinguish fire.

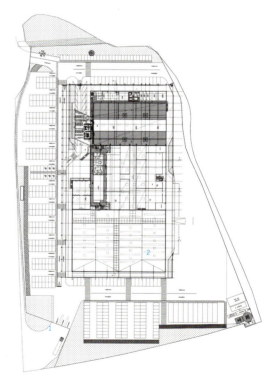

鲜花市场

鲜花市场的屋顶将市场内部三个独立的部分连接成为一个整体，每个部分根据其出售产品的不同拥有不同的物流和技术特征。插瓶花市场由于产品销售周期只有3天，采用了先进的工业制冷系统，室内温度被控制在2℃-15℃之间。

种植市场4,000平方米的辐射加热楼面是欧洲最大的加热楼面之一，被动式制冷系统保证了室内恒温在15℃-26℃之间，保证了15天左右的销售周期。这里不仅是一个贩售区，还可以作为仓储区和温室。装饰品市场出于中间位置，由于主要经营干花，也需要大面积的仓储区，因此防火问题不容忽视。这个部门的设计特别注重火灾的检查和灭火设施。

1. entrance
2. store

1．入口
2．店铺

Photo: Willy Müller Architects

Completion Date: 2008

Architect: Willy Müller Architects

Overall view 全景图

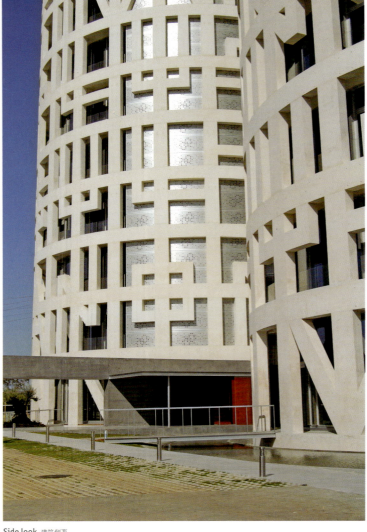

Side look 建筑侧面

Torres de Hércules

Located in the Bay of Algeciras (Cadiz), the new construction, which is surrounded by a man-made lake, is composed of two identical twenty-storey cylindrical towers, joined by a crystalline prism which houses the hallways connecting the two buildings. Its outer appearance is configured by the structure of the building, a gigantic lattice which completely surrounds the perimeter. There are the giant letters of the legend "Non Plus Ultra". Their job is to protect the inside of the building from excess solar radiation while providing panoramic views of the Bay of Algeciras, the Rock of Gibraltar, and the Serrania. This grid extends past the building's limits, as a "unique element" protecting the terrace roof-top deck, while at the same time acting as a base for possible energy collecting and telecommunications systems.

On the top floor, eighty metres up, will be a lookout restaurant. Above this, a panoramic roof-top deck will boast unique views over the Straight of Gibraltar, Mount Musa and the Alcornocales Natural Park. The building has a main entrance for pedestrians and cars which provides a clear view of the towers. The 200-spot ground-level parking lot is located on the other side of the towers and is organised around a landscaped area.

大力神大厦

大厦坐落在阿尔赫西拉斯湾，四周为人工湖围绕，由两幢相同的20层圆柱形塔楼构成，中间通过透明的棱镜状结构连通。大厦外观是巨大的格子结构，硕大的字母 "Non Plus Ultra" 镶嵌其上，格外醒目。设计的主要任务是防止室内太阳光线的过度照射，但同时确保视线清晰，便于欣赏外面的景致。格子结构将整幢建筑围裹起来，实现了设计的初衷。建筑顶层是一个观景餐厅，将直布罗陀海峡、穆萨山峰的壮丽美景尽收眼底。此外，行人及车辆通过的主入口也别具特色，200个车位的停车场位于大厦的另一侧，四周种植着各种植物。

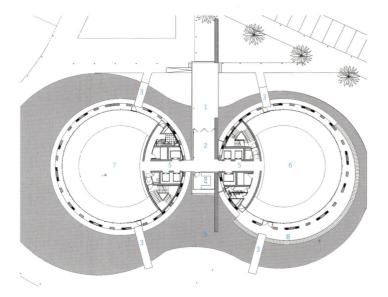

1. main entrance/cancpy	1. 主入口/华盖
2. entrance lobby	2. 门厅
3. side entrance/bridges	3. 侧门/廊桥
4. reception	4. 前台
5. lift lobby	5. 电梯大厅
6. cafeteria/bar	6. 餐厅/酒吧
7. commercials	7. 商业区
8. terrace	8. 平台
9. water	9. 水景

建筑细部 Details

老□ Stairs

建筑□部 nside view

Photo: Rafael de La-Hoz Castanys

Completion Date: 2009

Architect: Rafael de La-Hoz Castanys

Exterior 建筑外观

The Azahar Group Headquarters

The parent company originated from Castellón and, given its growth and expansion, was eager to have a corporate headquarters that would reflect its environmental and artistic commitment. With this as a framework and with the availability of a 5.6–hectare piece of land next to the N–340 highway, Half way between Castellón and Benicàssim, the project contemplates three interventions: the covered greenhouses and exterior nursery plantations; a building for services complementary to the activities developed by the company; and the group's corporate headquarters.

The headquarters is erected as an icon building maintaining a close relationship with the landscape. To both the north and west, the topography of the mountains serves as a backdrop for the building, against which the geometrical roofs repeatedly stand out. From a distance their facetted shape and outline help situate the building in the landscape. Orientated on the east–west axis, the building is structured as two wings united by a central body around two open patios of a very different sort. The first as a "parade ground" or external reception area for users and visitors, and the rear one landscaped and for more private use. The four wings that accommodate the company's different departments converge in a main hall which, as well as acting as a distributor, is a large exhibition space.

An important environmental feature of the building is the channelling of water from all the roofs and the outside areas to a cistern–reservoir, the latter being used in the watering of exteriors and nursery plantings on the plot of land.

艾资哈尔集团总部

鉴于集团业务的不断增长与扩张，公司急需打造一个总部，并彰显出其对于环保及艺术的执着追求。总部选址在临近N340公路一块面积达56000平方米的地块上。整个工程包括三部分：带顶温室及室外苗圃林、服务大厦及总部大厦。

总部大厦与周围景观密切联系——北侧和西侧的山脉提供完美的背景，使得几何形状分明的屋顶格外突出。整幢建筑以东西朝向为主，两个"翅膀"通过中央结构连结起来。两个风格各异的天井更为吸引眼球，其一命名为"操场"（室外接待区，供来访者使用），另一被景观覆盖，更为私密。另外四个"翅膀"内分别设置着公司的不同部门，它们在中央大厅（展厅）处交汇。

尤为一提的是，设计中专门打造了储水系统——水从屋顶及室外区域流入到储水槽中，然后用于浇灌植物。

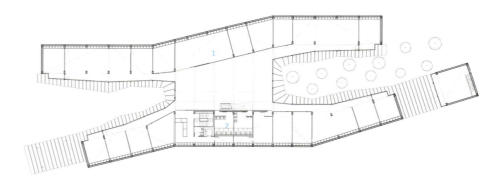

Ground floor plan   一楼平面图
1. office   1. 办公室
2. toilet   2. 洗手间

Exterior 建筑外观

建筑夕观夜景 Night view of the exterior

外面观内部办公室 Outside view of the offices

大厅 Interior lobby

Photo: Alejo Bagué

Corporate

Completion Date: 2009

Architect: Carlos Ferrater – Núria Ayala

General night view 夜间全景图

Abertis Headquarters

The brand-new building for Abertis, the second phase of Barcelona's "Logistic Parc", a 11,000-square-metre office building accommodates in its five floors the different business areas of the company.

The Abertis building is a part of a whole new-edge landscape, and three buildings with a continuous façade to the Ronda Litoral, form an interior plaza that conforms a patio protected from the highway. In this new phase three buildings conform an angled façade towards the highway, a dynamic, solid and transparent façade, in an attempt to capture the paradoxes of a new urban landscape.

It is an optimised ground plan with a feature, central greek cross-shaped open space. This space allows access from the lobby to the flexible working zones. It has some elements in common. The efficiency ratios exist from the beginning, reflecting the relationship between the gross and the usable space as ratio, the result of studies on economical and architectural functionality for a building.

There is a strong tendency to reduce price and quality. Of course, ask some of them, for example, Dior, Cartier, Rochas, Shiseido, Parisbas, Swift, JP Morgan, AXA and Chateau Lafitte Rothschild: all these companies have multiplied the final value of their buildings. The double skin façade was proposed some years ago, and has been used in many projects. Even though it was inspired, only ideally, in the great Egyptian or Medieval historical walls, the double glass façade has in common with these walls the element of thickness, acting as container, as barrier. We don't find these components in the single glass walls.

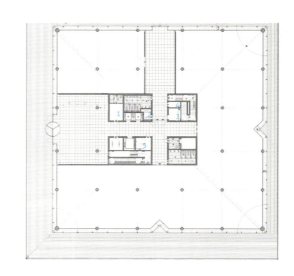

1. local control system
2. installation mechanism
3. installation electricity
4. local maintenance
5. installation telecommunications

1. 区域控制系统
2. 机械设施
3. 电力设施
4. 区域维护系统
5. 通讯设施

阿伯提斯总部

阿伯提斯总部大楼是巴塞罗那物流园的二期工程，总面积为11,000平方米，五层楼里设置着阿伯提斯公司的各个部门。

阿伯提斯总部大楼是先锋景观区域的一部分，三座建筑被包裹在一面朝向环形公路的连续外墙里，形成了一个室内广场。朝向高速公路的外墙呈一定角度倾斜，动感、坚实而透明，打造出一个城市新景观。

项目的布局十分合理，中间是特色希腊十字形开放区域，这个区域连接着门厅和灵活办公区域。项目的总面积和可用面积之间的效率比例被进行了优化，是建筑师对经济性和建筑功能性进行综合评估的结果。

日门全景图 General daytime view

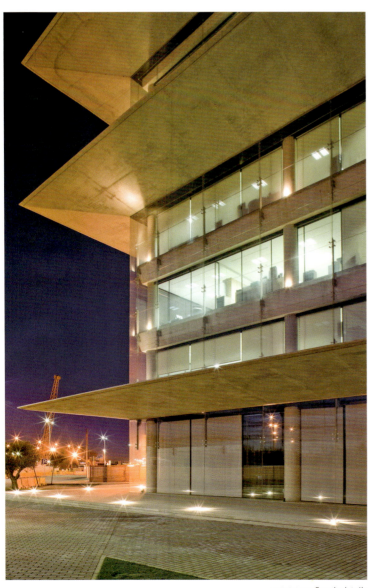

建筑外墙细部 Façade details

室内 Interior view

Photo: Carlos Casariego

Corporate

Completion Date: 2007

Architect: Ricardo Bofill Taller De Arquitectura

Exterior 建筑外观

Orense Swimming Pools for Vigo University

比戈大学欧伦榭游泳馆

The plot chosen is at the highest point of the university campus, a location that gives it a special value in the relationship between the city of Orense and the campus, currently encumbered by the presence of a busy road between them. Opting for a programme orientated towards leisure, and that may thereby serve both the university community and the people of Orense, it will help to heighten the value of the building as an element of urban interaction.

This special location has inspired the main conceptual and formal decisions of the project. The proposal is drawn up as a vast platform that looks onto the campus, and the whole organisation of functions is indebted to this idea. The users will be able to see the campus and its buildings from this raised platform containing the swimming pools. The platform is designed with bold projections supported by a strong base that, aside from spanning the existing drops, shapes the pools and contains all the water treatment services and necessary systems for the correct performance of the programme. The contrast between this massive, topographic base and the cantilevering light glass surfaces surrounding the public level of swimming pools is one the basic formal arguments of the project.

游泳馆建在校园的最高点，与欧伦榭城之间仅隔一条繁忙的马路。项目作为一个娱乐休闲设施，可以为学校和欧伦榭市民同时服务，与城市具有紧密的联系。

项目独特的地理位置决定了它的主要设计理念。项目被设计为一个俯瞰校园的巨大平台，其内部所有功能都围绕这一理念展开。游泳者可以在游泳池里俯瞰校园的景色。平台被设计为大胆的凸出结构，由一个结实的底座支撑。游泳馆内设有游泳所需的全套水处理设备和服务设施。巨大而结实的底座和游泳池所在的轻质玻璃悬臂结构形成了鲜明的对比，是项目设计的焦点之一。

1. 游泳池a
2. 游泳池b
3. 洗手间
4. 楼梯
5. 个人休息室
6. 行政办公室
7. 男子浴室
8. 女子浴室
9. 休息大厅

1. swimming pool a
2. swimming pool b
3. rest room
4. stairs
5. personnel room
6. administration office
7. men's room and showers
8. women's room and showers
9. rest hall

建築全景 General view

建築外景 Exterior

泳池 Swimming pool

泳池 Swimming pool

Photo: Pedro Pegenaute, Roland Halbe

Recreational

Completion Date: 2008

Architect: Francisco José Mangado Beloqui

Archeological Museum of Álava

The building adjoins the Palace of Bendaña, today the Naipes Fournier museum. Access to the building is through the same courtyard that leads to the Palace and conveys the full scope of the project. The proposal includes extending the courtyard surface area in order to upgrade the access area. This does not encroach on the whole court, however, taking only a narrow strip for an appendix perpendicular to the main building. As well as housing auxiliary programmes, this addition provides a more attractive access façade than the current party walls of the neighbouring constructions. Thanks to the sloping terrain, the courtyard is reached through a bridge over a garden that allows light to penetrate to the lower areas that otherwise would be permanently in shadow.

The areas housing the various activities, including the library and workshops, are located on the ground level orientated towards the street, and have an independent access. The assembly hall and galleries for temporary exhibitions are on the same level as the public entrance shared with the Naipes Fournier museum. The permanent exhibition halls are on the upper levels. The stairs linking the different levels define part of the façade onto the access courtyard. The outer walls comprise a series of different layers. The façade facing the access courtyard is bronze grilles, a material with clear archaeological references. In the middle, a double–layered wall of silkscreen printed glass contains the stairs that offer visitors views of the courtyards. In contrast, the façade fronting the street is more hermetic, comprising an outer layer of opaque prefabricated bronze louvres, with openings where needed, and an inner layer formed by a thick wall containing the display cabinets and systems. In this way the internal exhibition spaces are unencumbered and only traversed by translucent light prisms.

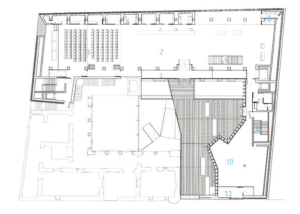

1. entrance foyer
2. entrance –reception
3. mul–tipurpose hall–auditorium
4. coat check
5. simultaneous translation room
6. circulation
7. toilets
8. stair shaft
9. hvac
10. temporary exhibition hall
11. foyer

1. 门厅
2. 接待处
3. 多动能厅/礼堂
4. 存衣处
5. 同声传译室
6. 流通区域
7. 洗手间
8. 楼梯间
9. 空调机械室
10. 临时展厅
11. 大厅

阿拉瓦考古博物馆

阿拉瓦考古博物馆与本达纳宫殿（现今的Naipes Fournier博物馆）相邻，并可通过同一庭院进入。设计的任务包括延展庭院大小，从而扩大入口的面积。当然这并不意味着将整个庭院改为入口，只是将一长条的空地改造成通往主建筑的入口结构，使其在外观上更具吸引力。通过横架在花园之上的小桥方可进入庭院，这样也使得光线可以照射到以前被阴影笼罩的区域。

博物馆一层包括图书馆、工作区以及临时展览区，永久展览区则位于上层。建筑外墙由多层结构打造，朝向入口庭院的外墙由黄铜格栅构成，让人不禁联想到考古。中间部分由双层丝网印花玻璃材质构成，可以望见内部的楼梯。朝向街道的外墙由不透明的窗户覆盖，突显出神秘的气息。

General view 全景图

建筑入口的外墙由铸造青铜制成　Cast bronze pieces configure the outer façades that define the access court

入口 Access

室内 Interior view

Photo: Pedro Pegenaute, César San Millán

Architect: Francisco Mangado / Mangado y Asociados

Cultural

Completion Date: 2009

Insular Athletics Stadium

海岛体育场

The sports building's location produces a large entrance square under which the indoor facilities are housed. The Elite Performance Centre is placed beneath a large horizontal platform set midway up the slope, using the existing incline. A repetitive construction system of concrete screens reduces the budget for a project in which the structural component is extremely important. High-resistance materials will permit intense public usage of the facilities with minimal upkeep. The Elite Performance Centre will have natural ventilation and lighting through patios and skylights. The athletics track meets the dimensions and guidelines required by the National Sports Council, with perimeter training belts. The inner track will have flexible usage, including partial usage for gymnastics. A linear network of services and dressing rooms separates the athletics field from the EPC, permitting alternate usage by both. An amenities block and a small canteen with independent access are placed in line in the top section of the tiers. The living quarters set in the southeast part of the allotment make the most of the slopes, aspects and views. There are four distinct entrances: the main public entrance, the athletes entrance, the direct entrance for marathons and external events and the service entrances.

体育场的地理位置使其下方入口处有很大一片空地,里面设置着室内体育设施。精英活动中心设置在看台的下方、斜坡的中央。反复的混凝土挡板建造系统缩减了项目的预算,高性能的建材既能保证设施的公共使用强度,又能将保养费用最小化。精英活动中心将采用自然通风系统,并通过天井和天窗进行自然采光。体育场跑道的规格完全符合国际标准,外围还有一圈训练场地。跑道内侧可以灵活运用,如进行体操项目。呈线形排列的服务设施和更衣室将体育场和精英活动中心分开,可供二者共同使用。看台的最上方有一个小型便利店和餐厅。东南侧的居住区最大限度地利用了地势的坡度和周边的风景。体育场共有四个入口:主公共入口、运动员入口、马拉松及外部活动入口和服务设施入口。

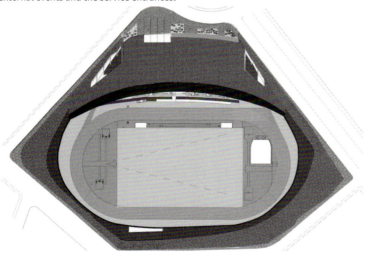

Photo: AMP Arquitectos

Completion Date: 2007

Architect: AMP Arquitectos

Private House in Menorca

Being a summer house, the main idea is not only to create the interior spaces of the house, but to distribute all outer space. The interior spaces seek good relations with the outside world, colonising their surroundings and their views.

Inspired by the typical Menorca "tanca", stone walls divisions of the realm, the plot is organised from a space frame, fully passable, based on a trace orthogonal, combining floors, platforms, water, trees, plants, tanca, pergolas, walls and the house itself. By combining these elements the architects are encountering this approach in which each piece is delimited and acquires its own identity within a harmonious whole. The diversity of outdoor stays provides the site with a space balanced richness.

The house is situated in the centre of the outer solar stays divided in two, front and rear. The hall of the house with two large openings on each side operates as a mixed external–internal transition. Falls outside the pavement creat a passage that connects the backyard with the front porch.

美诺卡岛私人住宅

作为一座夏日别墅，项目的设计重点并不在室内，而是分散在户外空间的各个部分。室内空间的设计寻求与外界之间建立良好的关系，将外界的景观纳入别墅之中。

受到美诺卡岛传统的tanca（一种边界的石墙）的影响，住宅的设计依循一个空间框架展开。项目以直角路径为基础，融合了地面、平台、水景、树木、植物、石墙、绿廊和住宅本身。通过将这些元素融合在一起，建筑师在住宅中划分出明确的界限，使每一处都具有极高的辨识度，同时又不破坏整体的和谐性。户外空间的多样性让住宅空间平衡而丰富。

住宅分为前后两部分，能够获得充分的阳光。住宅的门厅两侧是巨大的落地门窗，连接了室内外空间。小路外侧的下降区域连接了后院和前廊。

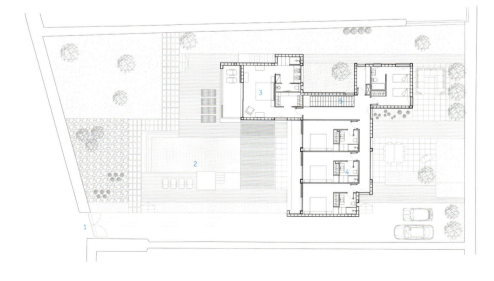

1. 入口
2. 客厅
3. 卧室
4. 洗手间
5. 楼梯

1. entrance
2. living room
3. bedroom
4. WC
5. stair

Residential

Completion Date: 2009

Architect: Pablo Serrano Elorduy

Bird view 鸟瞰图

Terminal 1 of Barcelona Airport

巴塞罗那机场一号航站楼

Designed by Ricardo Bofill, Barcelona Airport's new Terminal 1 is one of the best projects in his long career history.

The impressive infrastructure created stands out for its spaciousness and functional nature, the spaces inside it interrelating in a clear and orderly manner, offering the visitor great ease of movement. The general design, simple and minimalist, is the ideal framework for Nu benches, located in several of the terminal's spaces. The Bicilínea bicycle racks (designed by Beth Galí), situated outside the terminal, are also based on a design with well-defined lines.

Forged anchors galvanised steel ten-millimetre-thick dimensional control equipped and provided with expansive blocks M-16 and stainless hardware. Primary structure uprights and beams extruded aluminum, alloy 6063, adequately sized to withstand a wind load of 110 kg/m², bearing joints in both directions. Double glazed skin externally with twelve-millimetre tempered glass, restrained by structural silicone independent portavidrios frameworks for each hole. Inner skin consists of two glass sheets colourless practicable, which shall be operated and kept open with shock gas. The intermediate ventilation chamber shall have seventy-millimetre holes for inlet and outlet ventilation.

巴塞罗那机场一号航站楼由里卡多·波菲尔设计，这一项目是他职业生涯中的又一高峰。

航站楼宽敞的空间和便捷的功能而引人注目，其内部空间简洁而有序，为旅客提供了一个适宜的移动环境。项目整体设计简单而抽象，随处摆放着舒适Nu长椅。航站楼外，贝斯·嘉里所设计的脚踏车行李架同样拥有简洁、清晰的线条。

锻造锚点的镀锌钢板厚10毫米，同时还兼具M-16和不锈钢硬件。6063号铝合金的支柱和横梁所构成的原始结构比例适当，风力荷载可达110千克/平方米。双层玻璃外墙采用了12毫米的钢化玻璃，由硅树脂框架固定。内层壳板由两层无色玻璃构成，可以控制调节、任意开关。中间的通风室上有直径为70毫米的小孔，便于通风。

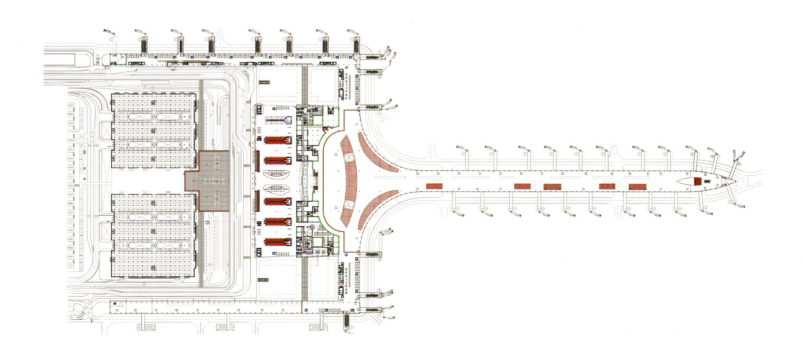

建筑正面 Front view

服务区 Service areas

候机大厅 Lounge of an airport

室内服务区 Inside view of service areas

Photo: Carlos Casariego

Transportation

Completion Date: 2007

Architect: Ricardo Bofill Taller De Arquitectura

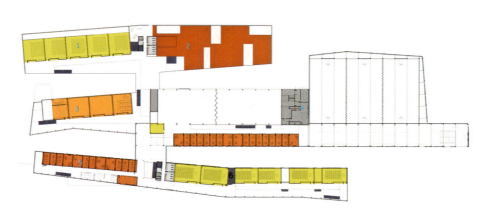

Façade 建筑外观

Pontevedra Campus

In the planning of the Campus of Pontevedra, the geometries shaping "local traces" and a nature still with potential, super put on the intention of creating an atmosphere that results from the permeability and presence of the environmental peculiarities, added to the decision to make the diverse activities of the university life take part among them, through two mechanisms: visual continuities in horizontal and vertical section, and conversion of circulation spaces in relation areas.

It leads to taking the most of reliving the river as organiser of the city, to deal and to propose the campus as community space of culture and leisure, and to recover the area of the campus and the river as park and as defined ecozone.

The central area of the campus is pedestrianised and creates a central-covered main square covered as a place that holds different events, which "sews" the buildings isolated. The landscape treatment joins the values of the natural space with the cultural place, coming together an ecozone of bank with a constructed garden, and the campus joins across entail with the river in the net bank path network that connect, with the city.

蓬特韦德拉学校

在蓬特韦德拉学校的规划中，具有区域特色的几何图案和自然的宁静感营造出一种亲和的氛围和独特的环境。校园设计通过两种机械系统——水平和垂直方向的视觉连续性与流通空间之间的相互转换实现了校园活动的多样性。

河流作为城市的规划者，将校园空间规划入社区文娱空间的一部分。校园和河流一起打造了一个生态公园。

校园的中心区域是步行区，有一个可以举行各种活动的中心广场。广场还将校园内分散的建筑串联了起来。校园的景观设计与自然空间结合在一起，与学校的文化空间一起在河岸形成了一个生态区。校园与河流、河岸一起形成了一个网络，与城市紧密联系在一起。

1. 教室
2. 大学图书馆
3. 工作室/实验室
4. 办公室
5. 楼梯/电梯

1. lecture room
2. university library
3. work room/lab
4. office
5. stairs/lifts

304

走廊 Corridor

门窗细部 Details of openings

体育馆 Gym

General view 全景图

Headquarters of the Environment Service and Public Spaces

The building houses the Forestry and Natural Areas Departments of Zaragoza City Council. Situated in the centre of the city, next to the Almozara bridge, the site has considerable differences in level, about five metres between the avenue and the square at the upper level and the Ebro riverbank park. The relationship with the adjacent urban spaces and the topographic features of the site become active conditions for the implementation of the building; it can clearly be seen in the project section that the extension of the public space of the entrance is also expresed, via the roof and the configuration of the building as a platform–viewpoint onto the Ebro River.

Given the differences in level of the site, the main entrance to the building is on the upper floor, which houses the administrative spaces and the environmental hall. The main hall joins and separates spaces, enabling them to be used independently. The restricted access offices of the Forestry and Natural Areas Departments are located below the entrance level, in the basement. This lower floor has pedestrian and vehicle access from the Riverbank Park.

环境服务和公共空间总部

建筑里设置着萨拉戈萨市政府的林业和自然区部门，位于城市中心，紧邻阿尔莫扎拉桥。该地区的地势差很大，在上方的街道、广场与埃布罗河畔公园之间有5米高的地势差。与周围城市空间和场地地形特征联系是建筑建造的活跃因素，这从建筑的剖面、公共空间的入口、屋顶和建筑的形态都得以体现。整个建筑即是可以俯瞰埃布罗河的景观平台。

由于场地的地势差，建筑的主入口被设在了建筑的上层，直通行政空间和环境大厅。主大厅连接着独立的空间，使它们可以被独立地使用。林业和自然区部门被设在下层地下室。行人和机动车可以从河畔公园直接进入下层空间。

1. garage	1. 车库
2. warehouse	2. 仓库
3. locker room	3. 更衣室
4. staff rooms	4. 办公室
5. installations	5. 基础设施
6. utility room	6. 杂物房

Elevation from Europe Square 从欧洲广场看建筑立面

从欧洲广场看建筑 View from Europe Square

办公室 Offices

Photo: Pedro Pegenaute

Corporate

Completion Date: 2009

Architect: Magén Arquitectos (Jaime Magén, Fco. Javier Magén)

General vew 全景图

Kindergarten in Rosales Del Canal

罗萨莱斯运河幼儿园

The Kindergarten in Rosales del Canal is located in an area of residential growth in the southwest of Zaragoza. The two main ideas that existed at the start of the project are based on children's special perception of the constructed environment. The first idea tries to combine the general volumetrics of the public facility with the domestic scale that must accompany the child. The second has to do with the sensorial relationship between children and architecture.

The basic unit of the school is the classroom. Its form responds both to the primary identification of the sloping roofing and to the advantages of height and additional lighting in the classrooms. The shape of the roofing of the classrooms is repeated to cover significant spaces that occupy a larger surface area such as the multipurpose hall and the dining–room.

The general configuration of the building clearly responds to organisational criteria, placing the classrooms around the patio, with service spaces situated between them and communicated on the inside with the corridor and on the outside with the patio, via the continuous exterior porch. The lobby, the multipurpose hall and an administrative area composed of the reception, the teachers' room and the administration complete the functional programme.

罗萨莱斯运河幼儿园位于萨拉戈萨西南部一片人口迅速增长的地区。项目最初的两个设计理念考虑了儿童对建筑环境的独特认知。一是试图将综合公共空间和家庭规模相结合，以适应儿童需求。二是建立儿童和建筑之间的感官联系。

学校的基本元素是教室。教室的造型由坡形屋顶、有利的屋高和教室的附加采光共同决定。教室屋顶的造型同样在其他重要的区域得以体现，如多功能大厅和食堂。

建筑的整体造型十分具有组织性，教室围绕着天井而建，服务区域则散落其间。室内空间由走廊连接；室外空间则通过连绵的门廊，由天井连接。门厅、多功能大厅、接待前台、教室办公室以及行政区域共同完善了幼儿园的功能。

庭院 Courtyard

建筑外墙 Wall

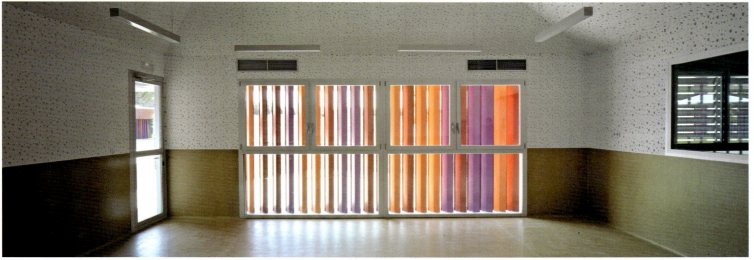

室内 Interior

教室 Classroom

Architect: Magén Arquitectos (Jaime Magén, Fco. Javier Magén)

Completion Date: 2009

Photo: Jesús Granada

Educational

44 Social Housing Tauste (Zaragoza)

The building is located at the south urban edge of Tauste, which is situated fifty kilometres away from Zaragoza. The project avoids the direct dialogue in a formal way with the closest out of context surrounding – the site borders at the north with a green space, at the south with housing buildings, at the east with an industry zone and at the west with the access road from Zaragoza and the municipal sports centre. The interior public space establishes connections with the adjacent spaces to get an appropriate integration in the place.

With the purpose of dealing with the urban character of the interior of the city block, two restrictions are proposed: the first one is to fit the garage of the semi-basement floor in the same width as the other floors, renouncing the allowed highest occupation, in order not to reduce the dimensions of the public space; the second one is to free a big porch at the north side, under the building, which relates the interior space with the pre-existence green space. This opening, together with the openings situated at the west and the south of the site, the trees and the urban furniture, helps understand the interior of the block as a public space.

塔西格44号社会住宅

建筑位于塔西格南郊，距离萨拉戈萨50千米。项目并没有直接与周边的环境相联系，而是以一种缓和的姿态融入环境。项目北面与绿地相连，南面与住宅楼相连，东面与工业区相连，西面与通往萨拉戈萨的公路和市政运动中心相连。

为了使建筑更具城市特色，项目设计增加了以下两个限制：一是在半地下楼层设置车库，减少建筑公共空间的规模；二是在建筑北侧设计一个大门廊，并利用门廊将室内环境与室外绿地连接起来。门廊与西、南两侧的门窗、树木、城市景观一起，营造了一个极具特色的公共空间。

Daytime view from the street 日间街景图

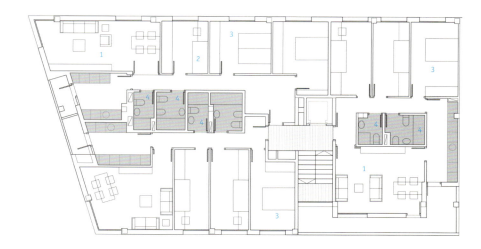

1. 客厅
2. 书房
3. 卧室
4. 洗手间

1. living room
2. study
3. bedroom
4. toilet

建筑外观 **Building exterior**

入口 **Entrance**

室内 **Interior**

Photo: Jesús Granada

Completion Date: 2010

Architect: Magén Arquitectos

Buildings with entrance 建筑入口

BTEK — Interpretation Centre for Technology

BTEK is an interpretation centre for new technologies, aimed at student visitors. The Centre's promoter, Parque Tecnológico, S.A., (Technology Park) set out the following as the most important guidelines:

The site's location, on one of the highest points of the Vizcaya Technology Park and close to the Bilbao airport's flight path for takeoffs and landings, helps with the aim of making the building a landmark in its landscape.

The building consists of two apparently uninterrupted pyramid-shaped volumes that connect below ground level.

+ The first is a heavy, black volume that emerges from the earth; it is enclosed by three metallic façades and completely covered with solar panels that form a patterned network.
+ The second volume, contrasting with the first, is formed by two façades of curtain walling with an artificial grass-covered roof that starts off as an extension of the terrain and continues on to cover the entire site.
+ Artificial grass also covers the below-ground-level connection, allowing it to merge with the site and its surroundings.

The five galleries are designed to be visited sequentially. In order to serve for a wide variety of possible exhibitions and contents, the galleries have been designed with very different characteristics: from those with ceilings at a conventional height to galleries with variable-height ceilings, reaching up to sixteen metres of clear height, and with or without natural lighting.

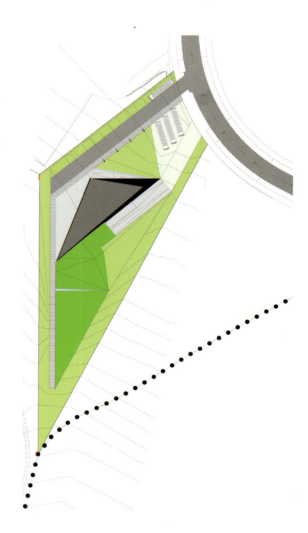

BTEK科技翻译中心

BTEK是一个新兴科技的翻译中心，以学生为参观目标。中心的承办方——S.A.科技公园——规定了下列指导原则：该中心的所在地是维兹卡亚科技公园最高地之一，而且邻近毕尔巴鄂机场飞机起飞和降落的跑道，这样的地理位置有利于实现使建筑成为该地地标的目标。这座建筑由两个看起来互不相连的金字塔形组成，二者在地下层相连。

第一座"金字塔"是一座比较重的黑色建筑物，从地面拔地而起；外立面三面是金属，全部用太阳能板覆盖，表面形成一种纹理。第二座"金字塔"跟第一座形成对比，其两个外立面采用幕墙，人造草坪屋顶从地面开始，延伸到覆盖整个场地。人造草坪也覆盖了地下层的连接空间，这样，这个部分就跟场地及其周围环境融为一体了。

5个展厅要按照顺序依次参观。为了满足各种展览和内容，这些展厅的设计各具特色：有的是普通高度的顶棚，有的是高度可变化的顶棚，能达到16米高，有的是自然采光，有的不是。

屋顶 Roof

建筑与景观 Building with landscape

建筑细部 Detail

夜景 Night view

Mora River Aquarium

Given the blazing Alentejo sun and the need to create shade, the building was devised as a compact and monolithic volume with a pitched shelter of thin white pre-cast concrete porticos with single spans of thirty-three metres, evoking the profile of the canonical Alentejo whitewash barns known as "montes". The shading and cross ventilation systems along with the water circuits foster the reduction of cooling energy, the sustainable increase of humidity and the wellbeing of animal and plant life.

Standing on a massive concrete plinth with a built-in stairway-cum-ramp entrance, the pitched shed veils a set of mute boxes that contain the programme, namely reception, ticketing and shop, cafeteria, changing exhibits hall, documentation centre, research and education, live exhibits, multimedia and a small auditorium. Inside, the exhibition spaces tend to be dark, in order to minimise UV impact on the live exhibits and allow visitors an in-depth viewing of the aquariums. The outdoor void between these programme boxes and the pitched shed generates not only accelerated viewpoints onto the outside but also a promenade that culminates in the passage through a bridge over the lake which, in itself, is also a live exhibit of animals and plants collected and nurtured in the region. The live exhibits, the main feature of an aquarium, reproduce, through complex life support systems, the habitat conditions of different regions allowing to exhibit side-by-side the various animals and plants. On the basement, these support systems guaranty stability of water temperature, pH, quality control and filtering for each habitat parameter, including a duct gallery below each exhibit to supply and monitor the water.

莫拉河水族馆

由于阿伦德如地区的日光十分强烈，建筑采用了紧凑的单体结构。由斜屋顶上延伸下来的白色预制混凝土柱廊长33米，与阿伦德如地区被粉刷为白色的谷仓十分相似。遮阳和对流通风系统与水流回路一起，减少了制冷所需的能源，也为水族馆内的动植物提供了适宜的湿度和温度。

水族馆的混凝土基座上设有嵌入式楼梯和坡道，倾斜的棚架结构内部是水族馆的各个功能区：接待处、售票处、商店、餐厅、展览厅、档案中心、研究和教学区、生物展区、多媒体区和小礼堂。室内展览区十分昏暗，以最小化生物展区的太阳辐射，让游客得以对水族馆进行深度观察。水族馆的室外区域不仅优化了室外景观，而且在湖上形成了一个室外步行桥，而湖本身也展示了当地的动植物。水族馆里的生物展览复制了不同区域的生态环境，让游客可以与各种动植物进行面对面的接触。地下室的支持系统保证了稳定的水温、pH值、水质量和过滤系统，每个展览区下方都有一个监控、供水的管道。

Completion Date: 2006

Architect: Promontorio Architects

Antas Educative Centre

The spatial and architectural design of the building of the new Antas Education Centre were formalised in several bodies, each containing part of the programme in accordance with principles of internal organisation, functionality, form and image, given the type of building and its specificity. This conception took into account the morphology of the terrain, solar orientation, access and links to surrounding bodies. It's a building consisting of several bodies expressed by a "simple architecture" that will build a close relationship with the exterior spaces. It was intended to create in the spaces between the various bodies the visual relationship between interior and exterior reducing relations with the urban surroundings.

There was an intention to turn into how the building relates to the exterior. However, there are some links to the outside also. The settlement found answers to a matrix that structuralise a functional organiasation of the school as a function of the planned programme and constraint imposed by various land levels.

安塔斯教育中心

安塔斯教育中心的空间设计分为几个部分，每个部分覆盖不同的项目。各个部分之间具有严密的组织结构，功能性和造型，令建筑的造型十分特别。这一设计考虑了地形、朝向、入口、与周边其他建筑的联系等诸多因素。这是一个由几部分构成的简单建筑，它与户外空间的联系十分紧密。

建筑本身的室内外视觉关系也是设计的重点。整个项目将学校的各个功能区组织结构化，在不同的地势高度上建立了不同的功能区。

Southeast view 学校东南侧

总平面图
1. 入口
2. 体育馆
3. 中庭
4. 运动场
5. 托儿所
6. 教室/厨房和餐厅
7. 户外游乐场
8. 教室/带顶操场

General plan
1. entrance
2. gym
3. atrium
4. playing field
5. kindergarten
6. classrooms / kitchen and dining
7. outdoor recreation
8. classrooms / covered playground

儿童户外游乐场 Children's outdoor play area

儿童户外游乐场夜景 Night view of children's outdoor play area

通往教室的通道 Circulation space to classrooms

Night view 夜景

Le Prisme

The building is a new venue for theatre, concerts, fairs and sports events. It contains retractable seating and demountable stage for versatility. The main space can accommodate up to 4,500 people during performances. As it is never permanently inhabited, the building is essentially a chamber for ephemeral events.

It is an instantly recognisable object. Three ribbons of concrete that vary in height and texture define the building. The ribbons delineate the different zones of the building: entrance, hall, storage and back of house facilities, and also create a series of residual spaces that contain services machinery and technical access. Inside, a twelve-metre-high ribbon defines a rectangle that is the hall. The hall is column free, measuring forty metres wide by sixty metres long. Externally the upper ribbon is made of prefabricated concrete panels with a regular grid of glass bricks.

During the day, the sunlight plays with the 25,000 bespoke pyramid-shaped glass bricks, producing glimmering effects and dramatic shadows. In the evening, the building awakens as the Fresnel lens-like surfaces of the glass bricks that amplify the intensity of the coloured lighting scheme, producing a glittering façade.

棱镜剧院

棱镜剧院内可以进行戏剧、音乐会、展会、运动会等活动。剧院内设有活动坐席和可拆卸舞台，最多可容纳4,500名观众，是一个举行临时性活动的主要场所。

建筑十分引人注目，由三个高低、材质不同的带状结构组成。这些带状结构划分了建筑的各个功能区：入口、大厅、贮藏室、后台设施。此外，剧院内还有一些剩余空间，安置机械、配电设施。大厅高12米，长60米，宽40米，呈矩形。里面没有任何廊柱。上层带状结构的外部由预制混凝土板构成，上面装饰有排列规则的方格玻璃砖。

白天，阳光照射在25,000块锥形玻璃砖上，闪闪发光，形成了奇妙的光影效果。晚上，建筑外表的玻璃砖使建筑产生了菲涅尔透镜的效果，扩大了彩色光线的光谱，让建筑外墙熠熠生辉。

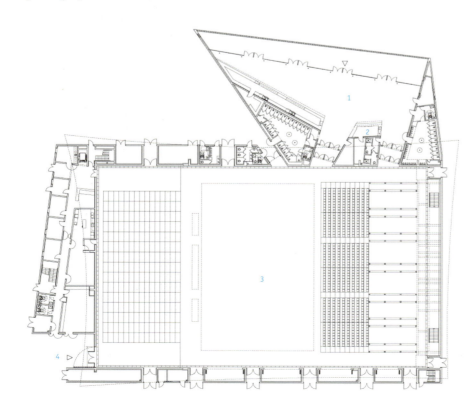

1. 门厅
2. 前台
3. 多功能厅
4. 配送室

1. entrance foyer
2. reception desk
3. multi-purpose hall
4. delivery

建筑正面 Front view

室内 Interior

入口 Enrance

Photo: Helene Binet, Brisac Gonzalez Architects

Completion Date: 2007

Architect: Brisac Gonzalez

319

CCVH–Gignac

At the entrance to the city of Gignac, the stake consists in positioning on this place a urban complex, a village around a square. Every building has an independent access to the garden, is autonomous and has its own identity. The traffic of vehicles is channelled around the entity allowing the heart to become only pedestrian.

A particular care is brought to the set-up of buildings. The main constructions are facing north/south to facilitate the natural ventilation. This choice of set-up has two other objectives: shelter the central garden by protecting it from the North wind, and then maintain views on the tower of Gignac, the Pic Saint Baudille and the low valley of Hérault.

Only the building used for the Communauté de Communes is perpendicular, it marks the city entrance and the access to the site, it closes the internal garden. The medical pole and the pole of services face each other around the central garden. Their composition is symmetric, unitarian, modular and evolutionary. This allows to answer both the climatic comfort and the respect for the surrounding built heritage. Zinc, stone and concrete are the essential materials of this composition and make echo for the traditional materials of the place, both for the houses and for the agricultural structures.

基格纳CCVH小区

CCVH小区位于基格纳市边缘，是一座围绕广场而建的城市综合体。每一座建筑都有通往花园的独立入口，各具特色。小区内部禁止机动车通行，只允许行人通行。

设计师对建筑的布局进行了精心设计。主要建筑结构全部采用南北朝向，保证了自然通风。这种布局实现了两个目标：保护中心花园免受北风侵袭；不遮挡建筑周围的自然人文景观。

只有社区行政楼是竖直朝向的，它标志着区域的入口，将花园封闭起来。小区内建筑的构成对称而统一、循序渐进、具有模块结构。这一设计既保证了气候舒适，又尊重了周边的已有建筑。锌板、石材和混凝土是建筑的基本材料，与当地的传统住宅和农业建筑有异曲同工之妙。

Central garden and community building 中心花园和社区楼

Community building 社区楼

社区楼 Community building

1-8. 社区楼

1-8. community building

夜景 Night view

Photo: Paul KOZLOWSKI

Architect: N+B Architectes

Complex

Completion Date: 2009

General view 建筑全景

Limoges Concert Hall

The Concert Hall in Limoges' outer envelope is made of wood arcs and translucent rigid polycarbonate sheets and the inner envelope of wood.

The use of wood was suggested by the location of the hall, in a clearing within a large forest surrounded by trees over 200 years old. The region also has an active timber industry. In addition, the soft translucency of the polycarbonate complemented the wooden frame by allowing light to filter in and out of the building. The strategy establishes reciprocity between concept and context.

The configuration of the double envelope with circulation in-between is a scheme that is advantageous for both acoustical and thermal reasons. The detached and fragmented envelope opens in two directions, towards both the forest and the road. Between the two envelopes are the movement vectors: two ramps, one extending downward toward the lower tiers of the auditorium, and the other upward toward the upper tiers. Additionally, two straight "flying" staircases extend directly toward the top row of seats.

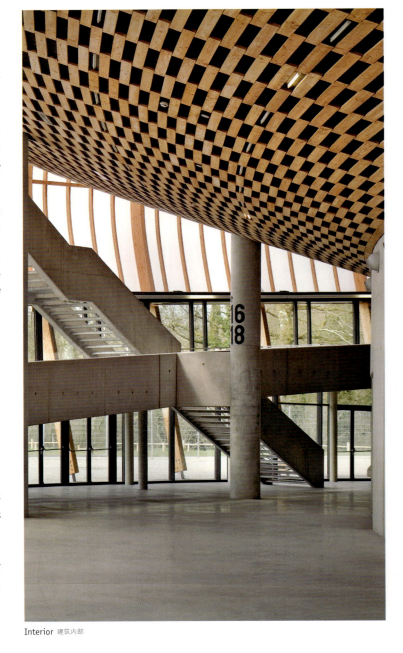

利摩日音乐厅

利摩日音乐厅的外层表皮采用了弧形木材和透明刚性聚碳酸酯板，内层则采用了木质结构。

项目对于木材的运用来自于它的地理位置——一块树林中的空地，该地区的木材制造业十分发达。此外，柔和而透明的聚碳酸酯板完善了木框架结构，让光线得以进入室内。整个设计策略将设计与环境有机结合在一起。

双层表皮结构和夹层流通区域既隔音又隔热。分离而零碎的表皮在森林和路面两个不同的方向开口。两层表皮之间是两个坡道——一个通往礼堂的下层、另一个通往礼堂的上层。此外，两个笔直的飞梯直通最上层的坐席。

Interior 建筑内部

夜景 Night view

建筑外观 Exterior view

Photo: Christian Richter

Completion Date: 2007

Architect: Bernard Tschumi Architects

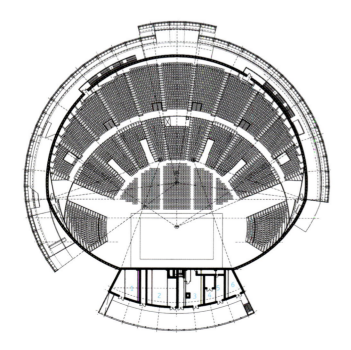

1. 区域发动机
2. 锅炉房
3. 区域适配器
4. 区域高压系统
5. 区域高压系统
6. 区域变压器

1. local engine
2. boiler room
3. local CTA
4. local TGBTS
5. local TGBTS
6. local transformer

Whole scene of the exterior 建筑全景图

Boulogne Billancourt

The first building within the Trapeze masterplan at Boulogne Billancourt, in the southeast of Paris, this seven-storey office development occupies a prime position within the mixed-use development. The building has an expressed white concrete structure, with a steel and glass façade. It is divided between two volumes, connected by a glazed, full-height atrium containing a café and two restaurants for staff. This dynamic space incorporates panoramic lifts and suspended walkways, which appear to float at double-height intervals to allow uninterrupted views out to the river or the park. Above the entrance, giant translucent brise-soleils filter daylight down into the atrium, while a wave-formed geometric façade on the east and west elevations comprise curved vertical panels of alternate translucent and fritted glass.

A rooftop pavilion to the east and the expansive south-facing balconies are accessible from the office floors and take advantage of the riverside location. The 22,000 square metres of interior floor space provides a mixture of cellular and open-plan configurations. The internal arrangement is flexible – the floor plates can be split either vertically or horizontally to accommodate multiple tenants. A fringe of shops enlivens the ground level and the boulevard between the river and the building has been landscaped to provide a waterside promenade.

布洛涅·比扬古办公楼

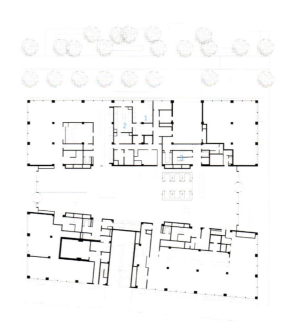

这座七层高的办公建筑是巴黎南部布洛涅·比扬古地区总体规划所建造的第一座建筑。建筑采用了明快的白色混凝土结构和钢铁玻璃复合外墙。建筑分为两个部分，中间由一个玻璃中庭相连，中庭里是咖啡厅和两间员工餐厅。这个充满活力的空间里将观景电梯和悬浮走道融合在一起。人们可以在每隔两层就出现的悬浮走道上俯瞰河流和公园的全景。建筑入口上方巨大的半透明遮阳帘过滤了射向中庭的阳光，东西侧的波纹型外墙则采用了由半透明和多孔玻璃组成的弯曲垂直面板。

从办公区可以直接前往东面的屋顶凉亭和朝南的阳台。22,000平方米的室内面积被设计成灵活的开放式布局，楼板可以根据不同租户的需求在水平和垂直方向进行调整。一楼的一圈店铺让空间充满了生气；建筑和河流之间的林荫道被美化成为了滨水人行道。

1. office
2. lounge
3. staircase

1. 办公室
2. 休息大厅
3. 楼梯

Photo: Foster + Partners

Completion Date: 2008

Architec: Foster + Partners

General view 全景图

Centre Dramatique National de Montreuil

The building complex, closed like a fist, has its correspondence in a series of spaces which expand and contract on the interior, reminiscent of the implausible spaces Alice in Wonderland traversed.

In the entrance area, the ceiling above the reception desk has multiple folds, similar to a geological zone of convergence, where tectonic plates are on a collision course but manage to slide past each other. In the stairway, gyrating concrete volumes are bathed in Caravaggioesque light. Dark and light, empty and full, this red space expands horizontally, ultimately leading to the auditorium, and then contracts until it becomes lost in the darkness of the foyer. Breaches in the building envelope occasionally offer up views to the city and allow daylight to penetrate deep into the space.

蒙特勒伊国家戏剧中心

戏剧中心的外墙几乎是完全封闭的，整体造型像一个拳头一样，室内空间时而扩张、时而收缩，让人感觉仿佛置身于爱丽丝的仙境。

入口处前台上方的天花板有许多褶皱，与大陆构造版块相互碰撞所产生的聚合地势相似。楼梯上的旋涡状混凝土结构沐浴在卡拉瓦乔式吊灯所投下的灯光中。光明与黑暗、充实与空虚在这个红色的空间里交错，延伸到礼堂，又消失于大厅的黑暗之中。建筑外墙上个别的开口为室内提供了城市风景和自然采光。

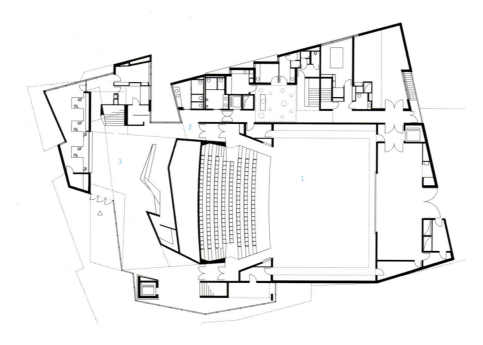

1. 大厅
2. 洗手间
3. 楼梯

1. salle
2. toilet
3. stairs

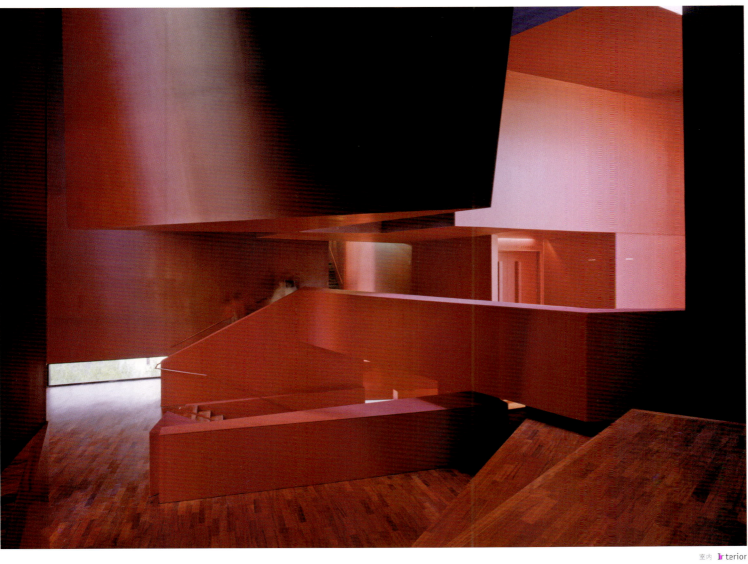

室内 **In**terior

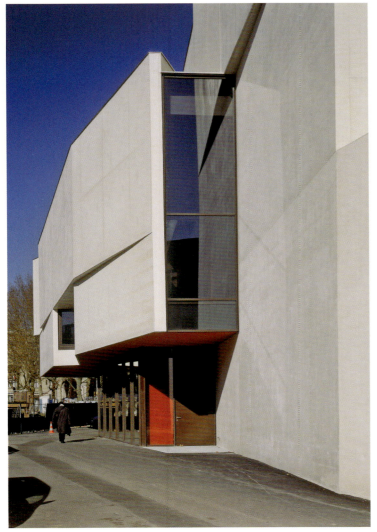

入口 Entrance

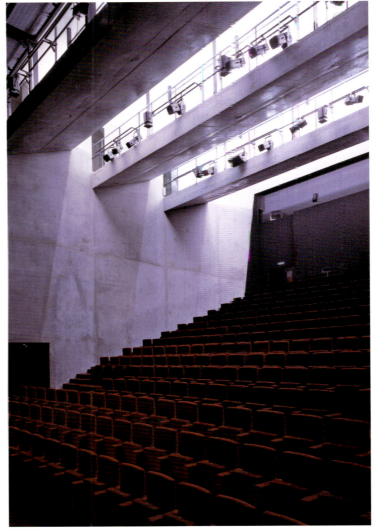

六= Salle

General view 全景图

Chapel Mussonville

马颂维勒教堂

Mussonville Chapel was built in 1880. The small theatre located in the chapel had to be improved and create new functions (reception, dressing rooms, equipment rooms, etc.). To ensure that these additions are not read as "warts" hung in the chapel, the extension has been made in the form of a "cloister" in which the various functions services take place.

A square grid, which consists of concrete columns of the equivalent sections in the foothills of the chapel, was developed from the chapel and deployed on three sides. In this constructive frame dictated by the chapel, the "brutality" of the constructed elements – joinery galvanised steel poles with coffered concrete planks of raw wood, compacted gravel floor, and chestnut logs provides consistency and a basis stronger in the park that surrounds Mussonville. The empty space created by this provision allows to extend the festivities outside. In the suburbs, a square enclosure formed by screen walls made of logs raw untreated fixed on a tubular frame, was galvanised to close or open space of the chapel, providing a "grid" vision on the spacious park.

马颂维勒教堂建于1880年，教堂内的小剧院需要整修并添加新的功能区，如接待处、更衣室、设备室等。为保证这些新添加的设施不影响教堂的整体效果，扩建工程被设计成一个回廊，所有的功能区都设在这个回廊里。

建筑师用水泥柱将方形的网格连接起来，从三面环绕着教堂。新建的结构由细木镀锌钢柱和原木方格厚木板组成，地面上铺装着碎石。这些元素简单而朴素，与环绕着教堂的公园十分相称。回廊和教堂之间所产生的空地可以举行庆典活动。外圈的原木围墙围成了可开合的方形大门，将公园的景色分割成一个个方块。

Wall 围墙

夜景 Night view

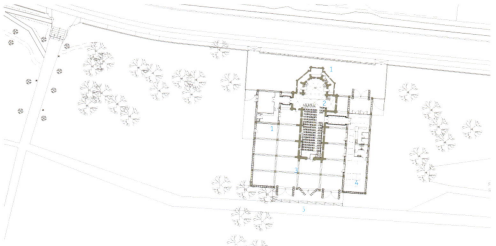

1. 蓬摊
2. 景色
3. 蓬摊
4. 天井
5. 接待处

1. loge
2. scene
3. loge
4. parvis
5. reception

庭院 Garden

Photo: Arthur Pequin

Completion Date: 2008

Architect: atelier d'architecture King Kong

European Investment Bank

The new headquarters building for the European Investment Bank (EIB), with its compelling 13,000-square–metre glass roof, extends Sir Denys Lasdun's existing buildings on Luxembourg's Kirchberg plateau. It provides 72,500 square metres of office space and other facilities for up to 750 employees.

The striking tubular glass roof spans the entire, 170–metre–long and 50–metre–wide structure. In combination with an extremely lightweight glass and steel superstructure, it offers a maximum of daylight and transparency. In addition, the building's zigzag plan encourages a non–hierarchical office layout that promotes interaction and communication. This unrivalled office environment is carried by an environmental programme that reflects a progressive approach towards sustainability in architecture.

Key to the new headquarters' ecological concept is the glass roof which curves around the floor plates to create the atriums in the V–shaped "gaps" of the building wings. The landscaped winter gardens on the valley side are unheated and act as climate buffers. In contrast, the atriums on the boulevard side serve as circulation spaces; hence temperatures have to be kept at a comfortable level. Both winter gardens and "warm" atriums are naturally ventilated through openable flaps in the shell to draw fresh air into the building and to reduce heat gain especially in the summer months.

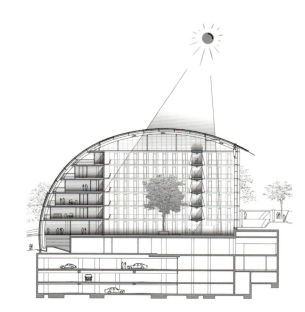

欧洲投资银行

欧洲投资银行新总部大楼13,000平方米的玻璃屋顶十人引人注目，将丹尼斯·兰斯顿先生所建造的建筑扩展到了卢森堡的教堂山高原上。新总部大楼为750余名员工提供了72,500平方米的办公空间和相关设施。

造型独特的管状玻璃屋顶跨越了长170米，宽50米的建筑物。屋顶的轻质玻璃和钢铁结构为建筑提供了充足的自然采光和通透性。此外，建筑内部的Z字形布局是办公室之间没有分级，促进了员工之间的互动和交流。这个无以伦比的办公环境在设计过程中运用了一系列的可持续和环保设计策略。

建筑生态设计的关键在于环绕玻璃屋顶，它包裹着整片楼板，在建筑两翼的V形空挡处形成了中庭。山谷一侧的景观冬日花园并没有空调设施，可以作为一个气候缓冲带。相反，临街一侧的中庭作为流通中转空间，一直保持着适宜的温度。冬日花园和温暖的中庭都通过建筑外壳的可控式风门实现了自然通风，保证了建筑内部的新鲜空气，降低了夏天的热吸收。

Architect: Ingenhoven Architects

Completion Date: 2008

Corporate

Photo: Ingenhoven Architects

General view of hotel 公园酒店全景

SOF Park Inn Hotel Complex

October 8th 2009 marks the opening of SOF, the new Hotel Park Inn, Krakow, Poland designed by the architects team of J. MAYER H. Architects, GD&K Consulting Sp. z o.o. and OVOTZ Design LAB. The hotel is being constructed in the city centre, at the intersection of important transport routes. The hotel is located in the vicinity of the planned Congress Centre as well as near to the Wawel Castel and the historical Jewish district. It offers splendid views onto the old city centre. The new porperty is characterised by clear horizontal lines, picking out the panorama view as a central theme. The façade is emphasised by black and white aluminium stripes, seperated by dark glass windows.

SOF公园酒店

SOF酒店由德国建筑师J·迈耶·H、GD&K咨询公司及OVOTZ设计工作室共同携手打造，并于2009年10月8日正式开业。酒店位于克拉科夫市中心区多条马路交界处，与国会中心紧邻，距华威城堡和犹太历史文化区仅有数步之遥。酒店建筑以清晰的水平线条为特色，将"城市全景"作为中心主题。外观由黑白相间的铝条打造，黑色的玻璃窗则起到分隔的作用，特色十足。

1. restaurant	1. 餐厅
2. kitchen	2. 厨房
3. bar	3. 酒吧
4. lobby	4. 大厅
5. reception	5. 接待处
6. canteen	6. 食堂
7. waste	7. 垃圾房
8. service	8. 服务区
9. delivery	9. 配送区
10. toilet	10. 洗手间
11. office	11. 办公区

麦气侧面 **Si_le view**

Photo: Jakub Kaczmarczyk, Ovotz design Lab (www.ovotz.pl)

Hotel

公园酒店外观 **Park Inn**

公寓酒店入口 **Par_Inn and_entrance**

Architect: J. MAYER H. Architects, GD&K Consulting Sp. z o.o., Ovotz Design Lab

Completion Date: 2009

View of the house from the driveway　从车道上看住宅

Aatrial House

One-hectare site near the forest, where the building is designed has only one weak point: south-west access. An obvious conflict developed between the driveway and the garden. The idea arose to lower the driveway in order to separate it from the garden. This prompted another idea of a driveway leading inside to the ground floor level, from underneath the building, which became possible thanks to the creation of an inner atrium with the driveway in it.

New type of the house
As a result, the building opens up onto all sides with its terraces in an unrestricted manner, and the only way to get into the garden is through the atrium and the house.

Structure and materials
The gateway is situated in the highest point of the site sloping to the east side. The ten-metre-wide driveway follows slope's declivity. The building was situated on the garden level. For the sake of neighbouring buildings, typical Polish "cube houses" arisen in 1970s, the structure of the house results from various transformations of a cube.

As a result of stretching and bending particular surfaces of the cube, all the walls, floors and ceilings were defined, together with inner atrium and terraces. This principle of formation has not only created the structure of the house, but also defined interior and exterior architecture, including use of materials. The building is a reinforced concrete monolith, and concrete is at the same time the finishing material of the transformed cube, while all additional elements are finished with dark ebony.

心房住宅

住宅选址在毗邻森林的地带，占地面积达10000平方米，其最为显著的弱点便是入口朝向西南方向设置。为使车道和花园之间互不干扰，车道被"拉低"，从而与花园分离开来。同时心房的设置，更使得车道一直上升通往一层。

新型住宅
住宅四面都设有露台，完全开放。通往花园的唯一通道便是穿过心房和住宅。

造型及材质
大门位于最高点，朝向东侧倾斜。10米宽的车道顺着斜坡打造，房子与花园位于同层。周围的建筑都采用70年代波兰典型的"立方盒子"结构，这一房子则突显出立方体的各种变化形式。

"盒子"表面蜿蜒伸展，奠定了墙壁、地面及天花的结构造型。钢筋混凝土打造了房子的框架，水泥及黑檀木作为主要的装饰材料。

Kitchen and dining room　厨房和餐厅

车道直通住宅的一楼 View of the driveway leading inside to the ground floor level

中庭 View of the atrium

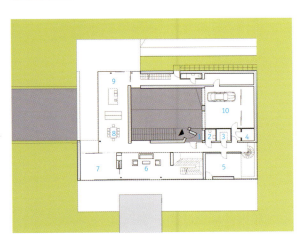

1. entrance	1. 入口
2. toilet	2. 洗手间
3. wardrobe	3. 衣帽间
4. storage	4. 贮藏室
5. study	5. 书房
6. television room	6. 电视房
7. living area	7. 起居室
8. dining area	8. 餐厅
9. kitchen	9. 厨房
10. garage	10. 车库

住宅背面和后院 Backyard Façade

Photo: Kwk Promes

Completion Date: 2007

Architect: Kwk Promes

Outrial House

The context and investor's expectations made designers to treat grass as a material. One part of the grassy plot is cut out and treated as the roof for basic functions. At the end of design work the investor wished to find a place for an orangery and a small recording studio. Due to "cutting a notch and bending" to inside, the roof became an atrium reached from the inside of the house. The smooth roof created space similar to bandstand that can be used to outdoor jam sessions. The very similar process was used to design the recording studio.

OUTrium – the new type of space
The green space smoothly penetrates into the interior which joins the merits of a traditional atrium with the outdoor garden. OUTrium – as it was called – lets inhabitants be pleased with the outside and stays the inside of the house strictly joined with the living room at the same time.

Technology
The selection of technologies was based on the investor's principles who wanted a simple house. The house is built in white plaster inside and outside walls. The green roof is a good isolation that reduces the loss of heat in winter and cools the house in summer. It assures a very positive microclimate inside the house.

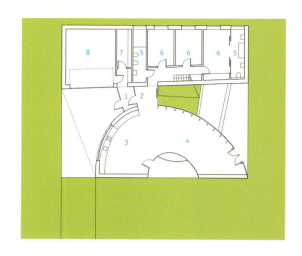

1. entrance
2. hall
3. kitchen with dining area
4. living area
5. bathroom
6. bedroom
7. storage
8. garage

1. 入口
2. 门厅
3. 厨房和餐厅
4. 客厅
5. 浴室
6. 卧室
7. 贮藏室
8. 车库

绿色屋顶住宅

设计师从其所处的环境背景以及业主要求出发，将绿草视为一种建筑材质。其中，草地一部分被用作屋顶。在工程接近尾声的时候，业主要求打造一个橘园和一个小录音室。为此，设计师构思了"旋转入室"的理念，将屋顶打造成心房。

新型空间
绿色空间一直延伸到室内，心房将其与室外花园连通。居住者生活在室内，但同时又有一种置身于室外的感觉。

工艺技术
工艺技术的选择完全依据业主的要求——他们需要一个简约的住宅。设计从这一点出发，室内外全部采用白色灰泥粉饰，绿色的屋顶构成了很好的屏障——冬天减少热量流失，夏天吸收热量，从而确保了室内的舒适环境。

The southwest façade 建筑西南立面

View of the house from the road 从路边看住宅

通往屋顶的台阶上长满了青草 The stairs leading to the roof with one mesh across which grows the grass

客厅通过楼梯与绿色屋顶相连 Living room with stairs leading to the green roof

作为入口 Main entrance to the house

Photo: KWK PROMES

Completion Date: 2007

Architect: Kwk Promes

Façade 建筑外观

Safe House

安全屋

Location

The house is situated in a small village at the outskirts of Warsaw. The surroundings are dominated with usual "Polish cubes" from the 1960s and old wooden barns.

Idea

The clients' top priority was to gain the feeling of maximum security in their future house, which determined the building's outlook and performance. The house took the form of a cuboid in which parts of the exterior walls are movable. When the house opens up to the garden, eastern and western side walls move towards the exterior fence creating a courtyard.

After crossing the gate, one has to wait in this safety zone before being let inside the house. In the same time, there is no risk of children escaping to the street area in an uncontrolled way while playing in the garden.

Movable elements interfering with the site layout

The innovation of this idea consists in the interference of the movable walls with the urban structure of the plot. Consequently, when the house is closed (at night for example), the safe zone is limited to the house's outline. In the daytime, as a result of the walls opening, it extends to the garden surrounding the house.

New type of building

The sliding walls are not dependent on the form of the building. That is why this patent can be applied to both modern and traditional, single- and multi-storeyed houses covered with roofs of different geometry. This universal solution gives a new type of building where not the form but the way of functioning is the most important. The name "safe house" gains a new meaning now.

地点
住宅坐落在华沙市郊的一个小村庄内，周围遍布着20世纪60年代修建的波兰风格"立方盒子"民房及小木屋。

理念
业主的首要要求即为最大限度确保安全性，这从而决定了建筑的外观及功能。房子呈现立方体造型，其中部分室外墙壁可自由移动——东西两侧的墙壁滑动到室外篱笆处，形成庭院，房子便完全朝向花园开放。来客走进大门之后，必须在安全区等候。

移动墙壁与地块格局互动
夜晚，房子关闭之后，安全区便完全被划分到外围；白天，墙壁移走，环绕在房子周围，并延伸到花园处。

建筑新模式
滑动墙壁并不依赖于建筑造型，可以广泛应用。相对于功能，形式便显得不再重要，这当然也赋予"安全屋"另一层含义。

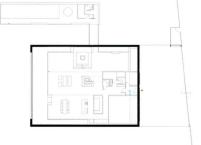

1. entrance	1. 入口
2. living area	2. 客厅
3. dining area	3. 餐厅
4. kitchen	4. 厨房
5. toilet	5. 洗手间
6. wardrobe	6. 衣帽间
7. television space	7. 电视房
8. fireplace	8. 壁炉
9. garage	9. 车库
10. storage	10. 贮藏室
11. swimming pool	11. 游泳池

开放式住宅 Open house

南朝平台 South terrace

通过敞开的百叶窗看室内空间 Open shutters show widely open inner space

Photo: Kwk Promes

Completion Date: 2009

Architect: Kwk Promes

Sustainable House at Lake Laka

拉卡湖可持续住宅

This simple sustainable house – like a chameleon – blends with its surrounding area on Laka Lake. Colourful planks within the timber façade reflect the tones of the landscape. The window reveals, clad in fibre cement, framed images of the countryside. Analogical to most creatures, the building is symmetrical outside, although the internal zones – according to function – are arranged asymmetrically.

The built form is designed to optimise the absorbance of solar energy. Approximately 80% of the building envelope is facing to the sun. The single–storey living space on the ground floor is externally clad with untreated larch boarding. Solar energy is gained there by the set–in glazed patio. Solar collection panels are located on the roof and a photovoltaic system is planed for the future. The dark façade of the "black box" – a three–storey structure clad with charcoal coloured fibre cement panels – is warmed by the sun, reducing heat loss to the environment. The passive and active solar energy concepts and a high standard of thermal insulation are enhanced by a ventilation plant with thermal recovery system.

The design of the project was determinated by the twin goals of low life cycle costs and a reduction in construction costs. All details are simple, but well thought out. The house did not cost more than a conventional one in Poland. Cost–saving was made by the application of traditional building techniques and the use of local materials and recycled building elements.

这幢简单的可持续住宅如同变色龙一样，融入了拉卡湖的环境之中，木墙面上的彩色木条反映了周边景观的基本色调。从纤维水泥窗框向外可以看到乡村的风景。与大多数生物相同，建筑的外观是对称的（尽管内部功能区的设置并不对称）。

建筑的造型充分利用了太阳能，约有80%的建筑外墙是朝向太阳的。一楼的生活空间外部包裹着未经处理的松木板，阳光可以通过嵌入式玻璃天井射进去。屋顶上装有太阳能电池板，可以用作未来的使用。由木炭纤维板所包裹的三层楼结构——"黑盒子"的外墙可以吸收阳光，减少对外热流失。具备热回收作用的通风系统进一步提高了主动和被动式太阳能策略与隔热系统的效用。

项目的设计具有两个目标：低生活成本和低建造成本。任何一个简单的细节都经过设计师的精心考虑。住宅的建造成本并不比普通的住宅高，传统建筑技巧和本地材料、回收建材的运用大大降低了建造成本。

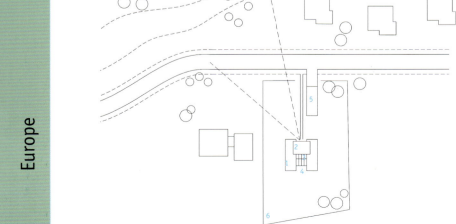

1. 绿色屋顶
2. 太阳能 "黑盒子"
3. 冬日花园
4. 平台
5. 停车场
6. 花园

1. green roofs
2. solar "black box"
3. winter garden
4. terrace
5. parking space
6. garden

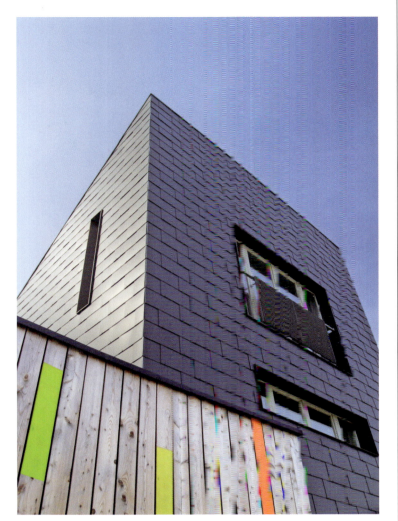

Photo: Tomek Pikula

Completion Date: 2008

Architect: Peter Piotr Kuczia

Front view 建筑正面

Andel's Hotel Lodz

The impressive weaving mill building was originally established in 1878 as part of a cotton factory complex. Due to subsequent turn of historical events, the building was forgotten and covered by the dust of history for many years. In 2009, the former cotton mill has been transformed into a sophisticated hotel by OP ARCHITEKTEN. It includes 278 designer style rooms and suites, 3,100 square metres of conference space, ballroom for 800 people and fine-dining restaurants and bars with seats for more than 450 people, swimming pool and wellness centre.

OP ARCHITEKTEN completely redesigned the roof area, proposing a wellness centre, event space and sun terrace. A lot of glass elements and skylights ensure that public space, including four-level atrium is properly lit by the great amount of natural light. Located on the roof of the building the big ballroom of 1,300 square metres is one of the largest hotel halls in Poland. To meet the needs of various occupants the place is fully customisable and acoustically separated from the rest of the building.

The most characteristic element of the roof area is a unique glass-enclosed pool, placed in a former fire water tank. The tank's unusual location allows for exceptional experience of the flow of spaces, where the pool-area ties together with the skyline of the city, impressive red-brick scenography of the façades and the elegant new landscape of the roof.

罗兹安德尔酒店

作为一家纺织厂，这座宏伟的建筑建于1878年，是棉花工厂的一部分。由于随后的一系列事件，这座建筑被历史遗忘了许久。2009年，OP建筑事务所将纺织厂改造成为一座精品酒店。酒店拥有278间设计客房和套房，3,100平方米的会议室，可容纳800人的舞厅以及450个坐席的精品餐厅和酒吧、一个游泳池和一个健身中心。

OP建筑事务所重新设计了建筑的屋顶区域，在上面建造了健身中心、活动中心和日光平台。大量的玻璃窗和天窗保证了公共空间（包括四层楼高的中庭）充足的采光。1,300平方米的舞厅位于建筑的屋顶，是波兰最大的酒店大厅之一。为了满足不同宾客的需求，这一空间完全采用个性化设计，并且具有良好的隔音性。

屋顶区域最具特色的元素就是一个罩在玻璃罩里的游泳池。游泳池是由废弃的消防水槽改造而成的。水槽的特殊位置使游泳池和城市的天际线、美丽的红砖墙和优雅的屋顶景观紧密相连。

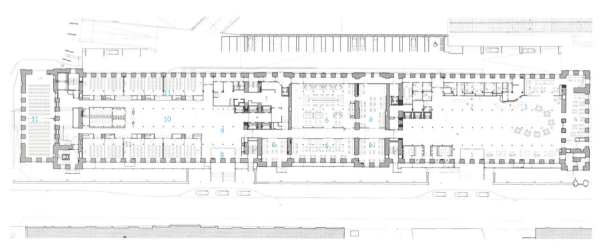

1. 主入口
2. 前台
3. 大堂吧
4. 咖啡吧
5. 办公区
6. 餐厅
7. 厨房
8. 会议室入口
9. 会议大厅
10. 前厅
11. 会议室

1. main entrance
2. reception
3. lobby bar
4. café, bar
5. offices
6. restaurants
7. kitchen
8. conference entrance
9. conference lobby
10. prefunction
11. conference rooms

屋顶 Rooftop

植多景观 Plant

会议厅 Ballroom

餐厅 Restaurant

Hotel

Photo: Op Architekten

Completion Date: 2009

Architect: Op Architekten

General view 全景图

Basalt Wine – Laposa Cellar

The wines of the Laposa Cellar following the millennium became well known amongst Hungarian wine drinkers under the brand name "Bazaltbor" or Basalt wine. The building is composed of connected panel elements, which were cast as monolithic visible concrete. The neutrality and rigidity of this are primarily detectable in the internal spaces and their relationship. There were two places where the ornamentation was necessary: when meeting the outside and at the cellar section for barrel maturation. For the former, following the principle of being like a model, the differentiation of the façades and the roof is missing; their homogenous covering is made up of prefabricated fine concrete facing panels, with a slightly transformed pattern of grapevines climbing and twining around them. If natural lighting needs to be provided in the inside spaces, the bands in the reinforced concrete model, following its geometry, were replaced by a light structure and glass cladding and the facing panel by a perforated metal sheet. Naturally the same grapevine pattern continues on this latter one, pulling it together into a unified surface.

The other ornament is to be found in the inside, in the deepest branch of the cellar. Although this space is of a longitudinal nature with a barrel shape, like a traditional cellar, its axis is broken several times and its structure is from reinforced concrete as part of the model. This bent–broken surface is covered by a brick layer characteristic of traditional cellars, but not according to the principles of tectonic order and brick binding, but diagonally, appearing as a woven fabric.

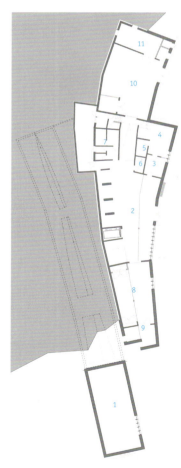

Basalt酒业——Laposa酒窖加工厂

Laposa酒窖所生产的Basalt酒在匈牙利颇负盛名。加工厂由连续的单块混凝土面板结构组成。这一结构的硬度和客观特点可以在建筑的内部空间和内外环境的联系中得以体现。有两个位置需要额外的装饰：建筑的入口和酒窖的酿酒处。前者的设计取消了外墙和屋顶的差别性，二者都由预制的细石混凝土面板组成，蜿蜒其上的葡萄藤图案可以作为点缀。当室内需要阳光时，钢筋混凝土结构可以被轻质玻璃罩所取代，而面板可以被网状金属板所取代。相同的葡萄藤图案也出现在后者的设计之中，使建筑达成了一种统一性。

另一个装饰出现在室内，在酒窖的最深处。尽管这一空间和传统的酒窖一样，像一个狭长的枪管，它的轴线有若干个转折点，结构也采用了钢筋混凝土。弯曲的表面由传统酒窖所采用的砖块铺装，但是并没有遵循普通的建筑模式，而是以对角线的方式铺装，就像一件纺织品一样。

1. airspace	1. 通风空间
2. manipulation area	2. 操作区
3. lab	3. 实验室
4. storage	4. 贮藏室
5. storage	5. 贮藏室
6. WC	6. 洗手间
7. changing room	7. 更衣室
8. bottling area	8. 装瓶室
9. technical room	9. 技术室
10. warehouse	10. 仓库
11. technical room	11. 技术室

建筑与景观 View with landscape

建筑细部 Details

外墙 Wall

门厅走廊 Passageway

Photo: Zsolt Batár

Completion Date: 2010

Architect: Plant-Atelier Peter Kis

Industrial

Exterior glass façade 玻璃幕墙

UNIQA Vital Business Centre

The new head office is located close to the previous one at the junction of two important traffic arteries in District XIII, Lehel út and Róbert Károly körút. The office building constructed by property developer Raiffeisen Evolution in twenty-two months, named the Vital Business Centre, has net usable floor space of 18,000 square metres. The nine–storey building houses all of UNIQA's divisions, while 40% of the usable floor space will be let to other companies. Companies renting the space so far include a pharmacy and a diagnostics centre. Árpád Ferdinánd explained that while designing the building he was guided by the principle of transparency. Transparency is one of the cornerstones of UNIQA's business philosophy, and is conveyed in the Vital Business Centre by many glass surfaces and expansive rooms, the architect said. Lighting up the night, a distinctive feature of the building is its LED façade, allowing colourful images and messages to be displayed on the street–facing side of the office building every evening. Although the idea is not new, it is unique in its scale: with some 80,000 pixels, the Vital Business Centre currently boasts Europe's largest LED façade.

UNIQA活力商务中心

UNIQA公司的新总部位于前总部的旁边，第十三区的主干线Lehel ut和Robert Karoly korut街的交叉点。UNIQA活力商务中心工程由房产开发商Raiffeisen Evolution开发，历时22个月，净使用面积为18,000平方米。这座9层楼高的大楼容纳了UNIQA公司的所有部门，还有40%的剩余空间出租给其他公司。目前已有一家药业公司和一个诊断中心使用了出租空间。设计师阿尔帕德·费迪南称，大楼的设计完全以透明原则为指导。透明是UNIQA公司经营理念的基础，活力商务中心的玻璃外墙和开阔空间都充分体现了透明原则。夜晚，大楼的LED外墙在临街的一面展示着多彩的影响和信息，使大楼十分抢眼。尽管LED概念并不新鲜，但是80,000像素的LED外墙是迄今为止全欧洲最大的。

1. 入口
2. 大厅
3. 前台
4. 管理室
5. 办公区

1. entrance
2. lounge
3. reception
4. management
5. office area

Photo: Istvan Oravecz

Completion Date: 2009

Architect: Ferdinand and Ferdinand Architects

Benozzo Gozzoli Museum

The new design of this building, which has a total area of about 400 square metres, closely follows the ground traces of the old building and it is luckily detached from the buildings around because it is located in an empty space which is a sort of square.

The building is rooted to the ground through a functional island-shaped base which solves the problem of urban furnishing meant in the classical sense (benches, flower pots, etc.). The curvilinear base takes over the space around the building and, at the same time, people take over a little of the museum's space: the base becomes a bench, a play area for children and adults, a theatre for small outdoor events. The building had to be entirely coated with cotto; in this way, it refers to materials and finishes of some of the local churches.

Inside, the museum is spread over four floors, three above ground and one underground. The ground floor is partly characterised by a low ceiling: a shaded area which quickly runs to the full-height space where the "Tabernacolo della Visitazione" is placed. This is illuminated by a cascade of natural light coming from the skylight in the ceiling. On the first floor, recessed into the corner-wall, we find the "Tabernacolo della Madonna della Tosse" which looks like a television screen. The staircase linking the floors, becomes a kind of visual path which frames the Tabernacles (now lacking their original context), according to new and constantly changing perspectives. It stops on the first floor, and then starts again on the opposite side thus reaching a room on the second floor, a naturally suitable space for small exhibitions and educational workshops.

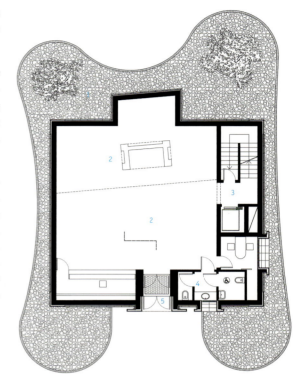

贝诺佐•戈佐利博物馆

建筑总面积为400平方米，与旧建筑的轨迹十分契合。由于地处一片类似广场的空地，它有效地和周边的建筑隔离开来。建筑通过一个小岛形状底座与地面相连，满足了典型的城市景观需求（如长椅、花盆等）。弯曲的底座占据了地面，而人类则占据了底座的空间：底座可以作为长椅、游乐场或是举行小型户外活动的场地。

博物馆的内部共分四层，三层在地上，一层在地下。一楼的天花板很低，穿过这片阴暗的区域就到了中庭——艺术品Tabernacolo della Visitazione的摆放区。日光透过天花板上的天窗倾泻而下。艺术品Tabernacolo della Madonna della Tosse摆放在二楼的墙角，看起来像一台电视机。楼梯将各个楼层相连，从视觉上环绕着"神龛"。楼梯在二楼略作停留，随后又转向三楼展览小型艺术品和工作室的房间。

1. landscape
2. exibition room
3. staircase
4. toilet
5. entrance

1. 景观区域
2. 展览室
3. 楼梯
4. 洗手间
5. 入口

Exterior 建筑外观

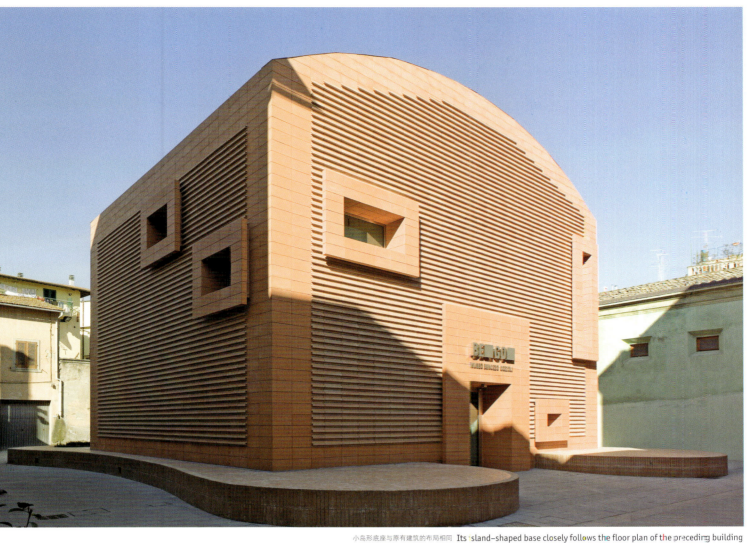

小岛形底座与原有建筑的布局相同 Its island-shaped base closely follows the floor plan of the preceding building

建筑外墙 Façade

夜景 Night view

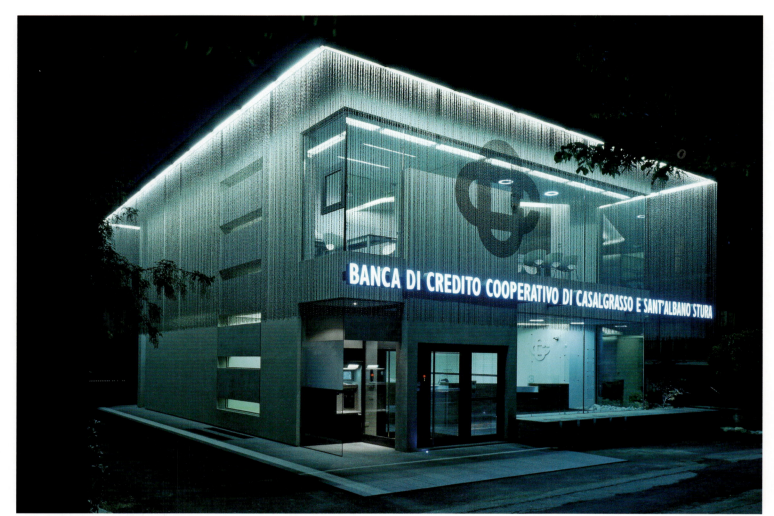

General view 建筑全景

New Branch of the Cooperative Credit Bank
11 Kilometres of Chain....

The architecture seems like an ode to clarity, transparency, to conceptual essence itself, while at the same time never losing sight of that direct functional relationship with the public, not just words from a bank worker's manual, but the real public; an old man, a parent a child, all with differing needs. There is a wide, luminous open informal space given over to a children's play area, quite a novelty in a bank, besides the obligatory comfortable sofas at the entrance. Wherever one goes from the counters to the meeting rooms one is breathing in that atmosphere of ordinary domesticity that never descends into the banal or predictable.

The inclusion of the chains on the façade, 11 linear kilometres, are no mere stylistic or aesthetic pretence, no external second skin. The idea is much more considered. The whole building, changes, shimmers and modifies its physical consistency. The curtain of chains hanging from on high fluctuates lightly letting off a gentle ringing, like Tibetan prayer bells. It captures the light bringing it inside, moulding it, and changing it, thus transforming the massive potentially heavy interior so that the whole building becomes a gigantic semitransparent lamp. In direct sunlight they act as protective sunshades.

合作信贷银行支行——由11千米长的链条装饰的建筑

建筑仿佛歌颂着简洁与通透，同时又没有缺失自身对于公众的功能特征，能够满足不同人群的服务需求。银行入口处摆放着舒适的沙发，沙发旁有一个宽敞明亮的儿童游乐区，这在银行设计中十分少见。从柜台走向会议室的人会体验到家一般的感受，而不是陷入平庸老套的银行氛围。

外墙上总长度为11千米的链条不仅仅是造型艺术或是第二外墙，这一设计是经过深思熟虑的。整个建筑在链条的围绕中变幻、闪烁，形成一个统一的主体。链条随风飘动，发出悦耳的响声，仿佛祈祷铃一般。链条捕捉阳光，将阳光引入室内，塑造它、改变它，厚重的室内随之变形，从而让整个建筑变成了一个巨大的半透明灯笼。日光直射时，链条还能作为防护遮阳网。

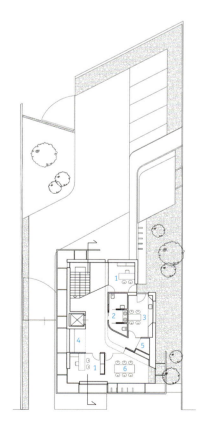

1. office	1. 办公室
2. toilet	2. 洗手间
3. kitchen	3. 厨房
4. waiting area	4. 等候区
5. storage	5. 贮藏室
6. meeting room	6. 会议室

建筑细部 **Details**

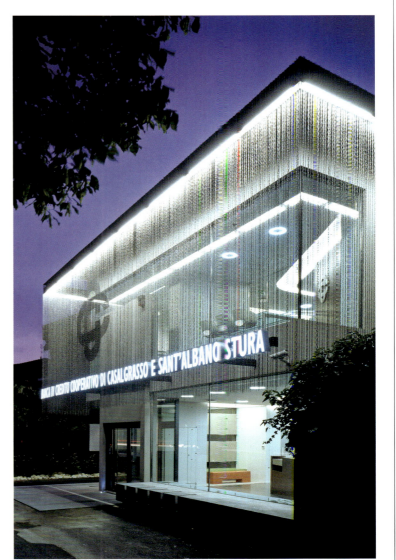

建筑侧面 **Profile**

室内细部 **Interior**

景观平台 **Landscape deck**

Photo: Alberto Piovano

Corporate

Completion Date: 2007

Architect: Studio KUADRA (www.kuadra.it)

Schreckbichl

The existing façade of the winery "Kellerei Schreckbichl" remains unaltered in its shape and its windowcases. A new structure will be suspended in front of this at a horizontal distance of fifty centimetres – a timber and metal structure.

The façade is a playful mix of spacings, colours, openings and shadows. This results in a rhythmic assemblage of forty-five metres total length and nine metres height. The black and red steel plates accentuate the window openings through their tilts and depths, focusing the views on production facilities, entrance, offices, reception and the sky. In contrast with the previous flat appearance of the façade, the differently sized openings produce a surprising and vivacious plasticity. The timber surfaces are inspired by the oak barrels, recreating a theme which adds a warm, inviting character to the building. At night the dynamic façade is immersed in light, again in a playful combination of colour and shadow.

The central element is a tilted plane, which represents the backbone of the staircase, the stairs themselves being integrated in this plane. A glass railing and the lighting from below create the impression that the staircase leads upward, beyond the printed image, in a seemingly weightless manner. A visitors' sofa has also been accommodated in the tilted plane, which bends and continues as a horizontal feature, thus rendering the reception and presentation area a unique visual entity.

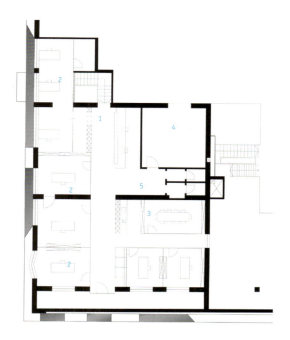

1. stairs   1. 楼梯
2. office   2. 办公室
3. meeting room   3. 会议室
4. storge   4. 仓库
5. entrance   5. 入口

施莱克比奇酒厂

酒厂原有的外观在造型及窗户设置上未经任何改变，设计师重新打造了一个由木材及金属构成的结构，并使其悬浮在与原有外观表面（相距50厘米）。新建表皮结构（长45米、高9米）是间距、色彩以及阴影等元素的趣味组合，节奏感十足。红色金属板通过倾斜度及纵深的变化使得窗户更加突出，从而将视线聚焦到生产设备、入口、办公区、接待台上。

规格不一的窗户带来了动感，木材表皮则增添了温暖友好的气息。在夜晚，整幢建筑沉浸在灯光里，玩味着光与影的变换。倾斜的结构形成了室内楼梯的龙骨，玻璃扶手和闪烁的光线营造了轻盈感。

External detail 建筑细部

External view 建筑外观

建筑立面 Front view

 室内 Interior ware shelf

Photo: Günter Richard Wett

Architect: bergmeisterwolf Architekten

Industrial

Completion Date: 2007

Matzneller

Three buildings are integrated among themselves. The areas of contact are forming the terraces. Different materials meet in a harmony of colours. The dihedral of the bodies is creating a tunnel-like atmosphere towards the green interior courtyard.

The plaster emphasises the individual volumes: brownish and strongly structured beneath, the admixed copper vitriol (bluestone) creating a bluish colouring, an association to the vine that used to grow here. A wooden slatted frame is connecting the individual buildings as well as the inside and outside. The glassy cube, separated from the living area, is planned to be the library, the office and the guestroom all in one. Windows in the shape of a belt allow the daylight to penetrate the underlying sports room.

The main bodies are connected by a staircase, the central element of this residence. A slot arises, a gorge made of glass panels, in which the wooden stairs are slotted and which are connecting everything from the upstairs until the cellar. The staircase is leading upstairs in a decorative way, right between the frameless glass panels and the OBS-panels painted with white varnish.

麦特辛乐住宅由三个单独建筑相互连结而成，连通区域形成了露台。不同的材质通过色调变换融合在一起。

设计师选用不同色调的灰泥材质赋予建筑各自的特色，褐色的底部结构让人不禁联想到葡萄园。木板框架将不同的建筑部分连结起来。图书室、办公区及客卧全部设置在一个"玻璃立方体"内，与客厅分离开来。光线透过条带状的窗户投射进来，一直照射到地下健身区内。

中央楼梯将内部空间连通，木质台阶镶嵌在玻璃栏杆内，格外吸引眼球。

Exterior corridor 室外走廊

自然环境中的建筑 Building in the natural environment

Photo: Jürgen Eheim

室内通道 Interior corridor

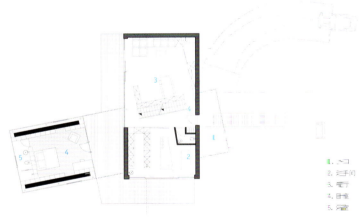

楼梯 Stair

1. entrance 2. toilet 3. dining 4. bedroom 5. bathroom

1. 入口
2. 洗手间
3. 餐厅
4. 卧室
5. 浴室

Completion Date: 2008

室内 Interior

Architect: bergmeisterwolf Architekten

External 建筑外观

Hi-tech Systems Headquarters

高新技术系统总部

The new directional building has been designed to abstract itself from the common building scheme, rejecting analogies and mimesis, while intended to assume technology and innovation as its essence.

The traditional building techniques make room to the use of innovative materials or the different exploitation of the old ones, and, at the same time, to the overcoming of conventional building schemes. Through the opposition between the sculptural concrete shapes and the lightness of the glass and steel structures, the designers have chosen to characterise the internal and external space, pointing out a clear and essential architectural language.

The façade is mainly conceived to go between the aesthetic and conceptual definition of the architecture. Built totally in steel and structural glass and closed among big concrete walls the front is both like an optical and material screen of the building, both its external projection; the aim is to symbolise the Company's talent to be always in search of innovative and new solutions.

高新技术系统总部新建建筑旨在从类似的结构中脱颖而出，摒弃模仿与相似，突出"技术"和"创新"的主题。传统建筑方式渐渐趋向于运用新材料或以现代方式诠释传统材质，从而打破古旧的建筑样式。设计师选择突出内外空间特色，彰显清晰、自然的建筑语言。

关于建筑外观的构思，他们寻求在美感和功能之间取得平衡。他们将两种对立的形式——水泥结构和玻璃钢筋外观融合在一起。大胆的设计手段进一步深化了主题。

Interior 建筑内部

建筑外观 **External**

室内 **Interior**

室内 **Interior**

1. entrance
2. access to upper floor
3. reception
4. waiting room
5. purchase office
6. office
7. toilet
8. café
9. delivery access
10. production area
11. technical office

1. 入口
2. 二层入口
3. 前台
4. 等候区
5. 采购办公室
6. 办公室
7. 洗手间
8. 咖啡厅
9. 配送入口
10. 生产区
11. 技术办公室

Elsa Morante Library

The pre-existing building for the new Lonate Ceppino Public Library already belonged to Lonate Ceppino's historical heritage. On a rectangular plan, the two-level buildings housed the Civic Library on the ground floor, while the first floor had been left unused. From the outside, the main entrance façade has a higher decorative part which is independent from the roof structure. The design of the fronts is organised in horizontal bands at different heights, while on the north, south and west fronts a system of vertical pilasters apportions the windows on both floors.

Besides the east front, a new well-balanced volume has been built. The new volume's architecture is marked by a narrower profile on its top, with a sloping side that restrains to give more space to the historical building pitches. The dialogue between the volumes is the key and main theme leading the whole intervention. The relationship between the two is nourished by juxtaposition between matterness and lightness, solidity and instability, opaque and reflecting materials. The highlighting of differences underlines the peculiarities of both volumes, in a mutual figure-background relationship. The two buildings are connected through a glazed roofed little volume. The entrance is on the left side and a further wooden connection goes to the first floor.

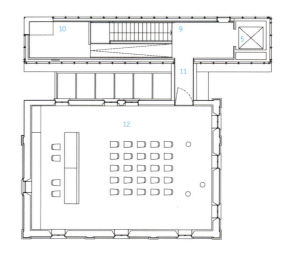

艾尔莎·莫兰黛图书馆

艾尔莎·莫兰黛图书馆位于瓦雷泽历史文化遗产建筑内——这一两层的建筑建在一个长方形地块上，一层用于图书馆，二层一直空着。远远望去，入口一侧外观与屋顶结构分离，好似塔楼一般，装饰性十足。建筑正面呈现高矮不等的水平条带造型，而另外三面则由垂直壁柱结构打造，窗户"穿插"其间。

新旧建筑之间的联系构成设计的主题。新建结构以狭窄的屋顶为主要特色，一侧专门设计成带有坡度，旨在不阻挡原有建筑。除此之外，设计师通过运用各种对比（轻与重、稳固与不稳固、通透与模糊）来进一步深化这两个结构之间的互补。强调不同点的设计方式使得各自特色更加突出。

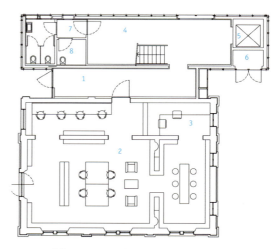

1. entrance hall	2. open-space library	3. reference office
4. stairscase	5. lift	6. technical room
7. vestibule	8. bathroom	9. hall
10. study hall	11. raised walk	12. flexible room
1. 门厅	5. 电梯	9. 大厅
2. 开放式图书馆	6. 机械室	10. 自修室
3. 资料室	7. 门廊	11. 高层走道
4. 楼梯	8. 浴室	12. 灵活运用空间

Front view 建筑正面

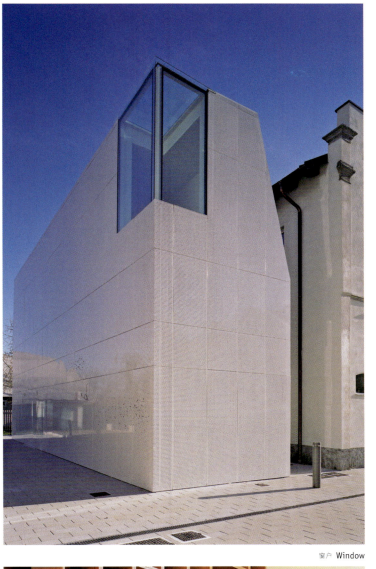

窗户 Window

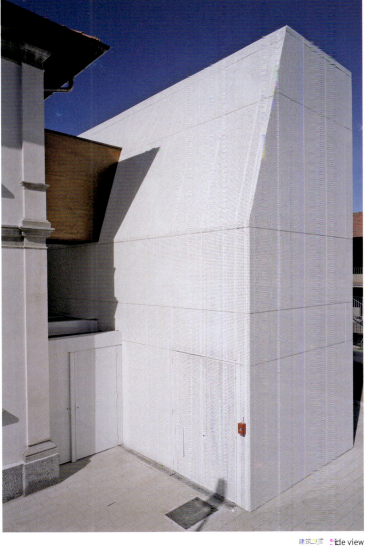

建筑一隅 Side view

开放式图书馆 Open-space library

楼梯 Staircase

Photo: Luigi Filetici

Completion Date: 2008

Architect: DAP Studio / Elena Sacco - Paolo Danelli

Educational

Building rooftop 建筑屋顶

Sloschek

The image of the concept is characterised by the landscape: three differently positioned cubes, connected by a levitating, protecting roof, are giving birth to an interplay of openings, insights and views and interspaces. Great stress is put on a good view of the steeples, the orchards in the valley and the mountain scenery. An excellent view towards the village, the vicinity and vice versa arises as a result of the spaces between the bodies and the roof. A "centre" emerges, a central location which opens itself towards the countryside and communicates with it. Different sizes of windows arise depending on the direction of view, which are defining the façade design.

Sloschek别墅

项目的设计特点主要体现在它的外形上：三个以不同方位摆放在一起的独立立方体结构由一个悬浮的屋顶联系起来，形成了一个从门窗、视野到内部环境都相互影响的整体。从别墅可以欣赏到尖塔、山谷中的果园和山脉三个方向的风景。别墅主体和屋顶之间的空间使得别墅具有极佳的视野。别墅中央出现了一片空地，使别墅和乡村风景紧密结合在一起。大小不一的窗户的位置和外墙设计都由外面的风景决定。

Interior lounge 室内休息区

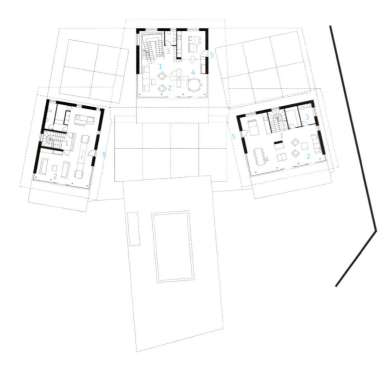

1. stairs
2. relaxing area
3. bathroom
4. dining table
5. entrance

1. 楼梯
2. 休息区
3. 浴室
4. 餐厅
5. 入口

建筑侧面 Side view of the exterior

屋顶上连接 Rooftop connection

From the road on the south, the skeleton of the roof gives evidence of the history of the building 屋顶的骨架结构展示了建筑的历史

San Giorgio Library

Apart from functional requirements, the building is designed to give shape to the urban space around it: the issue was to inject and express a sense of modern-day culture in this kind of context while making only minor alterations: how to provide a "library" while introducing in the new architecture a sense of both the old factory and the idea of being the research instrument inherent in a library; bringing together past and future, well aware that there is no conceptual difference between designing something new and recovering the past, just in the number of constraints to be faced.

The structure involves three vaulted aisles covering 4,000 square metres. The competition called for its conversion into a library containing 350,000 books, 600 readers, 100 multimedia stations, a conference room for 100 seats, a children's space with outdoor spaces, offices and a coffee bar. It is re-designed keeping the old vertical structures while introducing wide floors, changing the vaulted roof into laminated wooden ribbing, setting a compact image on the longitudinal fronts, stripped down at the north and south terminals. The result is an overall image of a skeleton which draws out the old features and breaks them down.

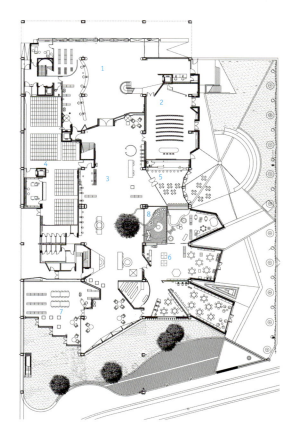

圣乔吉奥图书馆

除了具备图书馆的基本功能之外，这座建筑的设计还重塑了其周边的城市空间。它通过微小的改动为周边环境注入了当代文化感：在新图书馆的建设中融入旧工厂和研究中心的理念；连接过去与未来，使人们认识到进行新设计与恢复旧设计之间并没有概念性的区别，只是面对的约束条件不同而已。

建筑由三个圆拱形通道组成，总面积为4,000平方米。图书馆可容纳350,000本图书、600名读者、100个多媒体工作台、一个100个坐席的会议室、一个儿童中心、若干办公室和一个咖啡吧。设计保留了原有的纵向结构，拓宽了楼面，将拱形屋顶改造成了分层木纹屋顶，在水平立面上营造出紧凑的设计感，并且移除了南北两端的挡板。最终，建筑的整体形象是一个突破建筑原有特征的框架结构。

1. entrance
2. conference hall
3. central hall
4. book shop
5. coffee shop
6. children's library
7. mediateca
8. patio

1. 入口
2. 会议厅
3. 中心大厅
4. 书店
5. 咖啡店
6. 儿童图书馆
7. 媒体中心
8. 天井

图书馆、水景和发电厂之间的连接结构 Architectural relation between the library and the water and power plant

从一楼大厅看向二楼的演讲厅 The main hall on the ground floor, looking towards the large closed lecture hall on the second floor

儿童图书馆游乐空间 The space for the children's library

Photo: Pica Ciamarra Associati

Architect: Pica Ciamarra Associati, Franco Archidiacono, Federico Calabrese, Angelo Verderosa

Educational

Completion Date: 2007

Side façade 建筑侧面

The Central Library of the University of Molise

The scheme of the Library and the main lecture Theatre of the University of Molise includes the design of external spaces with a sheltered pedestrian path and new parking areas. The main hall has two entrance spaces, one from the public open connection system of the University and the other directly from the Faculty of Social Sciences and has been designed for 480 seats and is used by the whole university.

The scheme is grafted onto the morphology of the site with two main blocks, the first being used for the offices and the second for the library, connected by two service and security stairs.

Architectural unity, in relation to the existing buildings, is granted by typological features and the use of materials like bricks and iron windows, yet used for the main buildings.

The block is connected to the University complex and to the parking area through a grid of pedestrian paths and a future pedestrian bridge that overcomes the main street in the University area.

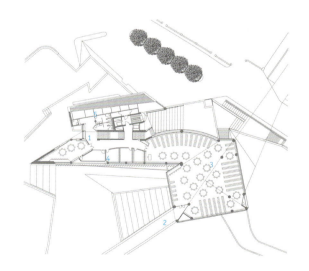

1. entrance stairs
2. entrance path
3. reading hall
4. study and research
5. administration

1. 入口楼梯
2. 入口走道
3. 阅读大厅
4. 学习和研究室
5. 行政办公室

莫利塞大学中心图书馆

莫利塞大学图书馆和主阶梯教室的设计还包括了户外人行道和新停车场的建设。主大厅有两个入口，一个通往学校的公共区域，另一个通往社会科学楼，里面共有480个坐席，可供全校使用。

建筑由两个主体部分组成，一部分里面是办公室，另一个里面是图书馆，二者由安全服务楼梯相连。

为了与校园内原有的建筑类型相一致，图书馆主楼的建设采用了红砖和铁窗结构。建筑与校园内的其他教学楼和停车场通过人行道相连，即将建造的人行天桥将覆盖校园里的主要街道。

主入口 The main entrance

建筑主立面 The main façade

室内空间 The inner space

Photo: Pica Ciamarra Associati

Architect: Pica Ciamarra Associati

Educational

Completion Date: 2009

General view 全景图

Officine Maccaferri and Seci Energia

The new headquarters of Officine Maccaferri and Seci Energia is a building designed from the inside out. One of the unique features of this project was bringing two of the group's companies to share one building and finding the right balance between the independence of the individual companies and the group as a whole. The interior space was divided with fitted mobile walls that are partly glazed and partly opaque with a wood finish. This made it possible to make open spaces, closed offices, shared areas, meeting rooms and toilet areas, depending on the needs of the various work teams and different kinds of activities. Progetto CMR designed brightly lit, invigorating offices for about 160 people, focusing on the functional aspects and flexible solutions in anticipation of possible future changes. It optimised the division, arrangement and fitting out of each area and work station to ensure that every user has the highest level of comfort and wellbeing as well as providing optimal efficiency.

Maccaferri制药和Seci能源办公总部

Maccaferri制药和Seci能源办公总部是一座从内向外设计的建筑。这一项目的特别之处在于它为一个集团的两家公司提供了共用的办公楼。因此,在公司的独立性和集团的统一性之间找到一种平衡便是设计的焦点。大楼的内部空间由木纹装饰的半透明可移动墙壁分割。这种设计让内部空间更灵活:开放空间、密闭办公、共享区、会议室、洗手间等设施的位置可以根据不同工作团队和不同类型活动的需要而调整。Progetto CMR为公司的160名员工设计了令人精力充沛的明亮的办公室。办公室的设计充分考虑了功能性和灵活性,易于调整。建筑师优化了各个区域和工作台的分配和布局,以保证每个员工都能享受高品质的舒适办公环境。

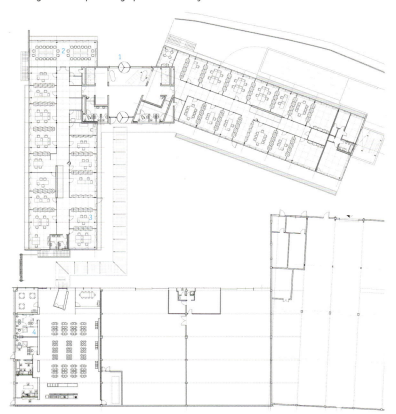

1. 入口
2. 会议室
3. 办公室
4. 卫生间

1. entrance
2. conference hall
3. office
4. WC

入口 Entrance

建筑侧面 Side view

入口和楼梯 Entrance and stairs

楼梯和走廊 Stairs and corridor

Photo: Oscar Ferrari

Corporate

Completion Date: 2007

Architect: Progetto CMR Massimo Roj Architects

Birken-elementary School and School "am Grüngürtel": Nursery Building

In the borough of Spandau, Berlin, the existing schools, Birken-elementary school and school "am Grüngürtel" will be extended to open-all-day schools with a capacity of 155 pupils within a national funding scheme. The project is cofinanced by EFRE-funds. The spacious sports-and playgrounds in the heart of a heterogeneous urban block will be reorganised within the design concept.

Under a firm protecting roof made of exposed concrete the new building offers plenty of space for the extension of the Birken elementary school and the school "am Grüngürtel" to a open all day school with a capacity of 155 pupils. The new building with two floors in the Berlin borough of Spandau (total construction costs about 2.2 millions) was financed from funds of the programme "future, education and mentoring" (IZBB) and the "European fund for regional development"(EFRE).

Placed in the extension of a picturesque Berlin "Gründerzeit" street-scape, the building mediates between the classical block housing on the south side and the open urban structure with four-storey-high housing slabs from the 1950s in the north. Within the building scheme the extensive sport and recreation areas were rearranged inside the large heterogeneous urban block.

伯克恩小学与绿带学校扩建楼

伯克恩小学与绿带学校扩建楼可容纳155名小学生。双层高度的拱廊界定了主入口的位置（介于两所学校连结通道处），同时这里摆放的长木椅为等候的家长以及孩子、老师提供了休憩之所。此外，拱廊内，开敞的格局更是加强了悬臂式屋顶的构想，混凝土等冷性材质更是与散发着温暖气息的木板（从地面一直延伸到屋顶横梁处）构成鲜明对比。

新建筑内包括食堂（带有厨房）、一个多功能区、四个团队交流区以及四间教室。室内，高大的木窗朝向四周的花园开放，通透而风景优美。

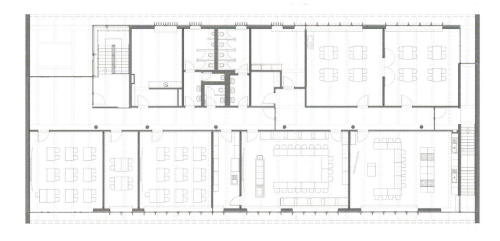

Photo: Werner Huthmacher

Architect: Huber Staudt Architekten

Completion Date: 2008

Educational

Hedwigshöhe Hospital

The intervention is based on the remodelling and extension of the existing hospital and on the design of the psychiatric units. In the open dialogue between the dense urban pattern and the spacious landscape, the new Hedwigshöhe Hospital is the turning point for Falkenberg. To contrast the existing structure, the two-storey pavilions of the psychiatric ward are playfully juxtaposed; the result is a small hospital city where the parataxis between old and new is evident.

The new psychiatric department is located north, on the slope of Flakenberg hill. Patients are welcomed in paired pavilions, each set as to create a slight crenelation toward the surrounding open landscape. The two levels follow the scale of close-by homes and their patios reach for a view over the landscape; the façades of the courts are wood and glass and wooden battens characterise the hospital. Along the external perimeter, the square windows are divided in three parts and slightly project from the wall. The scheme seems to support a positive contrast between the orderly layout of the existing structures and the vivacious layout of the psychiatric courts; the new sanitary structure establishes a conscious relationship with the landscape, the city and the final user.

Hedwigshöhe医院

这一工程以现有医院建筑为基础。紧凑的城市格局与开阔的自然景观共同存在，而新建的医院则恰好构成了两者之间切合点。为与原有结构形成对比，两层的精神科病房并排屹立，打造了一个小型的医院社区，新与旧的对比更加清晰。

新建精神治疗部选址在Flakenberg 山坡北侧，病人可通过两个庭院（如同周围景观的入口）进入。两层的结构依照周围的住宅样式打造，在天井内可以欣赏周围的景观景致；院子的外表采用木材和玻璃打造。方形的窗户被分成三部分，从墙壁处微微突出来。整体理念实现了原有结构规整的格局与新建空间动态的格局的有益对比，同时增强了景观、城市及使用者之间的联系。

Architect: Huber Staudt Architekten Bda, Berlin, Germany
Manuel Brullet, Albert De Pineda, Barcelona, Spain

Completion Date: 2008

Hospital

Photo: Werner Huthmacher, Jordi Bernadó, huber staudt architekten bda

Main entrance in the daytime 主入口日间景色

Media Centre Oberkirch

Three levels are connected with an open, organically formed stairwell. The centrally positioned open staircase is not only a movement and communication zone, but is an exposure element for the inside–recumbent zones of utilisation with the generously glazed upper light. The façades, with the large apertures, are understood like shop–windows, which permit varied and exciting views in the surrounding town space. The external, brighter window areas serve the reading zones and stay zones, partially furniture with an up–and–down movement is integrated which the visitor can use as a table or a bench. By the planning of the building it was respected beside the striking town planning architecture, particularly to create a high stay quality for the visitors in the building. In the whole building one finds the comfortable seating pieces of furniture, which also invite visitors in the free areas of the reading terraces for staying. As a sculptural architecture, which is modellised full of contrast, the building creates a sensuous experience space which should serve the citizens as a communicative centre where to all generations the varied media technologies and technologies of information are accessible.

欧伯克尔希媒体中心

建筑的三层楼由一个结构特殊的开放式楼梯连接起来。位于建筑中心的开放式楼梯不仅是移动和交流的空间，而且是休闲区的重要采光元素，楼梯顶部大块的玻璃提供了大量的自然采光。建筑外墙上巨大的玻璃窗看起来像商场的橱窗一样，从内部可以看到周边街区各种各样的风景。离窗较近的明亮区域是阅读和休息区，摆放着高低错落的家具，既可以作为桌子，又可以作为长椅。媒体中心的设计旨在为访客提供一个高品质的空间，大楼阅读平台的自由区随处摆放着舒适的座椅。这座雕塑般的建筑充满了对比和美感，为市民提供了一个具备多样化媒体信息科技的交流中心。

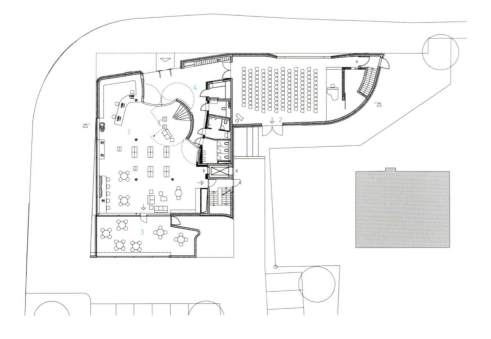

1. 前台
2. 阅览区
3. 平台
4. 走廊

1. reception
2. festival room
3. terrace
4. passageway

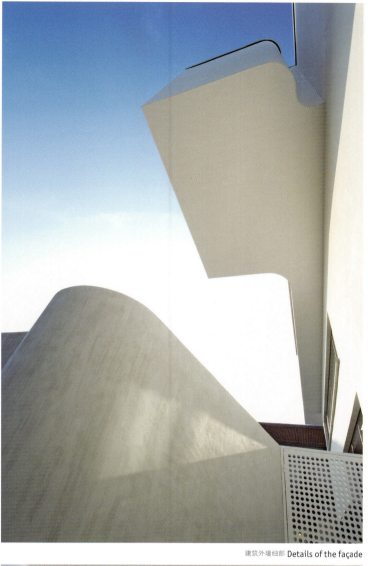

建筑外墙细部 Details of the façade

一楼走廊细部 Details of the passage way on the ground floor

屋顶采光细部 Details of the toplight

主楼梯和屋顶采光 Main staircase and toplight third floor

Photo: Guido Gegg

Corporate

Completion Date: 2010

Architect: Wurm+Wurm Architekten Ingenieure Gmbh

Shopping Centre Stadtgalerie

Directly located at Heilbronn's pedestrian zone, the new shopping centre "Stadtgalerie" strengthens the inner-city location. The architectural sumptuously-designed building integrates itself well into the town-space as a two-split building presenting different body languages. The Stadtgalerie Heilbronn consists of three levels which invite to stroll, a sales area of 13,000 square metres, and an attractive offer of approximately seventy-five shops and stores, cafés, and restaurants. The underground car park offers 600 parking spaces.

The cubic building with its straight and hard edge and the rounded building with the curved and soft façade melt into each other and are well integrated in the existing town-space. The independence of the buildings is especially highlighted via the façade concept. Backlit light-slits are crossing the wall area in an erratic way and thereby open the building alignment in an optical way. In the darkness the alleyway suddenly becomes a light creation. The shape softens the bulk of the building, integrateing itself in a smooth way into the surrounding building structure. A penthouse with a filigree projecting roof marks the end of the arched building shell. Vertical coloured glass lamellae accentuate the outline. Due to the different day and night impressions, the building offers varying appearances.

Stadtgalerie购物中心

Stadtgalerie购物中心位于海尔布隆的步行街区，稳固了市中心的位置。购物中心华丽的建筑形态让它很好地融入了城市空间，两座不同的建筑呈现了不同的肢体语言。购物中心一共有三层，总面积为13,000平米，包含75间店铺、咖啡厅和餐厅。地下停车场有600个车位。

轮廓鲜明的立方体建筑和弯曲柔和的圆形建筑完美地结合在一起，与原有的城市空间形成一个整体。建筑外墙的理念强调了建筑的独立性。背光的缝隙以一种古怪的方式横跨墙壁，从视觉上打开了建筑。小巷在黑暗中一下子就被点亮了。建筑的造型柔化了自身，与周围的建筑流畅地结合了起来。拱形外壳的最上端是有金银丝装饰屋顶的阁楼。纵向的彩色玻璃片强调了建筑的轮廓。白天和黑夜，建筑会呈现不同的面貌。

The building at night 建筑夜景

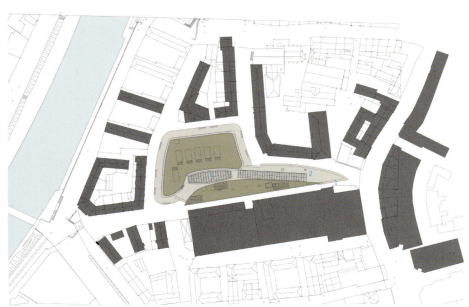

1. 入口
2. 餐厅
3. 屋顶

1. entrance
2. dining area
3. roof

室内 Interior

Photo: Blocher Blocher Partners, Stuttgart and Mannheim

Commercial

Completion Date: 2008

Architect: Blocher Blocher Partners

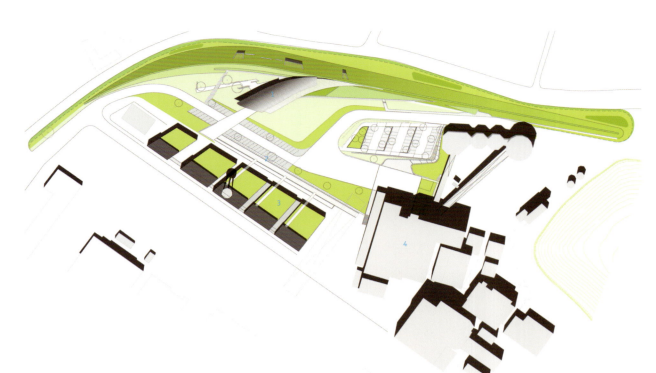

1. start in the lecture hall
2. historical trench
3. garages' views over the roof
4. waste bunker

1. 演讲厅
2. 历史纪念壕沟
3. 车库
4. 垃圾场

ZMS Administration Building

The task to design a new administration building, reorganise the power station compound and create a new noise protection barrier offers the chance to dissolve the dichotomy of landscape and building to realise the deconstruction of those categories into one designed environment, to be experienced in a dynamic and curious fashion. Four hundred and fifty metres long and up to thirteen metres high, the central part of a noise protection wall with a forty-five-degree incline simultaneously constitutes a new administration building for over 140 metres. The superimposition of building and earth wall allows one to explore and experience the landscape of this entire ensemble on various levels.

An auditorium with a visitor centre unfolds from this landscape and opens up towards the power station compound. It separates from the earth wall on the upper level, resting on two radial supporting walls and cantilevers up to twenty metres over the landscape. A long panoramic glass façade leans towards the power station. The administration building underneath is sculpted into the earth wall. Meeting rooms and common areas penetrate through the wall to establish a relationship with the bordering village.

The main structural challenges arose from the fact that they had to create an earth wall of up to thirteen metres high that is capable of supporting not only itself at a forty-five-degree slope, but also to accommodate a building within it, situated at about six metres above grade. It was very important to cover the building in landscape and vegetation that appear to be continuous, without indication of the shape of the building underneath.

ZMS行政楼

项目包含设计一座新行政楼、重组发电站和打造一个新的噪音防护墙。建筑师利用这一机会将景观与建筑结合在一起，打破了类别的界限，打造了动感而奇特的新建筑样式。噪音防护墙长450米，高13米，其中心部分的行政楼延伸了140米。建筑与挡土墙的叠加使人们能够全方位地欣赏场地的景观。

一个带有游客中心的礼堂从这个景观区域展开，并朝向发电站。它的上层与挡土墙相分离，依托于两面径向承重墙之上，悬在20米的高空之中，一面全景玻璃外墙直接朝向发电站。下方的行政楼则与挡土墙合为一体。会议室和公共区域穿透了围墙，与边界的村庄建立了联系。

Main entrance 主入口

庭院 Courtyard

一楼大厅和楼梯 Lower foyer and stairs

走廊 Corridor

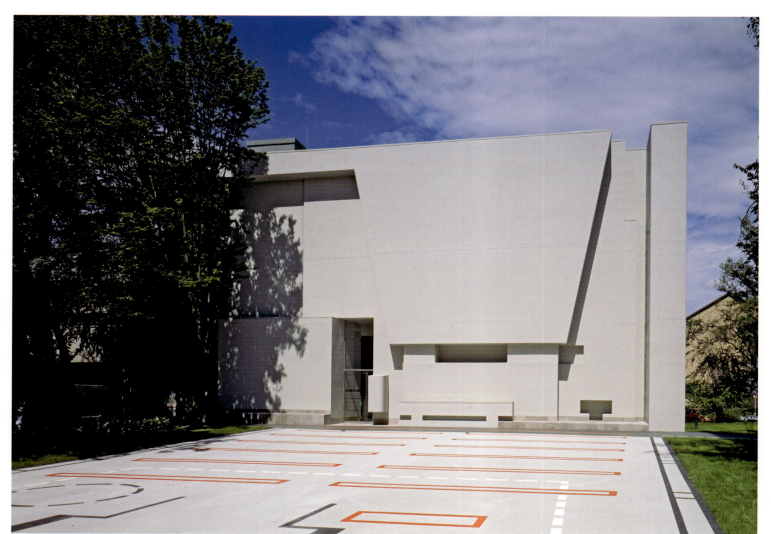

New churchyard with new façade 新庭院和新建筑外墙

Dornbusch Church

Dornbusch church is situated in a residential area in the north of Frankfurt. Due to the poor condition of this sixty–year–old church and the extreme decline in attendance at church services, a complete demolition and the new construction of a small "prayer room" as a replacement were under discussion. Planning studies were able to show, however, that the best choice is only a partial demolition. From a town planning point of view, a spatially and functionally intact ensemble remains – consisting of a community centre, "residual church" and tower: a new churchyard, with an attractive potential for public use, is created. The spacious area round the altar and the choir remains as the old / new church.

Reconstruction / New construction
The open side of the building, caused by the demolition work, is closed with a new wall or façade. The special nature of this location and the reduction process are made evident in that this new wall is marked with outlines and moulds of the "old" church, i.e. the structures which have been removed – such as the old entrance façade, altar and gallery – now form a sculpted structure out of the flat wall surface. Further factors contributing to the final form are room light exposure, completing construction features, the remaining building and the access to it.

The new wall is a mixed construction (reinforced concrete, masonry); the plasterwork surface corresponds with the plasticity of the concept. The new wall / entrance façade and the tranquil north wall as a counterpart are contrasted in light colours.

1. sitting area   1. 坐席
2. reception   2. 前台
3. platform   3. 讲台

多恩布什教堂

多恩布什教堂位于法兰克福北部一个住宅区内，拥有60年的历史。由于年久失修以及使用人数减少，考虑对其彻底拆除并新建一个小的"祈祷室"。经过研究之后，决定拆除其中的部分。从城镇规划角度出发，原有的主体部分（包括社区中心、教堂以及塔楼）被保留下来，祭坛周围的开阔空间仍留作教堂之用，并在此基础上新建一个墓地。

重建及新建工程
经过拆迁之后，原有建筑便出现了缺口，设计师新建一道外墙将其封闭。建筑独有的特色以及还

原过程在这一面墙壁上清晰展现——原有教堂的轮廓及格局（移走的雕塑、原来的入口、祭坛以及画廊）全部呈现在这里。室内光线透射出来，更加完善了墙壁的特色。另外，新建墙壁与北侧墙壁全部采用淡色装饰，相互补充。

建筑立面 Façade

建筑外观 Exterior view

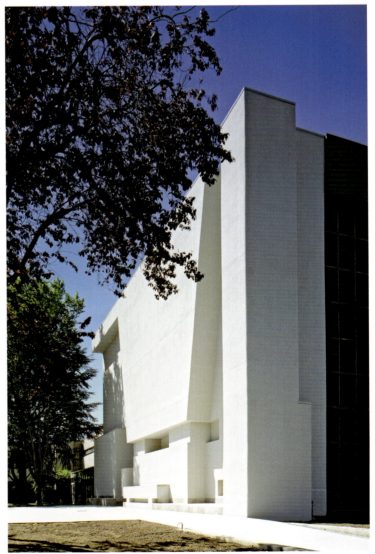

建筑侧面 Side view

Photo: Christoph Kraneburg

Cultural

Completion Date: 2006

Architect: MEIXNER SCHLÜTER WENDT Architecten

Façade 建筑外观

GFC Antriebssysteme GmbH Transfiguration and New Construction of a Production Hall with Service and Socialtract

GFC驱动系统公司产品生产大厅

GFC Antriebssysteme GmbH is looking back on more than 100 years history of developing and manufacturing of worm gear and hydraulic gear units at its enterprise location Coswig near Dresden. The existing production hall was established a faceless, simple box in standardised construction way from precast concrete parts. The existing building did no longer meet the requirements to modern jobs and the increased space requirement within the logistics area.

The designers decided to use the structure of armoured concrete, which was in a good condition, and to extend the original building with a two-storey administration building, logistics and stock area. The spatial connection between administration and production is made by wide glazings in both levels.

Different materials signify different functions. The administrative tract is coated by walls from exposed concrete, and production and logistics are characterised by the use of shingles from stainless steel and translucent boards from polycarbonate. The particular, large, calm surfaces stress the sculptural character of the whole body. The carrying structure of the office area is done as a grid from columns and beams from exposed concrete.

The support structure is slanting in two directions, forward and sidewise and thereby the façade along the railway line Dresden–Berlin creates a motion-impelling model. The movement arising from this skew is continued in the façade around the entire structure.

GFC驱动系统公司具有百年涡轮和液压装置制造历史，公司总部设在德累斯顿附近的斯维希。公司原有的生产大厅是一个简单的无外表装饰的预制混凝土盒结构，已经不能满足现代工作和物流区域日益增加的空间需求。

设计师决定保留原有的钢筋混凝土结构，并且在原有基础上增减一座两层的行政、物流、仓储楼。行政楼和生产大厅之间的连接结构采用了玻璃幕墙。

不同的材料象征着不同的功能。行政楼采用了露石混凝土外墙，生产和物流楼则以不锈钢板条和透明树脂板墙壁为特征。独特、巨大而冷静的外墙突出了建筑主体的雕塑效果。露石混凝土梁柱所形成的栅格支撑着办公空间。

由于辅助结构向前面和侧面所倾斜，朝向德累斯顿—柏林铁路沿线的外墙产生了一种前进的效果。这种由倾斜所带来的前进感一直延伸到整个建筑结构之中。

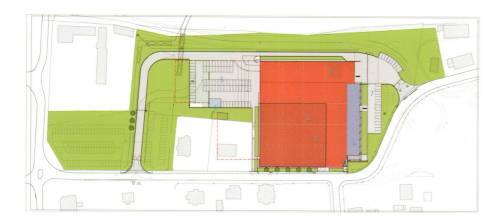

1. 高湾埠架和装卸区
2. 生产大厅
3. 原有建筑
4. 行政区
5. 78停车场
6. 自行车停车场
7. P3 – 23停车场

1. high-bay racking loading area
2. production hall
3. existing building
4. administration
5. 78 parkplatze
6. fahrrader
7. p3– 23 parkplatze

夜景 Night view

建筑背面 Back view

入口 Entrance

Photo: Ester Havlova

Completion Date: 2009

Architect: Wurm + Wurm Architekten Ingenieure Gmbh

Exterior view of the canteen and sports hall 餐厅和体育馆外部

Evangelical Grammar School

The project comprises a two-storey grammar school with refectory and sports hall as a full-time school. The site for the new Evangelical Grammar School is located in the northeast outskirts of Bad Marienberg. To the west it borders on an area with other schools such as a primary school, a secondary modern and a junior high school. To the east and north begins open land, while to the south there is a youth hostel and a special school.

Both sections of the building – the angled school building in the north and the second building with the arts rooms, the sports hall and the refectory in the south – circumscribe a schoolyard that is open to the west. The school thus has a relationship with the other schools, while on the other hand pupils are welcomed symbolically "with open arms" due to the positioning of the building.

In general, natural materials are used which radiate warmth, brightness and lightness while still remaining sturdy. The choice of colours and materials, openness, generosity, freshness and colourfulness form the basis of a relaxed atmosphere for learning.

福音教会语法学校

项目是一所两层楼的全日制语法学校，附带一间学生餐厅和一个体育馆。福音教会语法学校位于百得马瑞恩伯格的东北郊。学校西侧是一片学校群，有小学、普通现代中学和初中；东侧和北侧是空地；南侧是青年宿舍和一间特殊教育学校。

北面成一定角度的教学楼和南面包含艺术教室、体育馆和学生餐厅的副楼围城了一个朝向西面开口的操场。这样一来，校园展现了一种"敞开怀抱"的欢迎姿态，学校和周边其他学校的连接就更为密切了。

建筑所采用的材料以自然材料为主，注重保温性、透光性和坚固性。建筑装饰大方而新鲜，色彩鲜艳而活泼，营造出一个轻松的学习环境。

1. staff rooms	1. 办公室
2. meeting room	2. 会议室
3. entrance hall	3. 门厅
4. playground	4. 操场
5. classroom	5. 教室
6. arts rooms	6. 教室
7. artisanry room	7. 艺术教室
8. music room	8. 音乐教室
9. sports hall	9. 体育馆

呈角度弯折的教学楼 Exterior view on the angled school building

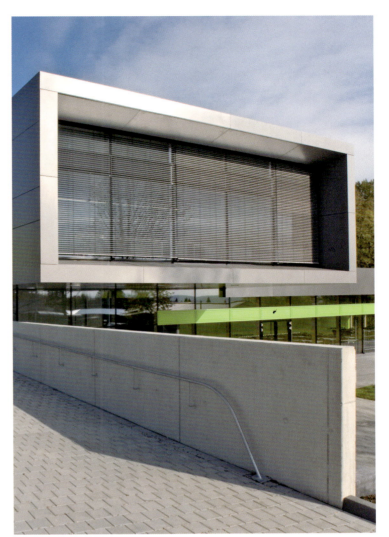

突出的翼楼 Striking wing

Photo: Guido Kasper, Constance, Germany

Educational

Architect: 4a Architekten GmbH, Stuttgart, Germany; Matthias Burkart, Alexander v. Salmuth, Ernst Ulrich Tillmanns; Project team: Andreas Behringer, Rike Hannes, Birgit Wäldin, Lars Goose

Completion Date: 2007

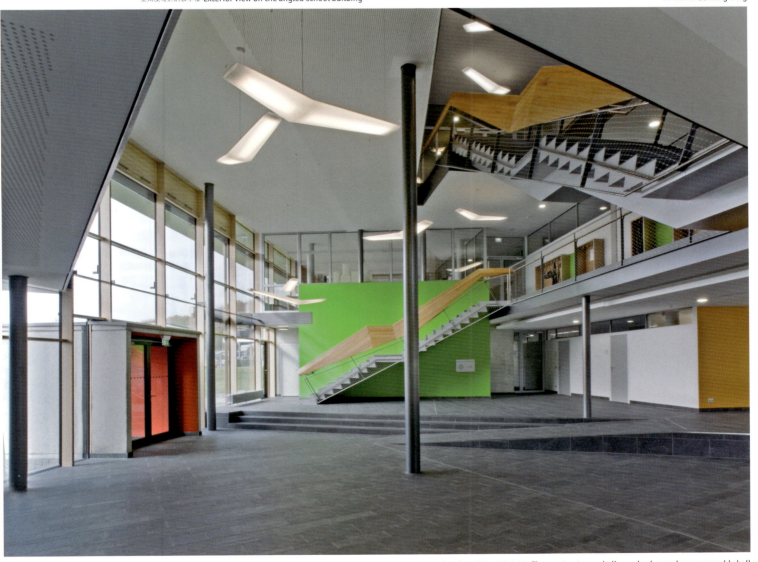

门厅也可以作为一个集会大厅 The great entrance hall can also be used as an assembly hall

Ener(gie)nger – Energy and Sustainability

The architect has named the project Ener(gie)nger – a coinage of the words Energie (Energy) and the name Gienger, the well established domestic engineering wholesale business. The geometry of a "band-spiral" creates a unique tunnel-like area, which by means of the side intersection surfaces exposed to light, throws a fascinating ray of light onto the slanting walls. The spiral begins at entrance of the Fa. Gienger showroom at the east of the building by means of a wide open projecting roof – opposite at the west side a similar opening is put to use for the cash-and-carry wholesale store.

The vibrant energy spiral stands for the energy within a fascinating area, with this area expressing the dynamics of new types of energy: The slanting external surfaces are fitted with both solar-thermal and photovoltaic panels – the latter of these could in the course of new developments, be fitted with products, which at the moment are still being devised. The outer shell is made up of solar panels, photovoltaic panels and metal and glass elements.

The core of the exhibition hall is not only the various bathrooms on display, but also the "educational energy trail" – an exhibition of current installations and equipment belonging to the energy-saving sector. Since its introduction, this "trail" has been accepted very quickly and intensively: it plays an important role in the understanding of technology, which in turn means that such technologies can actually be put to use.

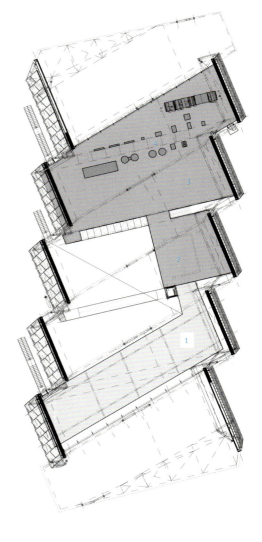

Ener(gie)nger能源与可持续发展展厅

建筑师将项目命名为Ener(gie)nger——是德语"能源"（Energie）和公司名称Gienger的结合。建筑的带状螺旋结构打造了一个独特的隧道般的区域。展厅的侧截面暴露在阳光之中，在倾斜的墙壁上打下一束迷人的光线。螺旋结构通过一个开阔的突起屋顶从Gienger公司的南侧展厅开始旋转，西侧与其类似的开口是批发商店的入口。

充满活力的能源螺旋象征着这块迷人的区域所展示的各类型新能源。倾斜的外墙上装备着太阳能热电板和光电板。后者可以被装置在新开发的产品上。整个建筑的外壳都由太阳能热电板、光电板和金属玻璃元素组成。

展厅不仅展示了多种多样的浴室套间，展览本身也是一个具有教育意义能源之旅——充分展示了节能的装置和设备。展览以一种简单易懂的方式让人们迅速理解能源科技的应用。

1. extension level	1. 扩建楼层
2. café	2. 咖啡厅
3. event	3. 活动大厅
4. energy trail	4. 能源展厅

Night view 夜景

入口 Entrance

建筑外观 Façade

室内 Interior

Photo: Roland Weegen

Architect: Peter Lorenz Ateliers

Cultural

Completion Date: 2009

Sipos Aktorik GmbH

西博思电动执行机构公司

The new building of the Sipos Aktorik GmbH is designed as a compact volume, embedded into the easily hilly landscape at Altdorf, which in its spatial effect, is more similar to a large mansion than a factory.

The building is divided into three parts: the assembly hall with attached automated store room, a two-storey administrative tract in front of that and a storage hall with goods receipt and goods issue afterwards to both ranges. The proximity of the individual operating ranges is to make possible a maximum of internal communication. The roof of the production hall is occupied with skylights shaped as a bar, so the lighting of the work stations reaches thereby daylight quality. This light impression on the inside is still strengthened by the bright colour of the construction.

On two sides ground–same, high window fronts with integrated glass doors enable contact wih the external space. The steel structure of the production hall is modular developed, whereby the manufacturing area is optionally expandable up to the double size. The functional separation between assembly area and the two-storey office area is abrogated by space–high glass walls. This conveys internal communication between these different working environments.

西博思电动执行机构公司新建的生产大厅结构紧凑、深嵌在阿尔道夫山景之中，比起工厂，它更像是一座巨大的宅邸。

建筑共分为三个部分：带有自动贮藏室的礼堂、前方两层高的办公楼和后方的收发货仓储大厅。操作区域彼此邻近的设计让内部交流更加容易。生产大厅的屋顶上装有条形天窗，保证了工作台上有充足的阳光。内部结构明亮的色彩加强了光亮的效果。

两侧的高窗和玻璃门增加了室内外空间的联系。生产大厅的钢结构采用了模块化设计，因此制造区域可以任意拓展，最高可到达原来的两倍大小。玻璃墙废除了装配区和办公区的隔断，使内部工作环境得以交流沟通。

1. 高湾垛架
2. 生产大厅
3. 行政区
4. 装卸区

1. high-bay racking
2. production hall
3. administration
4. loading area

Photo: Ester Havlova

Completion Date: 2009

Architect: Wurm + Wurm Architekten Ingenieure GmbH

Residential Area at Pelargonienweg

The lot of the residential area is characterised by a newly created gently-waved topography which through the redeployment of excavations on the site can be styled like dunes. So emerges a landscape on the outskirts of the town with a special identity: the houses are embedded in a space sequence of soft hills and troughs, in a grove of apple, pear, cherry and nut trees. There are also common spaces in this gentle hilly landscape which could be used, for example, for green houses, must cellars and playgrounds. The architects obtain the opportunity to develop their buildings "on the slope" through the newly created three-dimensional landscape. The construction "plays" with the slopes in the landscape and a number of configurations through various combinations of the two basic types, each bringing its own particular qualities. They thereby achieve a differentiated and rhythmical building typology which in a sequence of linked and free standing combinations of the buildings creates ever-changing townscapes. In doing so, particular attention is paid to a relaxed ground-floor zone, the contact areas of the building are reduced, and a visual connection of the transportation spaces with the garden landscape through the construction is made possible.

Pelargonienweg住宅区

Pelargonienweg住宅区所处位置呈现出柔和的波浪造型，犹如一个个小沙丘般。于是，整个建筑就恰似"嵌"在小山中，四周环绕着苹果树、梨树、樱桃树和坚果树，在小镇的边陲营造了一处完美的景观。其他的空地还可用于建造小木屋、酒窖和操场。建筑师们充分利用地形优势，建造了"斜坡上的结构"，周围的景观似乎在不断地变化着。休闲操场的设计吸引了更多的目光，建筑之间的连结区域被尽量减少，景观之间的联系在视觉上更加清晰。各部分结构之间强调变化，从而恰好增强了整体统一性。

   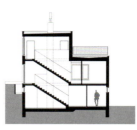

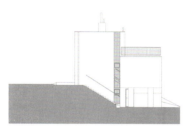 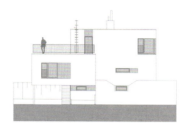  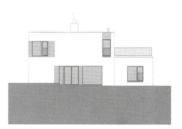

Architect: Josef Weichenberger architects

Completion Date: 2009

Photo: Lisa Rastl

Residential

University Mozarteum

The University was named after Wolfgang Amadeus Mozart and is a worldwide acknowledged training Centre for artists now for more than 160 years. Situated directly next to the Mirabell Gardens in Salzburg, the Mozarteum is part of the inner city ambience.

A stony solitaire is situated at the entrance and the Mirabellplatz Square is visually being led into the building. Entering the doorway of the hall is a special moment created by illumination effects through the floor. The hall is illuminated with down lighter from the shed-roof-construction. The lights are being arranged in bunches and can be activated as light isles in between which the students and concert guests might move animated by the light. A perron leads to the institute's areas. Open-access balconies with integrated light channels serve as attractive lounge zones. The walls appear to glow from cherry tree wood and therewith create a warm and festive character.

To make this outer illumination more visible, the façade of the solitaire is being shaded. The remaining three walls create an edged and bright solitaire. The light concept in context with the building makes a varied but calm and reserved impression. It wants to inspire and give room to the creativity of the occupants.

莫扎特大学

160多年来，莫扎特大学一直作为世界知名艺术培训中心，其名字源于著名音乐大师莫扎特。大学与米拉贝尔花园毗邻，已成为整座城市景观的一部分。

巨大的石雕坐落在入口处，莫扎特广场的风光"涌入"建筑内。走进大厅门口便会充分感觉到灯光营造的独特氛围——灯饰从屋顶上悬垂下来，一簇簇排列着，形成一座座"光岛"。经过一级台阶，便可达到其他区域，开放式的阳台可用作休息区，墙壁在灯光的照射下散发出温暖而喜悦的气息。

为增强建筑外部照明可见性，设计师专门将石墙一侧遮盖起来，另外三面墙壁则更加突显明亮感。独特的照明理念赋予整幢建筑淡雅沉稳的特色。

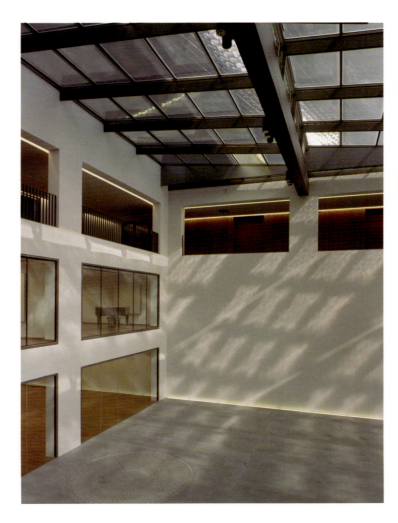

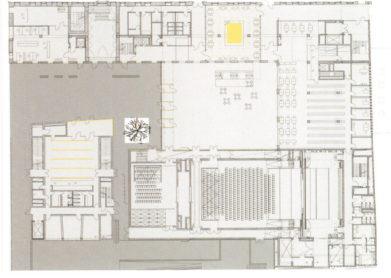

Photo: Andrew Phelps

Completion Date: 2006

Architect: Rechenauer Architects, Munich
Lighting Design: Gabriele Allendorf – light identity, Munich

Western elevation 建筑西立面

Main Railway Station Innsbruck

Although for the most part the trains travel on viaducts above street level, the station represents a real barrier across the main orientation of the valley. The reaction to this sharply defined urban situation that is further complicated by the long, narrow railway station forecourt was a relatively low and very long building with an extremely open and regular lattice–like façade that is placed at the east side of the forecourt but shifted six metres further back. This means that it moves out of the existing street line and – employing a classic way of shaping space – can assert itself as a liberated, freestanding building against the rest of the dense high–rise development on Südtiroler Platz. The building measures are grouped in detail around the theme of permeability.

All the important functions, such as travel centre, waiting and retail areas are placed in the central, lowered part of the railway station concourse. On the one hand this allows a direct approach from the underground car park to the concourse and thus to the trains, while on the other it also permits an uninterrupted view of the platforms from the city and vice versa.

因斯布鲁克火车站

尽管大部分时间火车都在高架桥上行驶，但火车站本身却也构成了实在的视觉障碍。为解决这一问题，设计师构思打造一幢低矮的长条状建筑，外观打造成开敞规整的条状样式，同时整幢建筑在原有场址上退后6米。这也就意味着，新建筑从原有的街道处退出来，确保其自成一体。所有的建筑细节全部围绕着"渗透性"这一理念展开。

所有的功能区：如游客中心、候车室以及零售区全部设置在中央空间内（车站大厅内沉下的部分）。一方面，这极大地方便了从地下停车场经由大厅去乘坐火车的旅客；另一方面，人们可以在其他地方一览无余地看到月台上的场景。

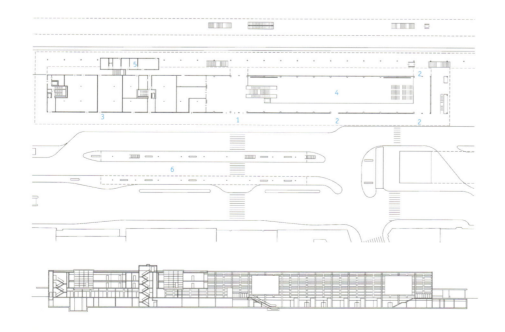

1. 主入口
2. 入口大厅
3. 入口商店
4. 大厅
5. 办公室
6. 汽车站

1. main entrance
2. access hall
3. access shops
4. hall
5. offices
6. bus stop

建筑南立面 Southern elevation

1号站台夜景 Platform 1 at night

站前空地夜景 Station forecourt at night

Transportation

Photo: Nikolaus Schletterer

Completion Date: 2004

Architect: Riegler Riewe Architects Pty. Ltd. / Florian Riegler, Roger Riewe

General view 全景图

Niederndorf Supermarket

Niederndorf is a small, rural village at the border between Austria and Germany. On the periphery of this village, there is a chance to realise a supermarket – urgently needed. The site lies between farmhouses, feedlots for cows and some residential areas. The client asked for a "cool marketplace" – protected from sun and heat.

Wood is the favourite building material in the Alps. Wood means homeland, cosiness and identity. The wooden stables do provide a clear local building identity. So the idea was to use decorticated pine trunks of the area: directly from the sawmill without any further treatment. This filter has several functions. It turned out to be the cheapest solution for sun protection, and is creating a public space and in some parts represents a curtain–wall façade.

The humanistic approach leads to the ideas of offering extra communicative areas for the clients. The "trunks" do separate an open space from the actual shopping area. Here the farmers can sell their X–mas trees, the local brass–band can play, and sometimes they offer some mulled wines, etc.

Niederndorf超市

Niederndorf是位于奥地利和德国边界的一个小村落，村落的边缘迫切需要建造一座超市。超市建在农舍、奶牛场和一些住宅之间，委托人希望超市能够免受日晒，保持凉爽。

在阿尔卑斯地区，木材是十分受欢迎的建筑材料，意味着故土、舒适和辨识度，例如木马厩就有清晰的区域辨识度。建筑直接采用了从当地伐木场运来的剥皮松木，并没有经过精加工。这层松木过滤网具有若干功能：它是最便宜的遮阳设施，营造了一个公共空间，也代表了一部分幕墙结构。

超市的设计具有人文尺度，为村民提供了一个交流的空间。松木柱在购物区和户外之间隔出了一片空地，村民可以在这里贩售圣诞树，铜管乐队可以在这里演奏，有时他们供应香料葡萄酒等。

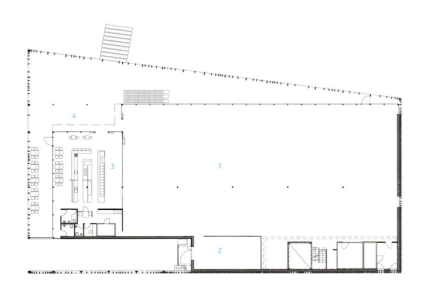

1. 食品超市/零售区
2. 面包房
3. 仓库
4. 带天棚的入口

1. food market/sales area
2. bakery
3. store
4. canopied access

带天棚的入口 Canopied access

零售区 Sales area

带天棚的入口 Canopied access

Photo: Thomas Jantscher

Commercial

Completion Date: 2008

Architect: Peter Lorenz Ateliers

Exterior plant 建筑外观和植物

Villa A

The property is about 5,000 square metres in a protected green zone with an old tree population. The architects orientate the living areas towards southwest. This side is entirely open to the panoramic view over the city and the garden. On the north side, the building façade is closed with natural stonework to provide intimacy towards the street. Following the topography of the site, the house is partly caved in the ground. A central hall with an open staircase gives access to various areas on different levels.

Garage, building services, fitness area, guestroom and office are allocated in the lower, carved-in storey. Living, dining and kitchen area are distributed on three different levels on the middle part of the building. The sleeping rooms are situated in the roof structure on the upper level.

The architect's intention was to differentiate formally and structurally the upper floor from the other two levels below. The purpose was to contrast the different function in the building. The lower levels are mostly designated for living, lounging and dining. These activities are embedded with different levels into the topography of the land and have direct access to the surrounding garden. The earthy relationship led the architects to opt for massive concrete structure that is cladded with stone. Above these, bedrooms are embedded into a roof structure carried by steel columns, totally detached from the massive walls below, almost flying over the massive walls.

A别墅

别墅位于一片绿树的环绕之中，面积约为5,000平方米。建筑师将生活区设置在西南侧，这个方位可以俯瞰城市全景和花园。北侧的墙壁由天然石材构成，保证了别墅的私密性。别墅依地势而建，有一部分结构嵌入地下。中央大厅的开放式楼梯可以通往不同楼层的各个区域。

车库、辅助设施、健身区、客房和办公室设在底层的嵌入式楼层中。起居室、餐厅和厨房分设在别墅中心的三个楼层中。卧室则设在顶层的阁楼。

建筑师试图将别墅的顶层和下面的两层区分，目的是明确建筑的各个功能区。较低的楼层基本是起居、休息和就餐空间，这些活动被分配在不同的楼层，可以直接通往花园。建筑与泥土的紧密联系让设计师决定采用覆盖石块的混凝土结构。顶楼的卧室由钢柱所支撑，与下方的墙体彻底分离，几乎是飞跃在墙体之上。

Dinner area 餐厅

建筑侧面的自然光 Side daylight

建筑外观 Exterior

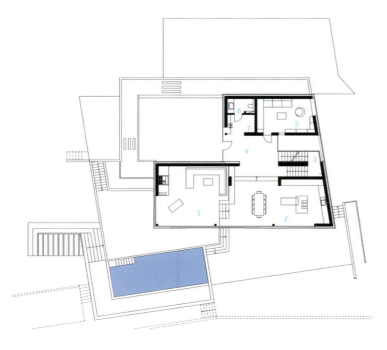

1. 起居室
2. 餐厅
3. 楼梯
4. 入口
5. 会客室
6. 洗手间
7. 保卫室

1. living area
2. dining area
3. staircase
4. entrance
5. saloon
6. WC
7. guard

Photo: Manfred Seidl (seidl.m.foto@aon.at)

Architect: Najjar-Najjar Architekten

Completion Date: 2008

Residential

Front view at night 建筑正面夜景

Oase Liezen Sustainability in the Shopping Town

利岑欧亚瑟购物中心

Over the past few years the major shopping town of Liezen has been developing with great energy. Now the new "Oase" on the main street offers a "haven of peace" to spend some time and linger: on a roofed "village square" in front of the centre, the square including a bistro as a café–restaurant. This complex is indeed a "small or refined" shopping centre with a real market hall. Due to its height and sense of space, it is one of the reminding historic examples in larger cities.

The centre is able to be accessed from two sides and has two parking facilities. At the pathway turning points, illuminated fountains provide a daylight effect similar to that of "forest clearings". Beyond the shelving the surrounding area's beautiful mountains are combined with the centre which make the customers feel free and at ease. H&M is spread out over two storeys – with a view around the market hall being provided. In between there is an attractive tobacconist's. Peter Lorenz finds that expressive architecture at this location is of importance: this is functionally important for a good identification of the customers with this "new marketplace" and also generates a relationship between customers and residents.

多年以来利岑的主要商业购物中心一直在蓬勃发展。现在，位于主要街道上的欧亚瑟购物中心将为人们带来一个集休闲娱乐于一身的购物天堂。购物中心的前方有一个带顶的小广场，里面有一个小酒馆。购物中心小巧而精致，里面有一个真正的贸易大厅。从空间和高度上来看，它和大型城市中的历史性建筑十分相似。

消费者可以通过两侧的门和两个停车场进入购物中心。道路的转角处，装有灯饰的喷泉营造出一种白昼的效果。购物中心周边美丽的山脉让消费者感到自由和放松。H&M品牌店占据了两层楼，环绕在贸易大厅的四周。二者中间还有一个引人注目的烟草店。彼得·劳伦兹建筑事务所认为具备表现力的建筑十分重要：这样消费者才能够找到适合自己的"新商场"（也就是欧亚瑟购物中心），购物中心和消费者与居民之间才能建立紧密的联系。

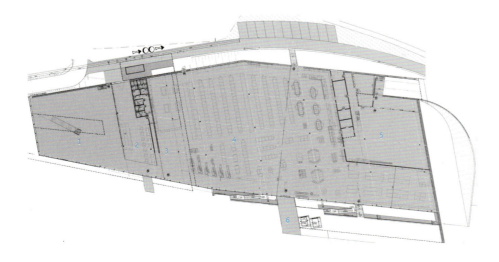

1. 商店
2. 咖啡厅
3. 交通空间
4. 食品超市/零售区
5. 仓库
6. 带天棚的入口

1. shop
2. café
3. traffic
4. food market/sales area
5. store
6. canopied access

从人行道看建筑侧面夜景　Side view at night, with light projection on the pavement

商场内部　Inside the mall

商场内部　View inside the mall

Photo: Thomas Jantscher

Completion Date: 2009

Architect: Peter Lorenz Ateliers

Façade 建筑外观

Q19

Metaphors must always be used carefully in architecture. However, approaching Peter Lorenz's new local shopping centre in Döbling quarter, in Vienna, from the north side, the associations of ideas spontaneously come out: a stranded ship, a rusted tanker. Actually, it is just with these images that the architect has introduced this work to his client: ships corroded by time, standing still and quiet, as waiting that something would happen.

This image is proper only when one observes the building from the north side and far off. Thus one can see the long and powerful hull, flanked by two protruding spiral ramps to the south and the north side, leading to the covered parkings on the upper floors. The "rusty" façade is made of corten steel. One can also see the domes diffusing the light, which are porthole–shaped elements of different dimensions.

On the entrance side, the scene is completely different. Here the continuum is created between the old and the new building. The old building is the Samum cigarettes paper factory. It is joined by the new one all along its length and, on the north side, even beyond. To the northwest there is the main entrance, with a large forecourt (which has been extended to include an urban public space too) and the glass front structure with uprights and crossbars. The façade is covered by a second skin, made of blinds, working either as solar control device or as dynamic advertising LED panel, which is 22 metres wide and 7.5 metres high.

Q19购物中心

在建筑设计中需谨慎使用比喻。但是，当人们从北面看到彼得·劳伦兹建筑事务所在维也纳Döbling区设计的Q19购物中心时，马上就会联想到搁浅的油轮。事实上，建筑师想表达的形象是：一艘历经时间侵蚀的轮船宁静地停泊在那里，默默地等待着一些事件发生。

只有从遥远的北面看去，商场才是一艘轮船。人们会看到一个长长的船体，船体的南北两侧是两条突出的螺旋坡道，坡道一直延伸到上层的停车场。商场锈迹斑斑的外墙是科尔顿钢板。商场的穹顶反射着阳光，就像是大小不一的舷窗一样。

入口处的风景则截然不同，新旧建筑在这里相连。旧建筑是热风卷烟厂，新建筑的长度和它相当。西北侧的主入口有一个巨大的前庭，可以拓展成为城市公共空间。入口的玻璃结构设有支架和横栏。外墙上增加了22米长、7.5米宽的百叶窗外壳，可以控制阳光直射和显示动态LED广告屏。

Inside q19 upper floor Q19购物中心二楼

夜景，入口连接了新旧购物中心 Night view – entrance, connecting the old and the new part of the shopping centre

建筑外观，LED墙和入口 Façade – LED façade, entrance

1. 总购物中心
2. 就餐区
3. 商店
4. 交通空间
5. 辅助区域

1. gross shop
2. dining area
3. shop
4. traffic zone
5. auxiliary area

Photo: Pia Odirizi

Completion Date: 2008

Architect: Peter Lorenz Ateliers

Commercial

Music Theatre

The mumuth theatre belongs to the university of music and performing arts graz which is a place where young musicians receive their instruction in the performing and musical arts.

The unit-based part of the organisation (the box) is situated on the right side, and the movement-based part (the blob) on the left side of the building as seen from the lichtenfelsgasse. There are two entrances: the everyday entrance on the park side which is used by students and staff, and the public entrance on the Lichtenfelsgasse which is used by the audience when there is a performance. On performance nights, the student entrance is transformed into a wardrobe using mobile closets. A removable ticketing desk and screen bulletin are placed underneath the staircase. The public ascends a wide staircase and enters a large foyer on the first floor. This foyer gives access to the multipurpose auditorium that can seat up to 450, and that is adaptable to a great variety of performances, ranging from solo instruments to opera to full orchestra.

音乐剧院

音乐剧院隶属于格拉兹音乐和表演艺术大学，是年轻的音乐人学习表演和音乐的地方。

从Lichtenfelsgasse街看去，盒结构位于建筑的右侧，而变形结构则位于建筑的左侧。共有两个入口：公园旁的普通入口由师生使用，Lichtenfelsgasse街上的公共入口由前来观看演出的观众使用。举办演出时，师生入口处可以使用移动衣柜，成为衣帽间。楼梯口摆放着可移动售票台和公告栏。观众通过一个宽大的楼梯进入二楼的大厅。从大厅可以进入拥有450个席位的多功能礼堂。礼堂可以举办多种演出活动，如独奏表演、歌剧、管弦乐队表演等。

First floor plan 二楼平面图

General view 全景图

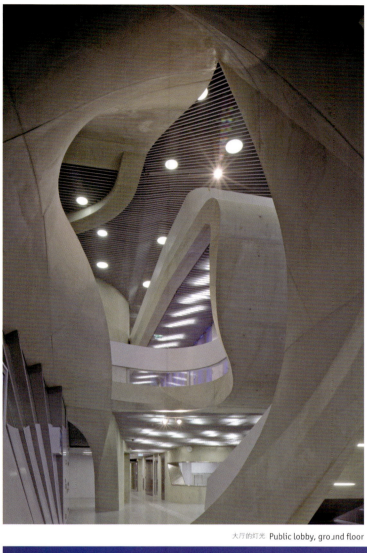

大厅的灯光 Public lobby, ground floor

楼梯 Stairs

Cultural

Completion Date: 2008

夜景 Night view

Architect: UNStudio

Buildings with entrance 建筑入口

Museum Liaunig

列奥尼格博物馆

The museum entrance zone is orientated towards both the centre of Neuhaus and the nearby historical castle. The substantial viewing storage depot, where visitors are accompanied by a "wine cellar of art", is one of the main areas of the museum, stretching the whole length of the gently sloping approach to the main exhibition hall. This underground volume offers the possibility to organise a variety of exhibitions by virtue of flexible screens and lighting arrangements. The building's core is a 160-metre-long, fully daylit exhibition hall, with protected terraces at each end. The continuous thirteen-metre-wide, seven-metre-high room is covered by a translucent curved skin – an industrial element permitting daylight. The hall is organised with mobile exhibition panels. In the exhibition hall are separately sited the graphic collection and the gold collection.

Besides, the museum claims efficiency and sustainability by its utility of the excavated soil for its construction. Industrial materials like concrete, glass and sheet metal dominate the visible portion of the building. Moreover, set into the hill, the building benefits from the constant temperature of the ground. Rooflight is adopted to substitute artificial light as much as possible.

博物馆入口朝向诺伊豪斯中心和附近的历史城堡。展示厅是博物馆的主要部分之一。游客可以沿着斜坡的通道走到主展示厅，参观这里的"艺术圣殿"。地下的巨大空间，配以活动屏风和灯光布景，适宜安排各种不同的展示活动。博物馆的中心是一个照明充沛的展厅，长160米，两端都有起保护作用的露台。考虑到日照因素，这个13米高、7米宽的房间采用半透明的波纹墙面。展厅里备有活动展台。博物馆展厅主要分为图画收藏和金属制品收藏，分别陈列在两个展厅里。

此外，这座博物馆以其高效能和可持续性理念著称，主要表现在从山体挖掘出的土壤不用运走，而是直接用作建筑材料。其他工业材料如水泥、玻璃、金属板等，主要应用于建筑外观可见部分。另外，博物馆因地制宜，对当地的恒温地热善加利用，并且最大限度地用屋顶采光取代人工照明。

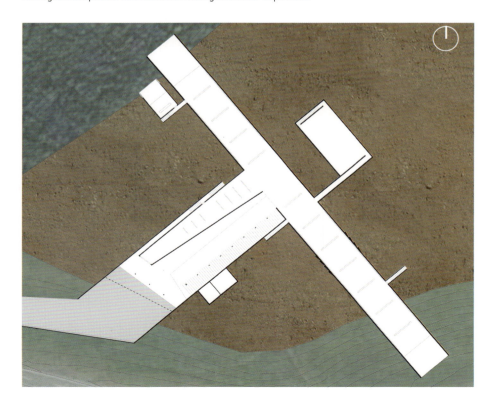

屋顶 Roof

建筑和景观 Building with landscape

建筑细部 Details

入口 Entrance

Photo: Erwin Stättner

Completion Date: 2008

Architect: Erwin Stättner

House in Canobbio

卡诺比奥之家

Nestled on the Alpine slopes north of Lugano, this house is characterised by a volumetric architecture that emerges from the terrain and follows the natural contour of the land. Its constructed volumes embrace the land in an organic and fluent sequence of spaces, each relating to another and to the surrounding landscape. In order to communicate an identity and a language to the inhabitants, the project has a strong and precise form, and its clearly identifiable geometric structure delimits an organised development of spaces. Carved in a clear square geometry, the spaces meet the slope and extend in a spiral, fluent movement that continuously changes the perception of the space and its relationship to the exterior, offering striking panoramic views across the hinterland and to Lake Lugano.

The succession of spaces and play of perception in the house are ideas derived from those principles of the Japanese garden. By their design and their nature, Japanese gardens offer varying levels of awareness of and responsiveness to space by offering an experiential sequence of different sceneries. As in the gardens, the organisation of the spaces in the house is condensed yet continuous, and is intended to take the experience of being in the house beyond the rooted and enclosed domestic scale. In addition, the house also works on both an intimate and vast scale. The clients, desire for a shell-like home has been met; the house is very private, protected and not overlooked, and the generated form and volume of the house also create an open, generous outlook, embracing its setting and taking quiet ownership of its prospect .

卡诺比奥之家"栖息"在卢加诺北部阿尔卑斯山一面斜坡上，拔地而起的巨大结构形成其主要特色，并与所处地块的造型完全吻合。各个结构之间相互联系，同时更与周围的景观紧密结合。显著而精细的方形造型向主人传达着它本身的特色，同时奠定了空间格局的划分基础。

连续的空间布局以及变换的视角设计源于日式花园建筑理念。房子室内空间简洁而流畅。需提到的一点是，这一房子本身具备双重特质——在主人眼里，它私密而亲切；在路人眼里，它开敞而大气。

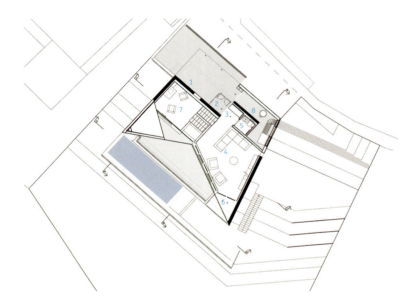

1. 停车场
2. 带顶的入口
3. 入口
4. 起居室
5. 浴室
6. 平台
7. 阅读室上方的空地
8. 热泵

1. parking
2. covered entrance
3. entrance
4. living room
5. bathroom
6. terrace
7. void over reading room
8. heat pump

External view up to house from southwest slope 从西南斜坡仰视住宅

住宅东侧的阶梯花园 East façade with terraced gardens

透过平台看餐厅和厨房 View across terrace into kitchen and dining space

从主入口看起居室 View into living space from main entrance

General view 全景图

Lienihof Residential and Business Complex

The Lienihof complex, situated between Albis and Heinrich-Federer streets in Zurich-Wollishofen, Switzerland, is incorporated into a virtually similar building site. Its complex sequence of twenty corners in the structure reflects respect for the building regulations in force as well as for the polygonal form of the plot. The building shape, determined by necessity, is balanced by the courtyard which opens up towards the south, thus constituting an architectural force of its own. Surrounding balustrades express the character of this meandering visually horizontal construction. The pine-wood panelling of the balustrades painted in dark red is a reference to the traditional carpenter's workshop which once stood on the site.

Following the sophisticated form of the building, the different apartments are designed as complex figures, closely interwoven with each other so that together they build but a single form. This organic adjustment of the ground plans to the outer structure generates inner spaces which offer a variety of perspectives, views and insights – an architectural approach which leads to interesting inner promenades continuously changing their orientation while moving through the apartments.

Each of the upper floors comprises eleven different apartments which are accessed by three stairscases. The ground floor accomodates both business premises and an apartment for persons with nursing needs. A long hall linking streets, courtyard and the three staircases, creates a community space.

Lienihof商住两用小区

Lienihof小区位于苏黎世阿尔比斯河和海因里希－费德勒街之间，看起来像是一个建筑楼盘。建筑共有20个转角，既符合建筑规则，又充分反映了多边形场地的特征。朝南的庭院也别具特色，与建筑的外形达成了一种平衡。建筑的围栏在水平方向装点了建筑，围栏的松木条被漆成了深红色，参照了场地上原有的传统木工作坊。

与精妙的建筑造型相似，公寓的设计也十分复杂，不同的公寓交错在一起，形成了一个统一的造型。外部楼面结构设计的有机调整让室内空间可以欣赏到多样化的室外风景。建筑设计让室内交通变得极为有趣，走廊在穿越公寓的时候不断地改变方向。

1. 卧室
2. 洗手间
3. 卧室
4. 凉廊
5. 客厅
6. 厨房
7. 卧室

1. bedroom
2. washing room
3. bedroom
4. loggia
5. living
6. kitchen
7. bedroom

走廊 Corridor

远景图 View from far away

内院 Inner courtyard

Photo: Adrian Streich Architekten AG, Zurich

Architect: Adrian Streich Architekten AG, Zurich, Roger Frei

Complex

Completion Date: 2007

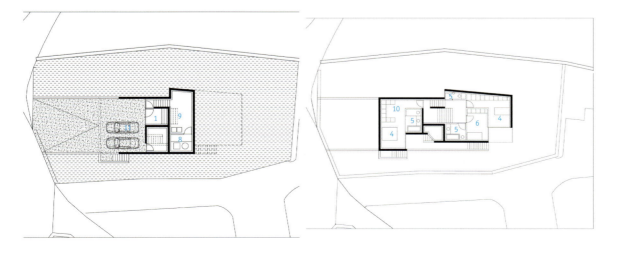

External view of the house in its Alpine context from the southeast 建筑西南侧与其周围的山脉景色

House in Lumino

Located in the Swiss Alpine village of Lumino, just north of Bellinzona, this house stands as a monolithic element, quietly complementing and echoing its context. The surrounding area is characterised by traditional stone built houses, many of which date back centuries and are marked by their use of this single construction material. The new house acts as a sort of bastion between the old core and the modern residential expansion.

In addition to the local scale references and material cues siphoned from the physical context, the concept and approach to the project was further influenced by the clients' expressed desire for a minimalist aesthetic. As such, the quality of the spaces in the house would be defined explicitly by the architecture and not by objects placed within it. The idea of the "minimalist monolith" was adopted as the conceptual generator of the project and became a principle applied to all elements of both the functional and construction programme, from the foundations up to the smallest finishing details.

The geometry of the plan is generated by two shifted parallelepipeds and follows the fall of the site. The double system of vertical connections, one internal and one external, relates all the spaces of the house in a spiral movement, and is in a constant play with its new inhabitants' perception of time and scale. What is interesting about the house is the ability of the spaces to expand and extend into the landscape, allowing the external become part of the composition. While the individual spaces may be defined geometrically, each space flows into the next and continues to the external.

卢米诺之家

卢米诺之家"屹立"在贝林佐纳北部小山村内，与四周环境相得益彰。村庄内多是几百年前的石头建筑，独特的背景使其更加突出，如同古老村落同现代环境之间的一座"堡垒"。

除地域及材质之外，这一设计更受到主人关于"简约美感"要求的影响。房子内部空间布局更应受到外观造型的影响，而并非室内摆设装饰。基于这一点，设计师构思了"简约独立结构"的理念，并将其运用到各个方面。

房子由两个错列平行六面体结构组成，顺着地势走向延伸。室内外分别修建楼梯，将不同的空间以螺旋形状连通，主人的视野更可以随时变换。更为有趣的一点是，房子室内空间可以延展，一直与室外景观"接轨"，各自空间之间又形成流畅的整体。

1. 入口
2. 起居室
3. 厨房
4. 卧室
5. 浴室
6. 客房
7. 平台
8. 洗衣房
9. 机械室
10. 贮藏室
11. 停车场

1. entrance
2. living
3. kitchen
4. bedroom
5. bathroom
6. guest bedroom
7. terrace
8. laundry
9. mechanical room
10. storage
11. parking

西侧的停车场和入口 View from west to parking and entrance

从厨房看餐厅和平台 View from kitchen through dining and out to terrace

起居室 View into living space

Photo: Enrico Cano

Completion Date: 2009

Architect: Davide Macullo Architects

New Industrial Development Centre

To the west of Neuchâtel, the landfill from the early 20th century spread over the lake is occupied by a centre of industrial production of Philip Morris. The project resulted in a cuboid of 106 metres long, thirty-three metres wide and twenty metres high. This particular configuration made of solid work on several levels is due to lack of space. The challenge of this realisation is simple: to provide the user with a tool capable of adapting to any new configuration of use and at a price of a conventional plant.

The proposal is based on a structural system capable of delivering significant ranges (sixteen metres) and integrating in its thickness and frame all the techniques necessary for the activity of this sophisticated development centre, namely the ventilation, electricity, gas, compressed air, dust removal related to the building, clean rooms and machinery.

Beyond its technical interest, this device offers a particular spatial quality and a quiet environment. The building envelope consists of a single material, namely the glass industry. This choice can optimally manage the confidentiality of the activity, quality natural light and indoor climate.

菲利普莫里斯工业发展中心坐落于纽沙特西部湖畔，原址是一处20世纪早期的垃圾填埋场。项目长106米，宽33米，高20米，是一个长方体。项目的目标十分简单：以传统工厂的造价为公司设计一个能够适应各种新装备配置的空间。

项目的结构系统可以运送长达16米的设备，里面各种技术设施十分齐全，如：通风设备、电气、煤气、压缩空气装置、除尘设备等与机械、清洁有关的设施。

除了机械特征之外，建筑还提供了一个独特的空间系统和极其安静的环境。建筑的外壳采用了单一的玻璃材料。这种设计保证了内部活动的私密性、提供了自然采光，并且营造了有利的室内环境。

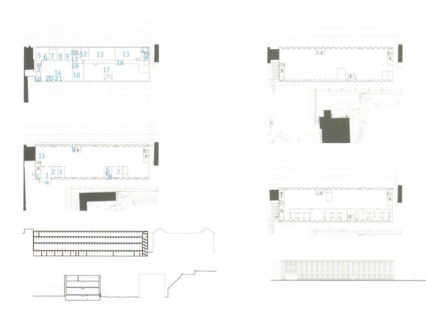

1. WC	1. 洗手间
2. lift	2. 电梯
3. airlock	3. 气闸
4. lift	4. 电梯
5. heating distribution	5. 热分布区
6. informatics	6. 信息工程区
7. power transformer	7. 电力变压器
8. power distribution	8. 配电室
9. airing +water treatment	9. 空气+水处理系统
10. technical	10. 机房
11. sprinkler	11. 洒水车
12. compressed air	12. 空气压缩机
13. stocking	13. 原料区
14. south–west workshop	14. 西南车间
15. meeting room	15. 会议室
16. CAD office	16. CAD办公室
17. prototype zone	17. 标准区
18. hallway + freezers	18. 走廊+冷柜
19. break room	19. 休息室
20. meeting room	20. 会议室
21. supervisor's office	21. 监管人办公室
22. cubbyhole	22. 文件室
23. emergency power	23. 紧急电源

South Façade, looking from the Neuchâtel Lake　从湖上看建筑南立面

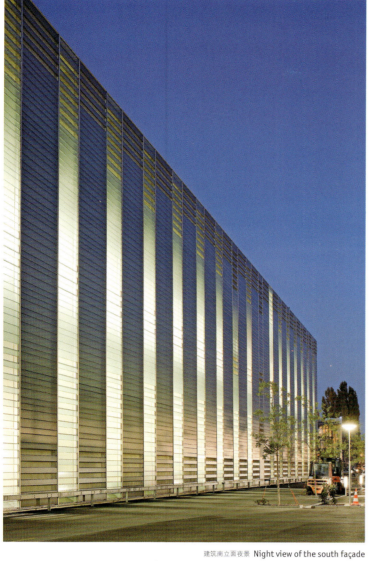

建筑南立面夜景 Night view of the south façade

工业玻璃墙细部 Details of the industrial glass

Photo: Thomas Jantscher

Completion Date: 2009

Architect: Geninasca Delefortrie SA

开放区域内部 Inside view of the open zone

Werdwies Residential Complex

The course of the river Limmat and the motorway represent two antipodes forming the boundaries of the insular microcosm of the Grünau Quarter in Zurich, Switzerland. The Werdwies Residential Complex provides the Quarter with an open centre characterised by the density of an inner space. The complex consists of seven rhythmically positioned prism-like constructions creating a sequence of built and open spaces. This leads to the emergence of several smaller and larger squares, each with its own character. A hard covering with integrated lawn areas forms the ground area of the exterior space and allows free movement within the residential complex. The planting of about 100 trees underlines the park-like character of the complex.

Each building comprises seven residential floors, having a total of 152 apartments in three different house types. The houses are distinguished by their staircases with loggias and stairwells. The spacious loggias create a dense relationship between interior and exterior spaces. At the same time, they mark the beginning of a principle of layers, generating a comprehensive modularity. Each apartment includes a sequence of rooms lighted from different directions, with areas for living, eating and sleeping. Simple layouts with long rooms and built-in cupboards extending over the whole length of the wall give the apartments a robust character.

Werdwies住宅区

利马特河的航道与其对面的高速公路共同充当了苏黎世 Grünau区的边界，形成了一片与世隔绝的微观环境。 Werdwies住宅区以其高密度内部环境为Grünau区营造了 一个开放的中心。Werdwies住宅区共有七座错落有致的 住宅楼，它们呈棱镜结构展开。这种设计在楼与楼之间 形成了大小不一、独具特色的广场。户外空间由硬板路 和草坪组成，人们可以在其间自由地穿梭。100余棵树木 的种植让住宅区看起来更像是一个公园。每座住宅楼有 七层，共有152间公寓，分为三种类型。不同的住宅楼的 楼梯拥有不同的凉廊和楼梯井。宽敞的凉廊在室内外之 间建立了紧密的联系。与此同时，它们也标志了住宅楼 的层次。每间公寓都有良好的采光，分为起居、餐饮和 睡眠三个区域。简单的长条形房间布局和整面墙的嵌入 式橱柜让公寓看起来十分稳定坚固。

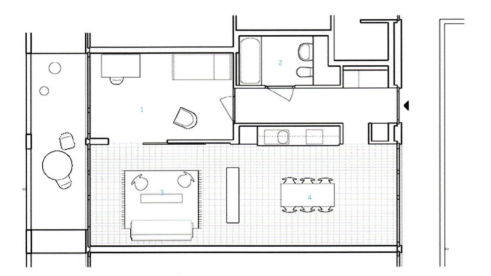

1. 卧室
2. 洗手间
3. 起居室
4. 厨房和餐厅

1. bedroom
2. washing room
3. living
4. kitchen and dining

Front view 建筑正面

景观设计 Landscape

庭院 Courtyard

餐厅 Dining

Photo: Adrian Streich Architekten AG, Zurich

Architect: Adrian Streich Architekten AG, Zurich

Residential

Completion Date: 2007

South view 建筑南立面

Sempachersee Golf Club

This is the largest golf course in Switzerland. The architectural office Smolenicky & Partner was commissioned to design two new buildings in connection with the enlargement of the course – the clubhouse and restaurant building and the new maintenance building.

The clubhouse is situated precisely on the topographical crest where the level plateau of the golfing green breaks into a steeply falling slope. At exactly this point, the vista opens out into a view over the lake and the Alps of inner Switzerland. The public footpath that transverses the golf club also runs along this topographical ridge.

Both the architecture and the interior design of the new building aim to combine two distinct atmospheric phenomena of the site into a single effect. This new manifestation is moulded, on the one hand out of the country character of the golfing culture of the Sempachersee course, and on the other hand out of its worldly sophistication. To this end the appearance oscillates between the rural warmth of a timber barn and the clear lines of a Masserati sports car. This is the attempt to embody both the reality of the dualism of the site and its potential within the building itself.

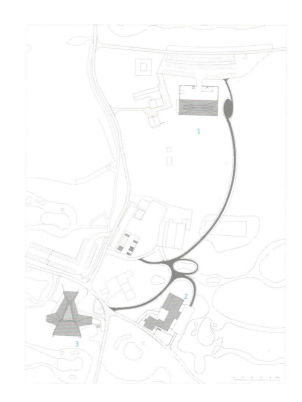

1. maintenance building – new building
2. reception – add-on
3. clubhouse – new building

1. 维护楼（新建）
2. 接待处
3. 俱乐部（新建）

Sempachersee高尔夫俱乐部

Sempachersee是瑞士最大的高尔夫球场。Smolenicky建筑事务所受托为球场设计两座新建筑，包括俱乐部会馆、餐厅和维护楼。

俱乐部会馆坐落在球场的最高点，背后是陡峭的山坡，可以俯瞰湖面和阿尔卑斯山的风景。通往俱乐部的小路沿着山坡顺势蜿蜒而上。

建筑的外观和室内设计都试图将高尔夫球场的两大特征融为一体。一方面，高尔夫文化带有显著的乡村风情；而另一方面，高尔夫具有世界级的精英受众。最终，俱乐部的外观是乡村风情浓郁的木板仓库和线条简洁的Masserati跑车结合体。这个尝试既体现了高尔夫的两面性，又具有合理的功能特征。

建筑西立面 West view

俱乐部会所 Veranda clubhouse

餐厅入口 Restaurant entrance

Photo: Walter Mair, Christoph Reinhard

Recreational

Architect: Smolenicky & Partner Architektur, Simon Krähenbühl, Dirk-Oliver Haid, Juan-Carlos Smolenicky-Munoz

Completion Date: 2007

Southwest　建筑西南立面

BSV College for Further Education

The new class building of the Viège/Visp College of Further Education achieves a transformation in terms of context and enables the linking of the neighbourhood's schools, the creation of a training campus, and the integration of the future workshops and gyms. Associated with the new building, the former school playground becomes the benchmark public space of the campus.

The new building's composition of volumes takes its cue from that of the schools in the area. The typology maker reference to that of the building with its central distribution hall, yet radicalises it into a plan featuring three strata, the central layer serving both as distributive and group-work space. The transparency of the lightweight partitions allows this space to benefit from natural light. The use of materials within the building is confined to the bare load-bearing concrete and to the aluminium combined with the coloured glass of the lightweight partitions. The materiality of the exterior confirms the volumetric character – the mirror-finish stainless steel frames of the glazing and the bare aluminium cladding applied to the section cut away from the base quadrilateral. The fragmentation of the reflections created by the bevelling of the façades generates a new context.

BSV继续教育学院

菲斯普继续教育学院的新教学楼实现了环境的转型，连接了邻近的其他学院，营造了新校园，并且整合了未来的工作室和健身房。与新教学楼相连的操场也成为了校园的公共焦点。

教学楼的结构与区域内的其他学院相仿，以中心大厅为核心。大厅分为三个部分，中间的部分是流通和集会空间。大厅的透明轻质隔墙让它能够得到充分的自然采光。教学楼采用了清水承重混凝土、铝材和彩色玻璃作为主要建筑材料，简单实用。外墙材料的运用强化了建筑的体量特征。镜面不锈钢框架与截面所使用的铝制面板相结合，清晰地刻画出建筑的棱角。斜切边外墙上支离破碎的倒影营造出一种全新的感觉。

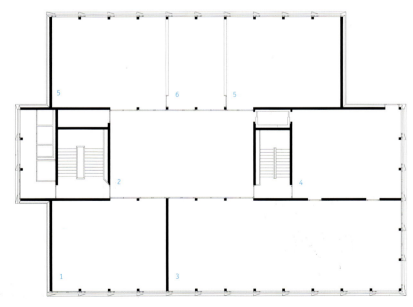

1. 入口
2. 门厅
3. 餐厅
4. 厨房
5. 教室
6. 准备室

1. entrance
2. foyer
3. refectory
4. kitchen
5. classroom
6. preparatory room

建筑东立面 East

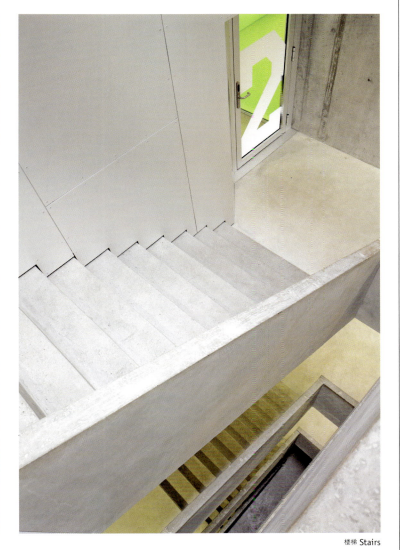

楼梯 Stairs

2楼大厅 Foyer level 2

Photo: Hannes Henz

Educational

Completion Date: 2009

Architect: Bonnard Woeffray Architectes

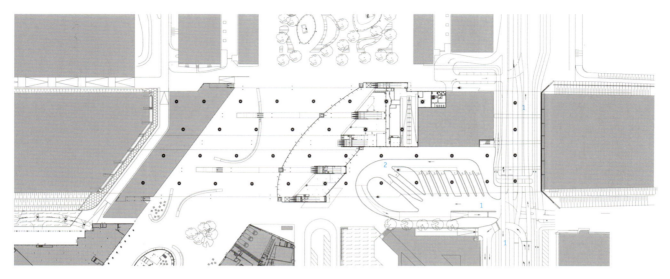

1. LUN bus
2. station

1. LUN巴士
2. 车站

Station Amsterdam Bijlmer ArenA

The station is designed to provide a high level of social security both during the day and at night. Long voids are cut into the platforms to break down the overshadowed sections of the 100-metre-wide area below the viaducts. These voids improve the sense of safety through visual contact and improved transparency between the platform and ground–level areas.

To avoid a dark 100-metre-long tunnel, the concrete structures were spaced apart. Each twenty-metre span was supported at each end on just one column via an integrated cantilevered saddle. Arrays of columns were then aligned on an axis with the boulevard to maximise visual connectivity from east to west. The base–element of the roof structure is a "V" shaped continuous hollow steel boom with steel arms cantilevered on either side to support all the roof glazing. The combined assembly is supported on a series of tubular "A" frames with only a single deep longitudinal stabiliser near the south end. Beyond their last supports these booms cantilever up to eighteen metres, thereby enhancing the sense of linearity and direction. The timber lined elements straddle each track–bed, and are open at ridge level to assist natural ventilation, and allow areas for pressure release in respect of 200k/h trains.

Bijlmer竞技场火车站

车站设计的主旨即为提供高水准的安全设施，长方形的空间被打造成月台，以使得高架桥下面的遮蔽空间得以充分运用。这些空间在增强了安全性的同时，也促进了月台同其他区域的联系。

为避免制造出100米长的隧道空间，设计师将其加以分割。每20米处便采用一个柱状结构支撑，一根根梁柱沿着轴线列成一排，并与大马路相接，在视觉上将东西两侧连通。屋顶结构呈现"V"型，钢材吊杆和悬臂结构分别位于两侧，支撑着整个屋顶。其他结构由"A"状框架支撑，仅在北侧安装有稳固装置。

Photo: Jan Schouten

Completion Date: 2007

Architect: Grimshaw Architects

Vuykpark, Capelle a/d IJssel

For the site of the former shipyard "Vuyk" in Capelle aan den IJssel, MIII designed a restaurant. The design of MIII was inspired by the history of this site, a shipyard where craftsmanship was at a highly competitive level. Marine lining, carpentrance, and wood constructions were starting point for the design in which lopside and vigour refer at navigation (shipping).

The outside, visible construction hints at the old craftsmanship of the ship's carpenter. The idea was to envelop the building, so to speak, in a wooden coat. Characteristic of these designs is the sharp contrast between the often introvert exterior and the open and transparant interior. A smooth wooden sheet around its outline makes a connection with the carpentrance of the past. The open–front coping has several advantages above a closed one. Most prominent is the feature that the wind can freely blow alongside the wooden front parts and by doing so, dehydrate these parts in a natural way. Thus, the backside can be ventilated.

Vuyk公园餐厅

Vuyk公园餐厅选址在一家旧船厂内，其设计充分受到所处地点的历史背景影响（技艺在造船领域占据极为重要的位置），因此木工手艺以及木质结构构成了这一设计的出发点。

餐厅的外观格外突出传统的造船工艺，完全采用木材"包裹"起来，犹如一艘木船。设计的特色更体现在低调的外壳与高雅通透的室内之间的对比。平滑的木条结构营造了建筑本身与古老木艺工艺的联系，开敞的屋顶相比较于封闭的结构更具优势。此外，尤为重要的特色即为风可以沿着木质结构表面自由通过，不仅起到干燥作用，同时增强了后侧的通风。

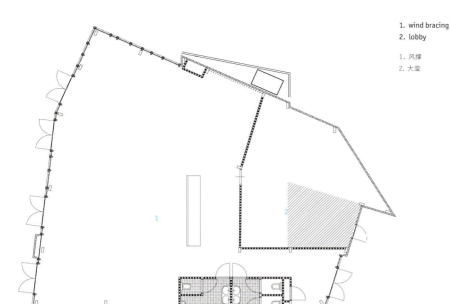

1. wind bracing
2. lobby

1. 风撑
2. 大堂

Photo: John Lewis Marshall

Architect: MLLL architecten

Front 建筑正面

Booster Station–South

Booster Station–South in Amsterdam is primarily a utility building which contains a technical plant and a pumping–engine for sluicing out sewage. The hidden technology of the sewage–system of the Dutch capital emerges in the public realm as an intriguing object. The location in the public realm asks for a careful design that will appeal to the public now as well as in the future, all the more because Booster is located next to a busy traffic junction where different types of transport intersect.

The programme for the pumping–station consists of a high and a low voltage space, an overhead crane, three pumps, a bypass and an entrance. The Booster Station–South can be seen as a metaphorical reference to a streamlined engine. The clinging, aerodynamic skin forms an envelope for the building's technical programme. It also emphasises the relation between form and function the building reveals. The constant stream of passengers by car, metro, train or bicycle perceive the Booster Station as a futuristic sculpture. With its cladding of stainless steel panels, it reflects the movements, shapes and colours of the environment. At night the illuminated seams in the steal skin make the building look like a mesh model.

布斯特泵站

布斯特泵站最初是一幢公用设施建筑，包括技术工厂及泵发动机（用于清理下水道污物）。荷兰首都的污物排放系统就这样出现在公众视野中，并以其独特的造型吸引着大家的关注。当然，这一地理位置（交通要道，多种交通工具交叉口）也决定了其精细的设计，必须达到吸引公众的标准。

泵站要求拥有高压及低压设备空间各一个，一台大型起重机、三台抽水泵、一条通道及一个入口。为此，设计师充分运用了一个比喻——将其打造成流线型发动机造型。黏附的空气动力材质表皮将整体"围和"起来，并充分强调了形式和功能的联系。不锈钢覆层突出了动态感，在过往的人群眼里，它更像是一座雕塑。

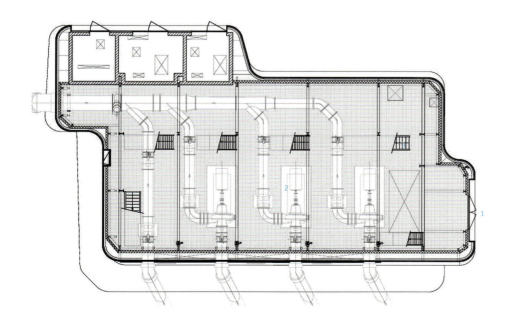

1. 入口
2. 泵
3. 楼梯

1. entrance
2. pumps
3. stairs

建筑内部 Interior

建筑细部 Detail

傍晚建筑一角 Evening side

Photo: Digidaan

Completion Date: 2006

Architect: Maarten van Bremen, Jaap van Dijk, Folkert van Hagen, Adam Visser, Jasper Hermans

School Piter Jelles

皮特尔·杰里斯中学

The building's shape metaphorically reflects the process of peeling a fruit. A transparent unity of theory and practice makes the fruit, the pupils are the seed. Like half-peeled paring spiralling around the fruit, the façade partially opens up towards its surroundings. The façade is transparent where contact with the outside world is encouraged and alternately open and closed where pupils are working independently.

As a token of appreciation and respect, pupils (and teachers) enter the building via a red carpet covering the stairs leading up to the building's main entrance.

The core of the building is home to practical subjects: mechanics and construction (ground level), kitchen and bakery (first level), electronics (second floor) and the building's technical facility room (third floor). It was a conscious decision to keep technical installations, such as pipes visible for pupils, in order to reveal the complexity of and to stimulate curiosity for the functioning of the building. Located around this core are the shops on the ground floor, with classrooms for theoretical lessons and offices above.

建筑的外形反映了削水果的过程。理论与实践简单易懂的结合是果实，学生是种子。建筑的外墙看起来像是被削到一半的水果皮，一部分向外部开放着。建筑的一部分外墙是透明的，当学生学习时，还可以自由选择关闭或开放。

学生和教师可以通过教学楼门口台阶上的红地毯进入，以示尊重。 教学楼的中心区域主要安排着一些实用科目：机械构造（一楼）、厨艺烘焙（二楼）、电工（三楼）。建筑的技术控制设备设在四楼。将建筑的机械装置（如管道）展示给学生，以此激发他们对建筑功能的好奇心是一个明智的决定。中心区域的四周是商店（一楼）、理论课教室和办公室（二、三楼）。

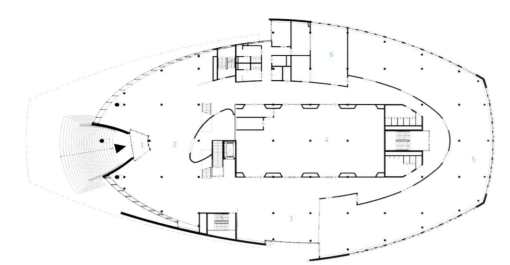

1. entrance
2. main lobby
3. restaurant
4. kitchen
5. classroom
6. administration

1. 入口
2. 主大厅
3. 餐厅
4. 厨房
5. 教室
6. 行政办公区

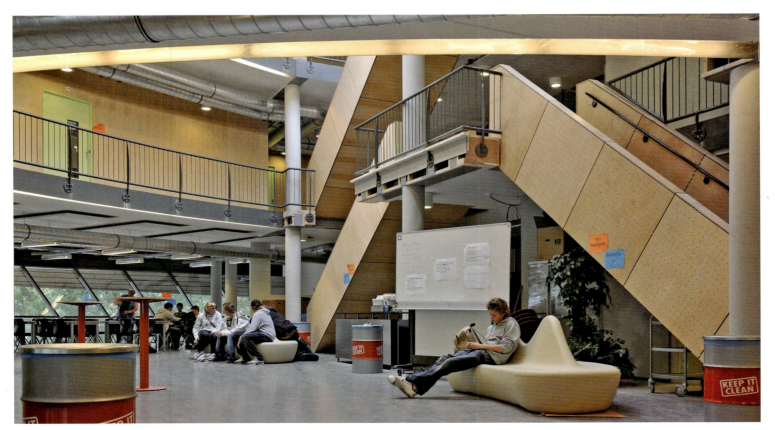

Photo: Bjorn Utpott

Completion Date: 2008

Architect: RAU

Exterior 建筑外观

Woonhuis VdB, Prinsenbeek

VdB住宅

Located in the outskirts of the city of Breda, on the edge of the urban and rural zone, this house is provided with great views. The clients preferred a maximum volume on this plot and asked the Dutch architects Grosfeld van der Velde to turn this desire into reality. They designed a house with a square floor plan of 15 x 15 metres, stacked up over three levels. The ground floor is raised above the surrounding land so that daylight can enter the half-sunken basement. By lifting the ground floor, the view from the living area of the natural environment is astonishing.

VdB住宅选址于布雷达市郊，四周景色迷人。客户邀请荷兰知名建筑师Grosfeld van der Velde为其设计并要求最大限度运用其所处地块面积。为此，设计师打造了一个15米x15米的方形结构，共为三层。一层比周围地势略微抬高一些，以确保半地下室空间内光线能够达到。与此同时，周围的美丽自然景色也被"吸纳"进来。

Exterior view 建筑外观

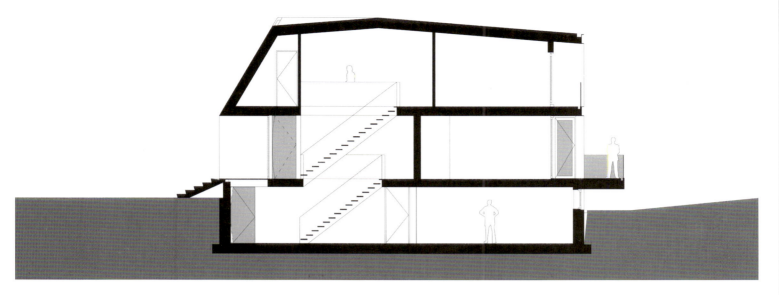

楼梯 Staircase

玻璃门 Glass door

客厅 Living room

Photo: Michel Kievits

Completion Date: 2008

Architect: Grosfeld van der Velde Architecten

Garden façade, northeast　建筑东北立面，花园

Villa BH

The villa is positioned on a rectangular plot of 35 x 50metres that is enclosed at three sides with similar plots and freestanding houses. On the back (Northeast) of the plot there's an old embankment with several tall trees. From the living programme; the kitchen, dining area and living are all orientated on this green scenery. Here the villa has a glass façade over twenty metres long.

Villa BH is inhabited by a couple of more than sixty years old. To optimise the accessibility of the house, the entire programme is situated on the ground floor around a patio. The specific form of the patio widens and narrows the interior space, making it a variety of areas. The façade of the patio is completely from glass panels, giving the villa great perspectives in its interior but also towards the context. The ceiling of the living area has an extra height in the shape of a sloped roof. The physical appearance of this area is very unique and highly qualitative.

The villa is designed as environmental friendly with extra insulated façades, roofs and floors. The roof is covered with sedum that regulates the distribution of the rainwater gently. On the flat roof are twenty solar panels for electricity. A heat pump warms the floors in the winter and cools them in the summer with natural temperature differences retrieved deep in the ground. As an extra heating there are two fireplaces for wood, one in the living and the other in the TV–room.

BH别墅

BH别墅坐落在一块面积为35米x50米的长方行地块上，其中三面被相似的地块或独立式住宅包围。后侧（东北方向）是一个古老的堤岸，高大的树木巍然屹立，厨房、餐厅及客厅全部朝向这一区域。别墅外观采用20多米长的玻璃材质打造。

别墅主人是一对60多岁的老年夫妇，为方便他们活动，所有的功能区全部设在一层，并围绕中央天井展开。其中，天井的独特造型使得室内形成一系列规格各异的空间，全玻璃的外观更在视觉上开阔了视野。客厅内，倾斜的屋顶结构使得天花稍微高出一些。

尤需提到的一点是，别墅的设计格外注重环保。屋顶上覆盖着植物，可以调控雨水的分布，而屋顶平台上安装的20块太阳能板则可用于发电。热泵可以调节室内地面的温度，客厅及电视房内的壁炉作为附加供暖设备。

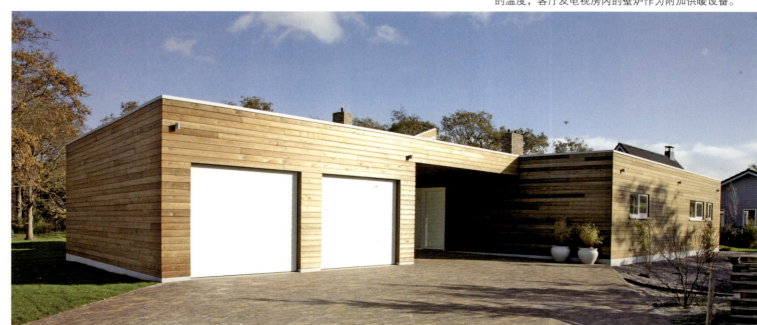

Entrance façade, southwest　建筑西南立面，入口

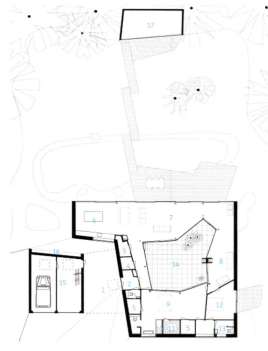

1. carport
2. entrance
3. toilet
4. installation room heat pump
5. closet
6. kitchen
7. living
8. TV room
9. main bedroom
10. bath
11. sauna
12. guestroom
13. guest bathroom
14. patio
15. garage
16. wood storage for fireplaces
17. garden house (still to be realised)

1. 车棚
2. 入口
3. 洗手间
4. 热电房
5. 壁橱
6. 厨房
7. 客厅
8. 电视房
9. 主卧室
10. 浴室
11. 桑拿房
12. 客房
13. 客房浴室
14. 天井
15. 车库
16. 木材仓库
17. 花园小屋（未建成）

建筑细部 Detail

从主卧看天井 Interior of the main bedroom with the view on the patio

Exterior view by day 日间建筑外景

Merry-Go-Round

Although ideas about recreation and the design of the landscape have changed over the years, the typology of the holiday cottage has hardly altered at all. Ever since the recreational outing of several days or more came into vogue in the 1960s, we have seen the same mini-version of the standard home.

Whereas the confined space of a boat or caravan has led to clever design solutions, the country cottage has never developed an identity of its own. The design of the Merry-Go-Round gives new meaning to the holiday cottage by taking the traditional floor plan with its rooms opening onto a central hallway and turning it inside out. The rooms are replaced by eight open alcoves in which furniture, colour, light, material, lines of sight and views of the outdoor surroundings are bundled into one compact, fixed interior. The alcoves are connected by a corridor that runs all around the perimeter of the dwelling and opens onto the landscape. By adjusting the façade, which is composed of shutters, the vacationing residents of the Merry-Go-Round can determine the view and their privacy themselves.

旋转木马

近年来，尽管休闲方式及景观设计一直在发生着变化，但是度假小屋的风格却几乎一直未变。自20世纪60年代休闲旅游开始盛行以来，度假小屋的样式标准就被制定了。

度假小屋的设计尽管受到小船或篷车式样的影响，但一直未形成自己独立的特色。"旋转木马"住宅运用传统的格局——所有空间全部朝向中央走廊，房间被8个壁龛结构取代，家具、色彩、光线、材质以及室外美景融合在一起。此外，这些"壁龛"通过回廊连接，分布在四周。房子外观安装有百叶窗，这样游客们就可以根据自己的需要调整空间的私密度。

Exterior by night 夜间外景

沙发 Sofas

走廊 Corridor

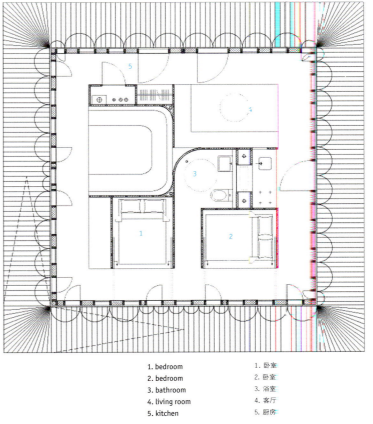

洗手间 Toilet

1. bedroom   1. 卧室
2. bedroom   2. 卧室
3. bathroom   3. 浴室
4. living room   4. 客厅
5. kitchen   5. 厨房

Photo: Bureau Ira Koers

Residential

Completion Date: 2009

Architect: Bureau Ira Koers (www.irakoers.nl)

Main entrance 主入口

WWF Netherlands Head Office

世界自然基金会总部

RAU transformed a former 1950s agricultural laboratory into the first CO$_2$–neutral and (almost entirely) self–sustaining office in the Netherlands. By breaking through the rigidity of the existing structure and adding an organic blob at the centre, the rejuvenated building got a friendly and inviting appearance. Natural materials have replaced bare concrete; what used to be grey and confronting is now in harmony with the surrounding natural reserve. RAU's intervention not only give a the building a new face, but fundamentally changed the user experience. Natural ventilation and the use of natural materials offer a balanced and healthy indoor environment.

Energy and Environment
The use of renewable energy is not the only environment–friendly aspect of the building. Energy and construction materials were saved by keeping the concrete skeleton of the former laboratory. All used wood is FSC–certified. The doormat is made of old car tires, and the flooring is made from recycled carpets. All used materials are child labour free. Bats have access to an especially prepared area of the basement and birds can nest in the façade.

Flexibility and Efficient Use of Space
The building is divided into two zones. All public functions such as the reception, the call centre, a shop and meeting rooms are grouped around the central staircase in the blob. Glass walls emphasise the open atmosphere in this part of the building.

The non–public functions are accommodated in the two wings of the complex. A smaller–scale floor plan creates a more calm environment, allowing employees to focus on their work without distraction. The first floor of the complex has a flexible layout so that it can partly or entirely be let to third parties.

RAU将建于1950年的农业实验室改建成荷兰首个碳中性及实现自我供给的办公建筑。原有建筑的规则性被彻底打破，并在中心处增建了新结构，赋予大楼友好而热情的外表。天然材质取代了混凝土，使得原来灰蒙蒙的结构完全融入到周围的自然环境中。这一改造不仅改变了建筑本身的"面孔"，更为使用者带来了全新的体验。

能源与环境
可持续能源的运用只是环保理念的一个方面，在材质选用上也遵循这一原则。使用的木材全部经过FSC（联邦科学委员会）认证，门垫由汽车旧轮胎改造而来，地面上铺设着回收地毯。蝙蝠可以自己找到地下室内的专属领域，小鸟更可以栖息在建筑表皮上。

灵活性及空间高效性
建筑分为两个区域，其中公共功能区——接待处、呼叫中心、商店及会议室全部"簇拥"在新建结构内楼梯周围。玻璃墙壁的运用增添了通透感。办公区则分别设在建筑两翼处，规模虽小但环境幽静，便于员工集中精力工作。二楼空间布局灵活，方便使用。

General view 建筑全景

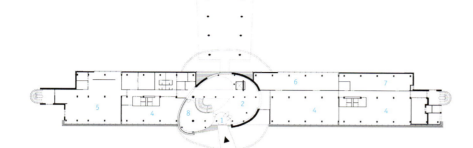

1. entrance
2. main lobby
3. call centre
4. office space
5. restaurant
6. messenger room
7. storage and archives
8. info centre

1. 入口
2. 主大厅
3. 呼叫中心
4. 办公区
5. 餐厅
6. 收发室
7. 仓库和档案室
8. 信息中心

三楼梯 Main stair

楼梯 Stairway

建筑外观 Exterior

Photo: Hans Lebbe, Kusters Fotografie, RAU

Corporate

Completion Date: 2006

Architect: RAU

Façade 建筑外观

The 4th Gymnasium

The façade has two effects that intensify the character of the building. The plinth is made of flat, coloured aluminium panels and continuously follows through into the façade of the courtyard; as a foretoken of the colour explosion in the court. The wooden façade has been developed more spatially and in depth and gives the building plasticity. This expressive modular built façade is hard to distinguish from a traditional façade because of a number of innovations, which prevents the monotonous picture of piled up units. By choosing a relative deep outside façade, it was possible to bring on relief. Under the frame, the façade withdraws twenty centimetres through which the image of two piled up arcades is created. And the seams between the modules are hidden; the wooden front parts are built from narrow planks, which are placed vertically and on small distance from each other. Through the number of artificial seams that arises, the real seams become invisible. At the plinth and the façade of the courtyard, the seams are hidden behind the rhythmically placed coloured aluminium boards of different widths. These creative solutions give the gymnasium a permanent and nevertheless dynamic charisma.

第四体育馆

第四体育馆的外墙极具特色。底部外墙的彩色铝板一直延伸到庭院的围墙，与庭院内部的色彩十分相似。而木质外墙则极具空间感，增加了建筑的可塑性。建筑的外墙与传统建筑大相径庭，一系列的创新设计使建筑免于单调。内陷的外墙结构让人感到安心。外墙在建筑框架中内陷了20厘米，形成了两块堆积起来的拱形图案。木制外墙由狭长的木板组成，木板垂直排列，中间留有缝隙。人造的缝隙将木板间真正的缝隙隐藏了起来。底部外墙和庭院外墙的缝隙隐藏在错落不齐的彩色铝板之间。这些创意设计方案为体育馆增添了永恒而动感的魅力。

Inner court 内院

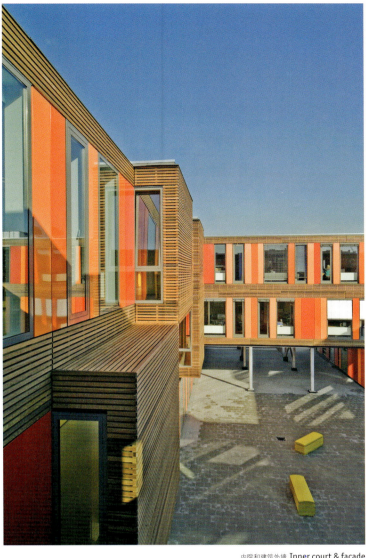

内院和建筑外墙 Inner court & façade

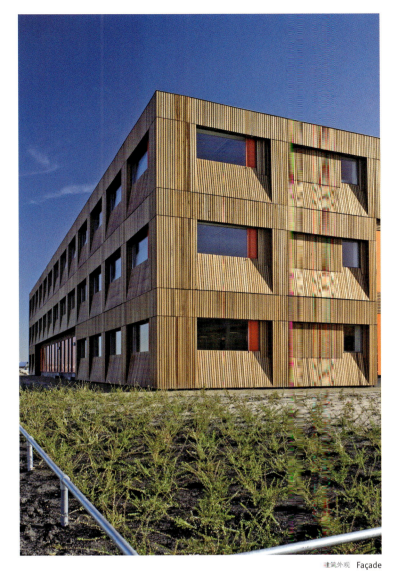

建筑外观 Façade

礼堂 Auditorium

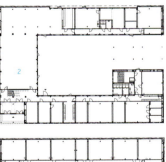

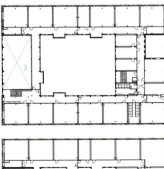

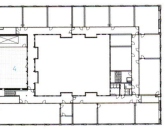

1. section A—A
2. ground floor
3. first floor
4. second floor

1. 剖面
2. 一层
3. 二层
4. 三层

Completion Date: 2008

Architect: HVDN Architecten

General view 建筑全景

Het Kasteel

Its location adjacent to the railway lines necessitates a high level of sound insulation and it is this that defines the external expression of the "Kasteel". The building is enveloped in a glazed skin that stands free from the apartment block behind. In order to give the skin a tactile quality, the panels are angled slightly to each other; this artifice lends the building the appearance of a gigantic crystal.

The "Kasteel" consists of a forty-five-metre-high tower standing on a four to five-storey base. It is surrounded by water and pedestrians and cyclists access the internal courtyard via a bridge. The car parking, storage spaces and some of the ground floor dwellings' living spaces are positioned underneath the courtyard's half-open wooden deck. The dwellings vary in size: those on the ground floor include a living space just above the water level while those above contain either a balcony or a terrace. The interaction between the apartment block's recessed elevation and the glazed panels of the building's skin ensures the entrance building acts as an icon for the Science Park.

城堡大厦

建筑建在铁路沿线，需要高强度的隔音设施，因此得名"城堡"。建筑被包裹在一个上釉玻璃外壳中，玻璃外壳与后方的公寓楼是隔开的。为了使外壳更具质感，上面的玻璃板呈角度倾斜排列，看起来像一块巨大的水晶。

城堡大厦由一座45米高的塔楼和一个4到5层高的底座组成，四周环水，行人和非机动车通过桥梁进入内院。停车场、仓库和一楼的起居空间位于庭院的半开放木板平台下方。公寓具有不同的尺寸：一楼的公寓有一个在水上的起居空间；其他的公寓则带有阳台或是平台。公寓楼凹进的立面和建筑的玻璃面板使它成为了科学公园的标志性建筑。

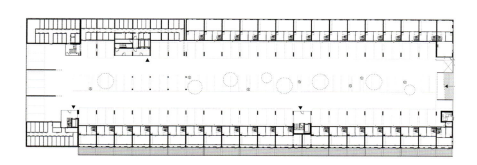

屋顶庭院 Courtyard on the roof

建筑细部 Details

室内 Interior

建筑外观 Exterior

Photo: Luuk Kramer, John Lewis Marshall

Completion Date: 2008

Architect: HVDN Architecten

Façade 建筑外观

Waldorpstraat

The building is more than 165 metres in length and its total surface area of more than 5,500 square metres is divided among the ground floor and the entresols. One of the units accommodates a wholesaler in paint accessories. Another houses an innovative, environment-friendly carwash, no less than fifty metres long, in which ten cars can be washed at the same time and waiting times are short. After the wash, customers can vacuum-clean their cars at one of the 38 indoor vacuuming points. Because the carwash is bounded by a side wall of perforated steel, there is always a light, sight and climate relationship with the outside world.

With its fixed grid pattern, the building has been fully designed according to the principles of industrial, flexible and sectional building, and it is easy to extend or to add extra floor surface on the inside. Because it is partly sunken in the railway embankment, the building is equipped with a curving roof that does not disrupt the view of the engine drivers and train passengers. At the front, the building is eight metres high, and at the rear it is only four. As the train occasionally spreads sparks and dust, the roof of the building is coated with sedum vegetation, which also offers a pleasant view to the residents of the neighbouring apartment block.

Waldorpstraat综合办公楼

建筑长165余米，总面积约5,500平方米，分为一楼和夹层楼面两个空间。建筑的一部分空间是油漆配件批发商店；另一部分是创新环保洗车场，不到50米长的空间内可以同时洗10辆车，无需长时间等候。车辆进行清洗以后，可以直接进行吸尘清洁。由于洗车场的侧墙是穿孔钢板，内部环境与户外世界联系在了一起。

建筑完全遵照了工业化、灵活化和剖面化规则，内部空间可以随意扩展或是增加楼层面积。由于建筑的一部分下沉到铁路路堤里，建筑的屋顶被设计为曲线形，不会影响火车司机和乘客所看到的风景。建筑的正面高8米，后面高4米。为了防护汽车所扬起的粉尘，建筑的屋顶上栽种了景天属植物，为附近公寓楼里的居民提供了优美的风景。

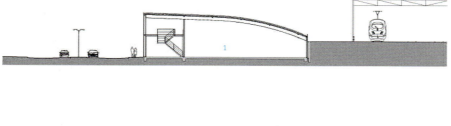

1. 建筑立面墙壁
2. 楼梯

1. elevation wall
2. stairs

主入口 Main entrance

鸟瞰图 Bird's-eye view

室内 Interior

Photo: Architectenbureau Cepezed

Completion Date: 2008

Architect: Architectenbureau Cepezed

Solar

Solar was initially intended as two separate buildings and later transformed into a single development in which the two halves of one building are connected by a full-height glazed atrium that facilitates all vertical and horizontal traffic movement.

The sturdy building was conceived as a rectangular concrete block in a strict grid pattern contrasted by a free-form shape of wire mesh. The grid pattern was then filled with window frames over the longitudinal direction of the building facing the adjacent square on one side and the street on the other. The image presented by the elevations on the cross direction with their brickwork infill reflects the area's nineteenth century roots. At the top, deflecting and refracting the sunlight, the wire mesh draped around the tough concrete edges and corners like a veil.

The interior remains loyal to the industrial character of both the area and the building in displaying stout metal constructions in the atrium and allowing the technique of the building to be in view in much of the office space. The harshness of it all is mediated by warm wood tones and smooth white surfaces.

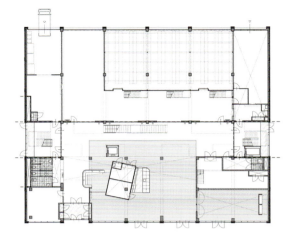

太阳楼

太阳楼原计划建成两座双子楼，但最终实行的时候改成一座楼。大楼的两个部分由一个透明的中庭相连，中庭里设有电梯和走廊。

太阳楼采用了规矩的长方形混凝土结构，而外面金属网的形状则十分随意。朝向广场和街道两侧的纵向楼面上装满了整齐的窗格。而横向立面上的砖砌结构则反映了该地区19世纪的历史特色。包裹着混凝土边角的金属网像面纱一样，折射和反射着太阳光。

室内设计保持了当地和建筑的工业特征，中庭采用了敦实的金属结构，办公区采用了与建筑造型相似的技术。温馨的木质材料和光滑的白色墙壁柔化了工业的粗糙感。

General view 建筑全景

造型随意的金属网 Freely-shaped wire mesh

金属网结构 Wire mesh construction

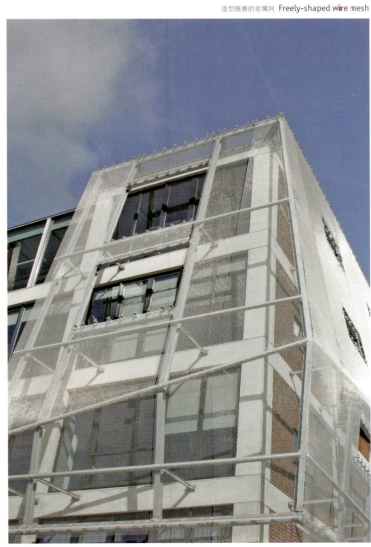

建筑细部 Detail

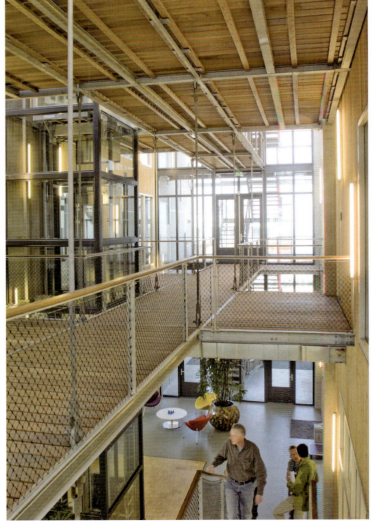

中庭和两楼的连接处 Atrium and connection between the two buildings

Corporate

Completion Date: 2009

Architect: MIII architecten

Shaken Office

Yushi Uehara search form that shines soul and energy, builds with user's dynamism, the Super Functionalism with the combination of subtle differences, and pursues the concept that innovate utilitarian aspects. Obviously, the intention to this "Super Functionalism" urged to stacked-up boxes of diverse dimensions, the minimalist floor plans with single cores, with which he created the workspaces with varied dimensions. He reached even to a building system. A light-body concrete on steel plates prefabricated timber sub construction façade are sealed by a sheet of black corrugated steel plates that shield the rain frills where façade panels are screwed on finely perforated, folded steel plates that let only the half of openings open. The monolith overshadows over these sensual openings. Lightness creates large impression: that is the rhetoric.

摇摆办公

建筑师上原侑士试图寻求一种具有灵魂和能量的建筑，他以使用者的需求为推动力，采用"超实用主义"，追求创新的实用设计。在这个项目中，尺寸不一的盒结构堆叠在一起，体现了"超实用主义"的应用。单核结构的平面布局将办公空间分割成不同的尺寸。建筑师创造了一个建筑系统。轻质混凝土墙面上嵌着钢板，预制木板建筑外墙上的黑色波纹钢板可以遮蔽风雨。遮雨板由精细的穿孔折叠钢板制成。这一结构下的窗户只能打开一半。建筑巨大的块状结构使这些窗口黯然失色。建筑的光影效果十分具有隐喻意味。

Photo: Jim Ernst

Completion Date: 2009

Architect: Yushi Uehara

Exterior view 建筑外观

Agora Theatre

埃格拉剧院

The Agora Theatre is an extremely colourful, determinedly upbeat place. The building is part of the masterplan for Lelystad, which aims to revitalise the pragmatic, sober town centre. The theatre responds to the ongoing mission of reviving and recovering the post-war Dutch new towns by focusing on the archetypal function of a theatre: that of creating a world of artifice and enchantment. Both inside and outside walls are faceted to reconstruct the kaleidoscopic experience of the world of the stage, where you can never be sure of what is real and what is not. In the Agora Theatre drama and performance are not restricted to the stage and to the evening, but are extended to the urban experience and to daytime.

Inside, the colourfulness of the outside increases in intensity. A handrail executed as a snaking pink ribbon cascades down the main staircase, winds itself all around the void at the centre of the large, open foyer space on the first floor and then extends up the wall towards the roof, optically changing colours all the while from violet, crimson and cherry to almost white.

埃格拉剧院是一个色彩鲜艳、充满活力的空间。剧院是莱蒂斯塔德总体规划的一部分，旨在复兴务实而严肃的市中心。剧院通过自身的功能来复兴这座战后的荷兰新城，创造了一个充满虚幻和迷幻色彩的世界。剧院的内外墙壁都被分割成小截面，重现了万花筒一般的舞台世界，让人迷失于真实与虚幻之间。在埃格拉剧院里，戏剧和表演不仅仅局限于舞台或是夜晚，而是延伸到了整个城市日间生活中。

室内设计的色彩强度更甚于室外。蜿蜒的粉红色丝带从主楼梯上倾泻下来，环绕二楼宽敞宏大的大厅，并且沿着墙壁一直延伸到屋顶，色彩从淡紫、深红、粉红，直到白色。

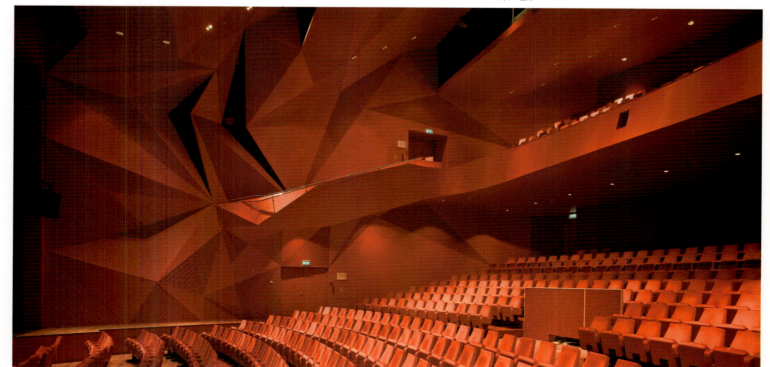

Stage main hall 舞台大厅

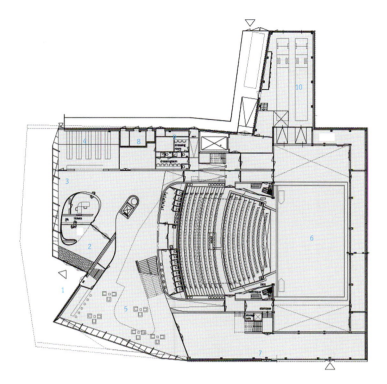

Ground floor plan

1. entrance
2. ticket desk
3. reception counter
4. cloakroom
5. foyer
6. stage main hall
7. restaurant
8. machinery
9. dressing rooms
10. loading bay

一楼平面图

1. 入口
2. 售票处
3. 前台
4. 衣帽间
5. 大厅
6. 舞台大厅
7. 餐厅
8. 机械室
9. 更衣室
10. 装卸处

大堂天窗 Foyer with skylight

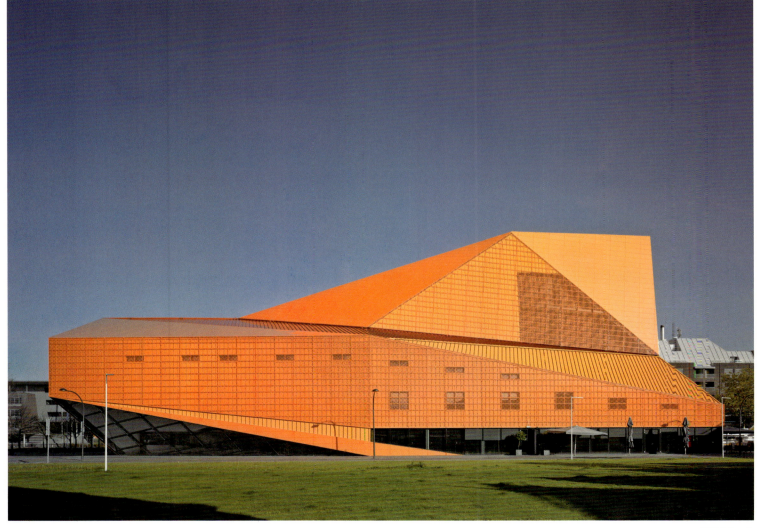

建筑外观 Façade

Photo: Christian Richters

Cultural

Completion Date: 2007

Architect: UNStudio, Amsterdam

General view 全景图

Betty Blue

As unambiguous as this shopping machine is lying here on its doorstep, waiting for visitors, as ambiguous it is in relation to its shape and colour, it is sometimes straight and other times round, from the one side purple and from the other side blue. In the shelter of this enormous lifted and stretched drop of water, an inner square with almost exotic conditions has been shaped. It is as if a whole life of its own has been able to develop itself inside this inner space, in which façade openings, bill boards, lampposts, wastebaskets, bicycle sheds and road markings have gone through a joint and balanced growth.

That exclusivity does not necessarily mean an extraordinary budget. In the task the designers set themselves by making something with a modular, and therefore efficient building system is specific and thus unique. Where modular systems usually result in all too predictable shapes, the designers managed, within the regime of recurring façade elements, to put up a system of façade openings with such variation that a seemingly much bigger variety of windows, shop windows and entrance doors can be made.

贝蒂蓝购物中心

正如购物中心不清楚什么时候会有顾客光临，它的造型和颜色也模糊不清——时而圆，时而方；时而蓝，时而紫。购物中心就像被拉伸、抬高的水滴，内部是一个造型奇特的广场。广场内部有建筑门窗、广告牌、灯柱、垃圾桶、自行车棚、路标等设施。

项目的独创性并不意味着额外的预算。设计师采用了模块结构，构造了一个高效而独特的建筑系统。虽然模块系统容易陷入既定的模式，设计师在相同的外墙元素中加入了不同的窗口、橱窗、大门等元素，丰富了建筑的内容。

1. supermarket	1. 超市
2. home electronics	2. 家电
3. dieren	3. 动物类
4. 5. home electronics	4.5. 家电
6. speelgoed	6. 玩具
7. PDV	7. 产品数据管理
8. outdoor	8. 户外
9. woonwarenhuis	9. 商店
10. 11. 13. PDV	10.11.13. 产品数据管理
12. Sport	12. 运动区
14. fitness	14. 健身中心
15. disco	15. 迪斯科舞厅
16. 17. fast food	16. 17. 快餐

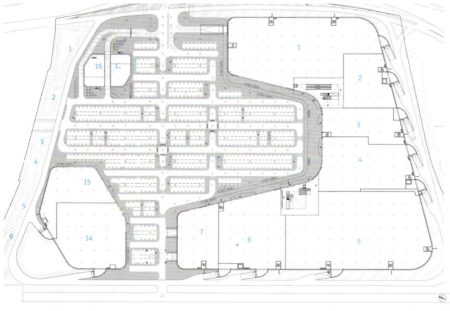

Entrance 入口

鸟瞰图 Birc's-eye view

夜景 Night view

室内 Interior

Photo: Van Pol beheer, Arjen Schmitz, Hans Pattist, Hennie Retera

Commercial

Completion Date: 2008

Architect: NIO architecten

Exterior glass façade 建筑的玻璃外墙

The New Martini Hospital

Flexibility is currently one of the most important factors in an ever–mutating healthcare environment where rapid developments in medical technology make it difficult to predict the future. That is why it is important to design a hospital now that will last for the next forty years and can easily adapt itself to an unknown future. It is for this reason that the concepts of the IFD (Industrial Flexible Demountable) programme spearheaded the design of The New Martini Hospital in Groningen, the Netherlands.

It is always difficult to predict what functionality must be accommodated for in a hospital building with a set lifespan of forty years. For this reason the designers chose a uniform building block which, in a general sense, complies with the demands of safety, natural daylight, structure, services and floor planning. The design brief was tested on a number of important frequently occurring departments such as general nursing and outpatient clinic. A uniform building block therefore acquired a useful aspect in that it could be functionally totally interchangeable in the design phase as well as later on once the building is being used. A nursing department can be converted to an outpatient clinic or offices.

新马尔蒂尼医院

在医疗技术日新月异，医疗环境迅速变化的今天，空间的灵活适应性是医院设计的主要考虑因素。因此，建造一个在未来40年间能够灵活调整自身环境的医院极为重要。在这样的背景下，位于荷兰格罗宁根的新马尔蒂尼医院的设计率先使用了IFD（工业化灵活可拆卸）规划。

很难预测未来的40年医院将配备何种功能、何种类型的设备，所以设计师为建筑选择了标准的统一造型，在设计中综合了安全、自然采光、结构、服务和平面布局等元素。设计纲要经过了一系列常规医疗部门的测试，如普通护理和门诊。标准医院大楼的特点是它的功能区可以在设计阶段、乃至使用阶段进行任意改变。例如：护理部完全可以被改造成门诊或是办公区。

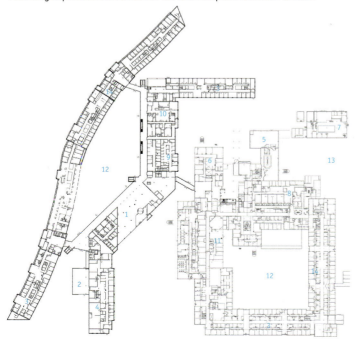

1. main entrance	1. 主入口
2. ambulance entrance	2. 急诊入口
3. outpatient clinic	3. 门诊
4. emergency	4. 紧急出口
5. personnel's restaurant	5. 员工餐厅
6. conference Centre	6. 会议中心
7. technical services	7. 技术服务设施
8. rentable floor area	8. 可出租楼层
9. phlebotomy	9. 静脉手术室
10. endoscopy	10. 内窥镜手术室
11. teaching and training	11. 培训室
12. inner courtyard	12. 内院
13. delivery	13. 收发室
14. lift core	14. 电梯

建筑的玻璃外墙 Exterior details

走廊 Passage

Photo: Mr. Rob Hoekstr and Mr. Derk Jan de Vries

Hospital

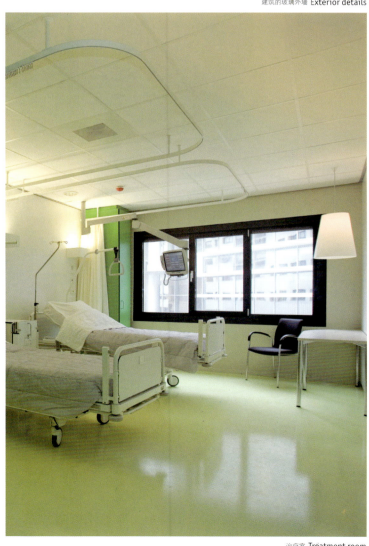

治疗室 Treatment room

彩色的走廊 Colourful corridor

Architect: Mr. Burger / SEED architects (before Burger Grunstra architecten adviseurs)

Completion Date: 2007

Tea House on Bunker

The project involves the reprogramming of a historical and derelict building through renovation and addition. The original bunker is part of an intricate water management system that enabled the inundation of land situated in a classic, Dutch polder landscape.

Stables and polo fields now surround the building and the new addition is intended as a large space with facilities to support a meeting space or business retreat. The existing 1936 bunker remains intact except for a portion of the concrete roof where the new structure connects whilst the new addition is like an umbrella, an addition that could be removed and does not damage or permanently influence the historic structure. The metallic addition appears to have grown out of the still visible concrete façades of the bunker, cantilevering out towards the sports fields with its large single window. In fact the space is designed with steel structures within its two main walls which act as one-storey-high beams. These beams are balanced off centre on two columns that land directly in front of the existing bunker. Stability is achieved by using the massive concrete shell of the bunker as a counterweight.

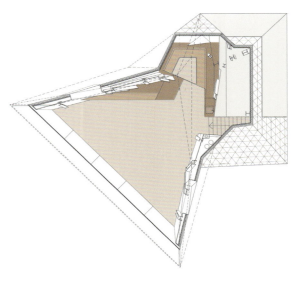

First floor plan 一层平面图

地堡茶室

该项目是对一处被遗弃的历史建筑进行翻新和扩建。原有的地堡是错杂的水处理系统的一部分。这个系统设在一片经典的荷兰圩田景观区域。

建筑四周环绕着马厩和马球场，新的扩建工程则被作为设施齐全的商业会所。建于1936年的地堡完好如初，扩建部分通过混凝土屋顶和地堡连接起来，宛如一把雨伞，丝毫没有影响历史建筑结构。扩建的金属结构看起来像是从地堡的混凝土墙面上生长出来一样，大型单玻璃窗高悬在马球场之上。整个空间采取了钢结构，其中的两面墙壁就是单层楼高的横梁。正对着地堡的两个圆柱平衡着这些横梁，地堡巨大的混凝土外壳作为一个平衡力，保证了建筑的稳定性。

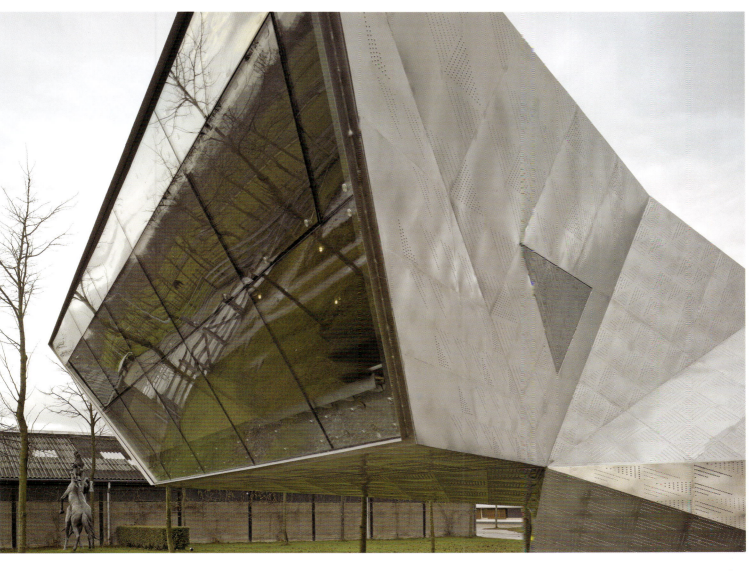

Night view 夜景

School 'tij49

'tij49号街区临时学校

Even though the time frame from conception to completion was less than six months, the client stipulated that this should not manifest itself in the building's appearance.

The proposal involves a three-storey building with a wide, double-loaded central corridor. By compartmentalising the building vertically to comply with the fire regulations, the stairwells and voids form part of this central space. The three entrances are located in the building's long elevation facing the schoolyard.

To harmonise the stacked prefabricated elements into a convincing building, the horizontal bands in the façade are strongly articulated. The cantilevered strips also function as effective sun screens and shelter for the entrances. They are finished with a sprayed rubber layer, white on the outside and with a different brightly coloured soffit per floor. The colouring corresponds with the school's internal colour scheme. By illuminating the bands at night, the building acts as a beacon in the neighbourhood.

项目从设计理念到竣工仅用了不到六个月的时间，而建筑的外观却也十分优秀，丝毫看不出来是仓促建成的。

三层的教学楼有一个宽大的双倍荷重的中心走廊。为了分割垂直空间，遵循防火规则，楼梯和缓步台都设在中心空间。教学楼的三个入口并列在朝向操场的一侧。

为了让叠加的预制结构形成一座令人安心的建筑，建筑外墙上的水平带状结构异常牢固。伸展出来的悬臂结构也可以有效地遮阳，并为入口遮蔽风雨。悬臂外面装饰着白色喷涂橡胶层。每层楼都有不同颜色的拱腹。外墙的色彩和教学楼内部的色彩相一致。夜晚点亮带状结构，整个建筑就成了街区的灯塔。

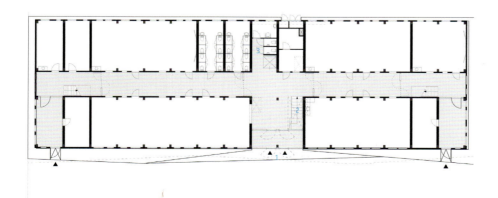

1. entrance
2. stairs
3. washing room

1. 入口
2. 楼梯
3. 卫生间

外墙上的图案 Wall with patterr

建筑细部 Details

建筑三面 Front view

Photo: Luuk Kramer, Jan Derwig

Completion Date: 2007

Architect: HVDN Architecten

View of the warehouse cargo unloading area　仓库和装卸区

Warehouse – Refrigerators & Office Building

冷藏仓库和办公楼

The issue in the architectural composition was to provide, in morphological terms, for the co-existence and unification of two main functions (office building and warehouse – preservation refrigerators) so that the structure was a uniform one. This objective was achieved by including the following morphological elements: The office building's sides were covered in Etalbond panels and passive metallic systems (blinds) because of the unsuitable orientation. The warehouse – refrigerators' sides were covered in aluminium panels and, in morphological terms, an attempt was made to include the functional areas of the warehouse, loading – unloading bay with their special requirements into the overall scheme. Metal frames in the warehouse were repeated, spaced at thirty-metre intervals, covering blank spaces and assisting in integration with the office building. The metallic decorative elements (in the form of netting) and partial covering with inclined panels demarcate the entrance area in the form of a niche.

The partition walls between offices are panels covered in plasterboard. The floors are covered in granite tiles. The suspended ceilings are made of mineral fibre tiles. All facilities including lighting and air-conditioning for the building are hidden behind the suspended ceiling.

项目要求将两个建筑元素——冷藏仓库和办公楼结合在一起，形成一个统一的结构。这一目标通过以下造型元素所实现：由于建筑朝向的不适合，办公楼的外墙上包裹着Etalbond面板和金属百叶窗系统；冷藏仓库包裹着铝板，并且还有一个功能齐全的装卸处。仓库的金属外框每隔30米就重复一次，遮蔽着空白区域，也间接与办公楼相连。网状金属装饰元素和局部斜屋顶标志着入口区域。

各个办公室之间的隔断采用了石膏板，地面上则铺设着花岗岩地砖。吊顶由矿物纤维瓷砖组成，照明、空调等辅助设备全部被隐藏在吊顶之上。

1. warehouse cargo unloading area
2. office building facility entrance
3. main building
4. warehouse loading area

1. 仓库卸货区
2. 办公楼入口
3. 主楼
4. 仓库装货区

建筑全景：仓库、冷库和办公楼　General view of the entire complex: warehouse, cold storage, and office facilities

仓库装卸区　Warehouse unloading area

办公楼入口　View of the office building facility entrance

办公楼入口　View of the office building facility entrance

办公楼一楼接待处　Reception area on the ground floor of the office building facility

Architect: Yanniotis Yannos, Yannioti Vasiliki Nilent Paul;
Civil Engineer: Karoukis Panagiotis–Polixronopoulos Kostas
Mechanical Engineer: Klissiounis Dimitris, Yanniotis Constantinos; Owner: Mevgal S.A. Construction: Terna S.A.

Completion Date: 2005

Photo: Psaros Vlassis

Industrial

A view of the stairwell leading to the swimming pool from the main yard 主庭院的台阶直通游泳池

Two Detached Holiday Residences in Paros Island

The compositional objective
Two residences to be integrated, in terms of volume and shape, into a single building but retaining their difference. In morphological terms, the composition had to refer to traditional Cycladic architecture in other words small sized one or two-floor volumes blending into the natural terrain, arrayed around courtyards, and a strong presence for the colour white.

Choices made
Uniformity in terms of volume was achieved by an intermediate semi-outdoor area which led from the parking area to a well-proportioned inner courtyard that allowed access to the residences and the pool area. The detached houses were laid out horizontally with one section extending to a second floor, and the roofs were flat in the style of local Cycladic architecture. Each was different because of the layout of the guest houses, one being located around the inner courtyard and the other just beside the pool and its grounds. All walls and ceilings were white. The floors were the same colour laid with squared-off artificial stone tiles in a light grey colour. In addition to natural light, lighting is provided with sconces or standard lamps.

帕罗斯岛度假屋

设计目标
将两幢独立的住宅连通，在造型及体积上营造统一性，但同时使其不失各自特色；从形态学角度来说，住宅模仿基克拉迪风格建筑，即小型的一层或两层结构——融入到周围的自然环境中，围绕庭院展开，装饰基调以白色为主。

最终方案
两幢住宅平行设置，其中一幢为两层，平顶结构充分展现了基克拉迪风格。住宅各有特色，一幢毗邻室内庭院，另一幢则与游泳池相接。设计师通过打造半户外结构将它们连结起来，这一结构从停车场延伸到室内庭院，从这里便可达到住宅室内及游泳池。墙壁、天花全部采用白色粉饰，地面铺设着浅灰色方块图案的人造石材瓷砖，简约而不失典雅。照明元素除自然光线外，以烛台及落地灯为主。

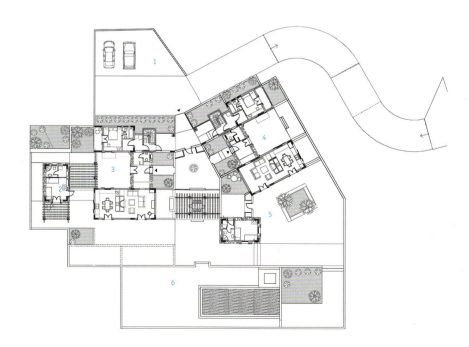

1. 车库
2. 门房
3. 房间1（餐厅和卧室）
4. 房间2（餐厅和卧室）
5. 花园
6. 游泳池

1. garage
2. gate house
3. room1 (dinner room and bedroom)
4. room2 (dinner room and bedroom)
5. garden
6. swimming pool

从公路上看住宅 View of the residence from the nearby road

从半露天区域看客月住宅入口和庭院 View of the guest house entrance and courtyard of the residence from the semihypaethral area

客厅 Living room

Residential

Photo: Psaros Vlassis

Completion Date: 2009

Architect: Yanniotis Yannos, Yanniotis & Associates; Civil Engineer: Retzepis Ioannis;
Mechanical Engineer: Klissouris Dimitris; Yanniotis Constantinos; Lighting Design: Simon Simos, Lighting Design SARL
Construction Firm: Structura Aete; Owner: R.E.

Bird's-eye view 鸟瞰图

Office Building in the Centre of Athens

The building is situated in the old centre of Athens on a main street. It has a total surface of 1,634 square metres on six floors with two basements for electromechanical equipment and parking area. The building makes use of all the depth of the site and the interference of two patios, and two gardens bring natural light and air to all the offices. All the façades to the patios are enclosed by glass partitions which give a visual connection between all the offices. The curtain wall of the main façade is divided in horizontal strips which recess into the building and project over the street providing a fluid border between the urban and private space. At the same time they offer a play of shadows during the day and a play of light at night.

On the ground floor the urban space penetrates into the private. Only a small part of the ground is covered for the building entrance, and the rest, the arcade, the two patios and a terrace are open to the city. Through the open patios on the ground floor, the urban space penetrates also into the building. The patios are planted with Mediterranean plants offering a garden to the city and a different value to the urban space.

雅典中心区办公楼

建筑位于雅典旧城区中心的主要街道上，总面积为1,634平方米，共有六层楼，还包括两层设置着停车场和机电设备地下室。建筑充分运用了场地的深度。两个天井和两座花园的结合为办公楼提供了自然采光和通风。天井的外墙全部采用了玻璃隔板，在视觉上与各个办公室相连。建筑正立面的幕墙被划分成横条状，一部分凸起，一部分凹进，划出一道流畅的曲线，分开了公共和私密空间。此外，这些条纹营造了特殊的光影效果。

一层的城市空间和私密空间相互融合在一起，只有少部分的空间属于建筑的入口，其余的空间——拱廊、天井和平台都像城市开放。城市空间透过开放的天井渗透到了建筑之中。天井中种植着地中海植物，形成了一个城市花园，别具风情。

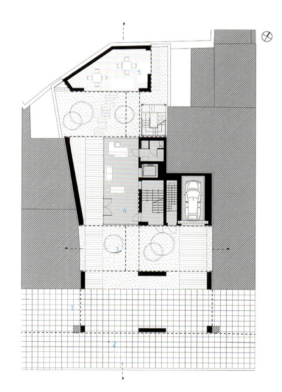

1. arcade	1. 拱廊
2. pavement	2. 通道
3. garden	3. 花园
4. entrance	4. 入口
5. outdoor sitting area	5. 户外休息区

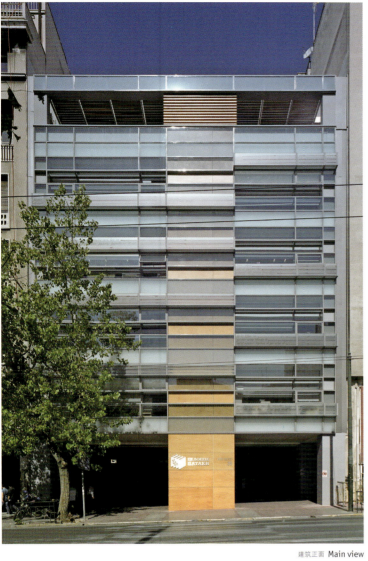

建筑正面 **Main view**

建筑细部 **Details of façade**

入口通道 Entrance road

入口 Entrance

Photo: Charalambos Louizidis

Corporate

Completion Date: 2008

Architect: Alexandra Kalliri and Associates Architects

Dot Envelope

The existing site is listed as an industrial historical area with buildings of an old butchery complex, which included the water-tower and old butcher hall. Demand of National heritage was to rebuild the tower as it was originally and to integrate the main façade portal of old hall in front of the planned new shopping mall.

The client's permission and expected plan was prefabricated concrete hall of 46x42x7 metres. After detailed calculation the budget for covering all elevations was at 60,000 EUR. The pattern was based on different stepped elevations in order to soften basic cube shell.

The surface which could fit into the budget for the façade used of basic metal sheets which painted in bronze structured colour. After cost evaluation only 20% of the concrete shell could be covered with the metal sheets. So the sheets were perforated with holes in different sizes. Furthermore, the cut metal circles from the sheets were used and arranged on the rest of the façade surface. The new shopping mall has parking facilities and customer approach on three sides of the building. Therefore it was important to cover three sides with final decorative finishing with the budget of one side only.

圆点商场

超市场址原为历史工业区，曾经用于水塔和屠宰场。为满足保护国家古老遗址的需求，需还原建筑原有的面貌同时赋予其全新的外观。

客户的原有计划是打造一个体积为46x42x7（米）的预制水泥建筑。覆层外观预算60000欧元。设计师决定采用金属片元素装饰，并将其漆成赤褐色，然而使用既定预算仅有20%的表皮可以实现覆层。为此，设计师将金属片打出小洞，将切割之后的剩余部分覆在其他部分，既节约了预算，又满足美观需求。

此外，超市设有停车场，顾客可以从三个不同的侧面进入。通过巧妙的设计，设计师运用装饰一面外观的预算实现了三面装饰。

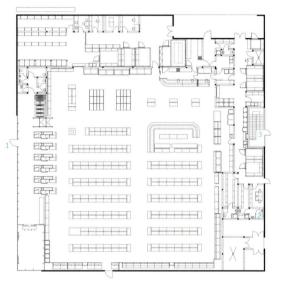
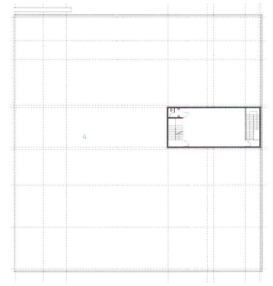

1. 入口
2. 洗手间
3. 楼梯
4. 屋顶

1. entrance
2. washing room
3. stairs
4. roof

Photo: Tomaž Gregoric

Commercial

Completion Date: 2008

Architect: OFIS Arhitekti

650 apartments, Ljubljana

To make planning and construction simple and to allow for the use of such prefabricated elements as bathrooms, windows and façade panels, the buildings were designed in module form. Each building is divided into four identical modules, each with its own vertical communications core. There are forty-two apartments in each module, varying from small thirty-square-metre studios to 1.5–bedroom sixty-square-metre apartments on four identical floors, and larger duplex apartments from 85 to 105 square metres on the top two floors. The module is repeated four times with slight variations at the far ends of the building.

The façade is designed in two layers, the inner façade and outdoor space being formed by items such as glazed loggias, balconies, terraces and verandas. The second skin is constructed with pre–formed wooden panels, glass and metal rails. The structure of the apartments is such that each apartment gets at least one balcony and loggia that connects outdoor and indoor spaces. Like the modules, the façade layer is also repeated four times, but given the different geometry of the elements and the repetition passes virtually unnoticed.

There are two parking levels beneath the site. The landscape provides a contrast to the geometrical façade through the use of gently curving rails and other features. It breaks up the sightlines through the complex and creates a difference between public and private spaces.

650公寓

为尽量简化规划及建造过程，设计中运用了大量的预制结构，如浴室、窗户及表皮结构。每栋公寓均为四个相同的模块结构打造，同时每栋公寓内包括42间住宅，下面四层面积为30平方米至60平方米不等，上面两层为85平方米至105平方米不等。

公寓表皮为双层结构，阳台及露台垂悬出来。预制木板、玻璃及金属栏杆构件作为主要的材质，被大量运用。每一间住宅都设有至少一个阳台和凉廊，借以将室内外空间连通。停车场位于地下一二层，周围的景观与几何图形拼接样式的外观形成鲜明的对比。

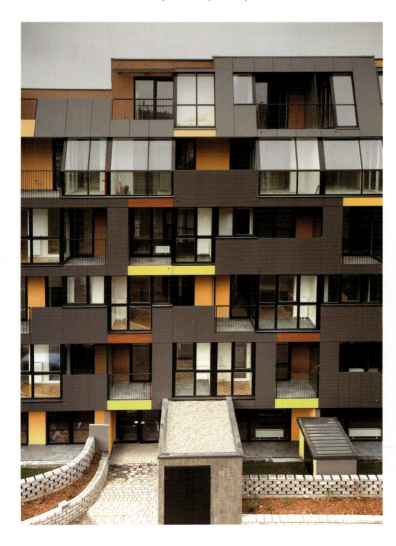

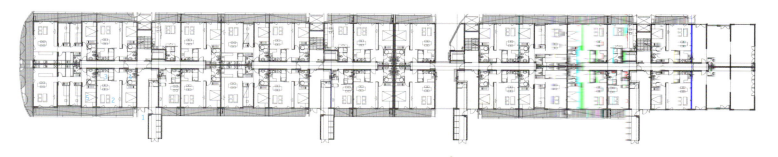

1. entrance
2. living room
3. kitchen
4. washing room
5. bedroom

1. 入口
2. 起居室
3. 厨房
4. 洗手间
5. 卧室

Photo: Tomaž Gregoric

Completion Date: 2006

Architect: OFIS Arhitekti

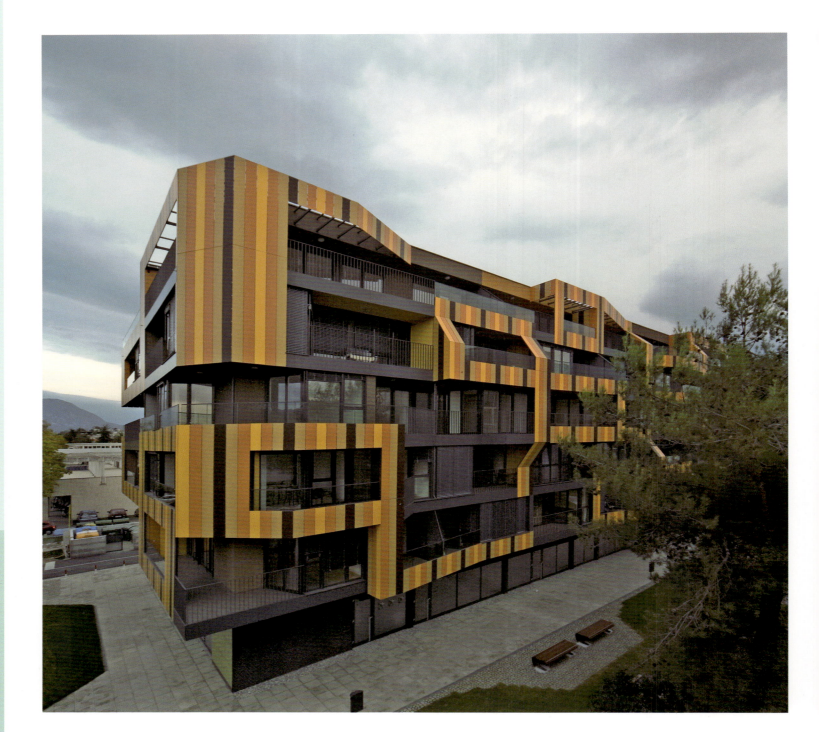

Lace Apartments

The Lace Apartments are located in Nova Gorica in the west of Slovenia, on the Slovene–Italian border. Situated ninety-two metres above sea level, the town is said to be the hottest town in summer, while in winter it suffers from very strong winds. The climate, vegetation and way of life of Nova Gorica are very Mediterranean, with a strong emphasis on outdoor living, making external shady areas an important feature of the town's architecture.

The client of this project requested rich external spaces with different characters. The client was also very specific about the apartments' size and typology, which needed to be simple and repeated. Because of the fixed urban plot, the building had to be an orthogonal block of forty-eight by sixteen metres and five levels. The architects studied the external spaces of the area's existing house and proposed balconies and terraces, which can both be opened and covered with a roof or pergola, and loggias that are closed from the side and fully or partly glazed, with different type of fences – transparent with glass or metal, full or of varying heights. This second skin of terraces gives each apartment a different character and allows the buyer to choose a space that responds to his/her lifestyle.

Though the façade's colour pattern is inspired by the area's typical colour elements, such as the valley's soil and the wine and brick roof tops, the locals soon nicknamed the building "pyjamas", as it reminded them of a pattern on a man's nightwear.

蕾丝公寓楼

蕾丝公寓楼位于斯洛文尼亚西部的新戈里察镇中部，靠近斯洛文尼亚与意大利的边界线。新戈里察海拔高92米，典型的地中海气候，夏天很热，冬天风又很强。独特的生活方式决定了户外生活区的重要性，带有凉棚的天台及阳台已成为该地建筑的一大特色。

除了要有大量风格各异的户外生活区，客户对公寓的大小及类型也十分挑剔，要求其风格简朴、易于模仿。由于地形的影响，公寓成直角结构，面积为48x16平方米，共为五层。设计师做了充分研究之后，增设了带有屋顶或凉棚的天台及阳台、侧面开门的凉廊（全部或部分使用玻璃材料，并带有高低不一篱笆墙）。这一设计赋予了每幢公寓独特风格的同时，也给客户带来了更多的选择——根据自己的生活购买喜欢的房子。

公寓的外观使用了当地典型的元素装饰，如沙土、砖瓦等。由于其独特的造型让人不禁想到睡衣上的图案，因此被当地人形象地称作"睡袍建筑"。

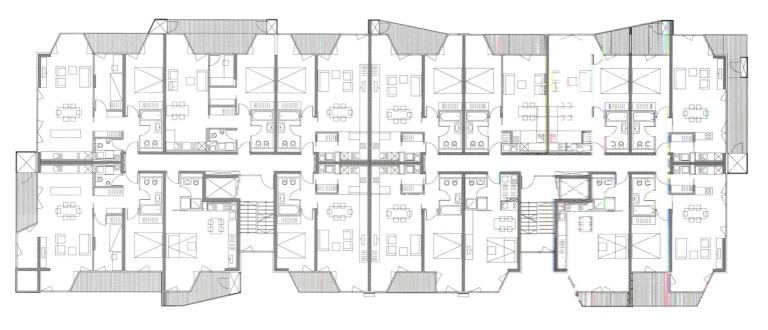

Photo: Tomaž Gregoric

Completion Date: 2008

Architect: OFIS Arhitekti

Tetris Apartments

The building stands on the edge of the "650 Apartments" development which was finished a year ago. By urban regulation the block is sixty-five metres long, in width fifteen metres and three floors high. Since the orientation of the building is towards the busy highway, the apartment opening together with balconies is shifted as thirty degrees window-wings towards the quieter and south orientated side. Long after the elevations were planned, many people associated them to Tetris game, and so the building got its name.

"俄罗斯方块" 公寓

这一项目与一年前刚刚竣工的650公寓相邻，由于建筑朝向繁忙的车道，因此阳台便稍微向南侧倾斜，呈现30度角。建筑立面规划之后，让人不禁联想到俄罗斯方块游戏，公寓也因此得名。

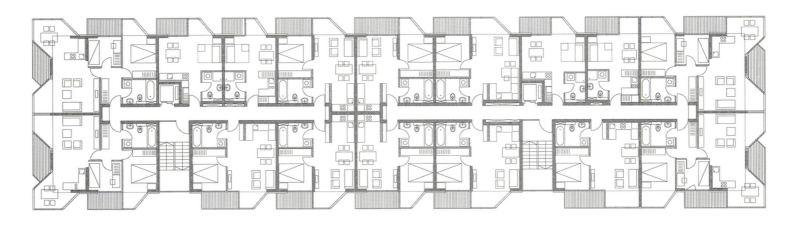

Photo: Tomaž Gregoric

Residential

Completion Date: 2007

Architect: OFIS Arhitekti

Hayrack Apartments

观景公寓

The site is the edge of Alpine town Cerklje (near the Ljubljana Airport) with beautiful views to surrounding fields and mountains. On the site there is a beautifully protected 300 year old lime-tree. The plan of the building therefore is L-shaped and embraces a green area around the tree. And mountain views are opening from this courtyard, therefore most of apartments have beautiful views.

The landscape and villages in the area remained unspoiled with many examples of traditional architecture such as old farms, barns and hayracks. The concept of the façade is taken from the hayrack system, wooden beams follow traditional details and patterns. Traditionally farmers store grass and corn on beams, and on the housing one can store flowers or other balcony decoration.

The balcony layer runs all around the block. The wooden ornamental construction elements in front of the balconies and loggias are designed in the same sense as traditional hayracks, wooden objects in function of storing and drying the grass. They provide first entrance temperature zone to the main living and sleeping areas and also create shading for the balconies. Additional aluminum shading panels are placed on the outer sides of the winter loggias and balconies. The service and communication spaces are reduced to minimum thus the daylight is provided on the shafts. The monthly basic energetic and service costs are very low, so they are also economic for the habitants since the apartments are social type.

公寓位于阿尔卑斯山小镇边陲（临近卢布尔雅那机场），四周环绕着一望无际的田野和连绵起伏的山峰，300年酸橙树在这里郁郁葱葱的生长着。公寓呈现L形，将绿树环绕其间，庭院内更是可以饱览山间的别样景致。

这一地区的景观和村庄一直保留着原始的模样，农场、马厩、干草架这些古老的建筑样式一应俱全。公寓外观的设计正是从干草架结构中获得灵感——木质梁柱呈现出古老的细节和样式。在古代，农民们经常在梁柱上存放干草和谷物，现今住户们可以在上面摆放鲜花或饰品。

阳台贯穿整幢公寓，凉廊以及阳台前面的木质装饰结构则起到了干草架的作用。凉廊外侧同时采用铝板结构覆盖，起到遮蔽作用。

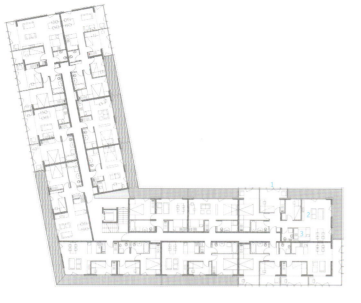

1. 入口
2. 起居室
3. 厨房

1. entrance
2. living room
3. kitchen

Residential

Photo: Tomaž Gregoric

Completion Date: 2007

Architect: OFIS Arhitekti

Social Housing on the Coast

This project located on the Adriatic coast in the south-western tip of Slovenia is the winner of a competition organised by the Slovenia Housing Fund, a government-run programme which provides low-cost apartments for young families. The winning points of this project were economic, rational and functional issues, but more importantly, the ratio between gross versus saleable surface area and the flexibility of floor plans.

The residential blocks are built on a hill with a view of Izola Bay on the one side and the surrounding hills on the other. Each block sits on a sixty by twenty-eight metres plot. The brief required thirty apartments of different sizes and structures, ranging from studio flats to three-bedroom apartments. There are no structural elements inside the small apartments in order to provide plenty of flexibility and the possibility of reorganising the space.

Considering the Mediterranean climate, each apartment has a veranda, partly connected to the interior, which provides an outdoor space for the tenants as well as shading and natural ventilation inside thanks to perforated side panels, which allow the summer breeze to ventilate the space. Semi-transparent textile shades block direct sunlight and help accumulate an "air buffer" zone. In the summer the hot air is naturally ventilated through the ten-centimetre holes in the side panels, while in the winter the warm air provides additional heating for the apartments.

沿海公寓

住宅位于斯洛文尼亚南端的亚得里亚海岸，由政府出资兴建，旨在为年轻夫妇提供廉价居住空间。该项目的竞标由斯洛文尼亚住房基金会承办，最终胜出的方案因其经济、合理以及实用的特性而获批。还有最重要的一点，总面积同销售面积之间的合理比例以及空间的灵活性。

住宅区建于山上，一侧被群山环绕，另一侧遥望伊佐拉海湾。每栋住宅占地60x28平方米，包括30间大小、形状各异的公寓，从一居室到三居室不等。此外，面积较小的房间内无任何装饰，便于灵活运用。

考虑到当地地中海气候的影响，设计师在每间公寓内打造了一个阳台。通风板隔断的运用既确保室内空气流通，又可遮阳。半透明的板材（上面带有织品图案）围和了一个如同"空气缓冲器"的区域。夏季的热空气穿过10厘米厚的通风板后冷却下来，冬季的暖湿空气透进来提高室内的温度。

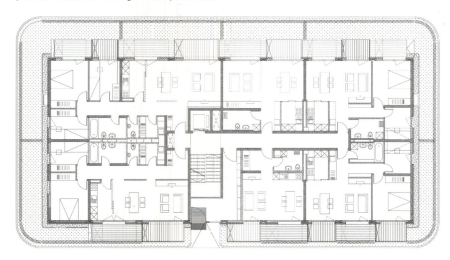

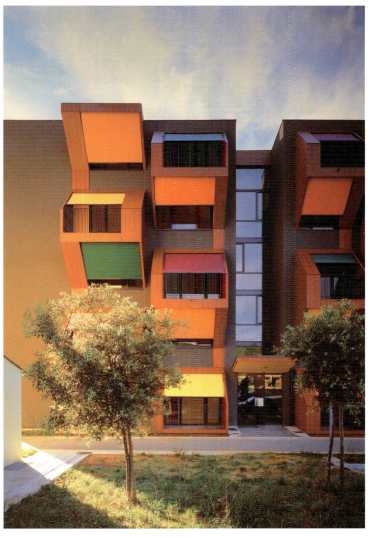

Photo: Tomaž Gregoric

Completion Date: 2006

Architect: OFIS Arhitekti

Shopping Roof Apartments

The initial task from the client was to build a new shopping mall on the plot of the existing one. Furthermore, the new project proposed use of the shopping roof for additional volume as new apartments.

The stepped volume of the building follows the silhouette of surrounding landscape. On top of the shopping mall apartments are set in the form of stepped L-volume. From the west where strong wind and snow arrives, the façade is opened only towards enclosed balconies and its material is grey slate-it is designed as a vertical roof. L-shape volume encloses inner communal garden that is the roof of the shopping mall. The front and courtyard façade is warm and open, made of wooden verticals with different rhythm.

The organisation of the housing and the envelope of the apartments open towards mountain views and the sun. Therefore the front, wooden façade is mostly transparent with panoramic windows. From side windows views also open to the mountains. Local larch is used and slates in diagonal pattern are traditional materials used for roof and façade. Play of transparency formed by wooden verticals that form balcony fences, façade panels or mask characterizes the north and south part of the building. On the east and west, pitched rhomboid-textured roof interpolates into vertical surfaces that protect apartments from snow and wind. Shopping mall façade is combination of steel and glass panels.

"购物中心屋顶上"的公寓

设计的初衷是在原有的场址上打造一个全新的购物中心，随后决定在屋顶上加盖公寓结构。整幢建筑在造型上与周围的景观相协调，屋顶上的公寓设计成L形。西侧立面设有封闭阳台，防止风雪来袭。小花园环绕其中，正面及花园一面采用木质结构排列而成，营造出温暖的气息。此外，公寓整体朝阳，侧面封闭。当地落叶松被大量运用到屋顶及表面结构上，南北两侧整齐排列的板条韵律十足，之间的缝隙更增强了通透感。购物中心的外观主要采用钢材和玻璃板打造，与公寓外观形成鲜明对比。

Architect: OFIS Arhitekti

Completion Date: 2007

Photo: Tomaž Gregoric

Commercial

Office, Store & Shop Concrete Container

The building dimensions are 35 x 22.5metres and 11.5metres in height. Furthermore, contract included executive Construction Company for entire industrial zone with their system of prefabricated concrete system with ready-made openings on each elevation.

The project task was to merge a programme inside the given volume and redefine the existing elevations. The existing sections had to remain the same. A client's company produce and merchandise safety equipment and devices which had to be stored in the 2/3 of the volume.

The elevation cuts break the functional façade grid and reinstate flowing concrete elements in between translucent screens. Offices are made up with transparent double-glazed façade, storage spaces with semi-translucent polycarbonate elements, and two openings on the back are used as loading dock doors.

The roof is ready-made functional pitched system that is incorporated into façade boards in a way that the exterior seems a rectangular block.

Materials of exterior are prefabricated concrete, glass, metal and polycarbonate plates. Interior is functional, flexible and simple. Storage has industrial durable reinforced floor, and wall finishing is concrete. Offices and shop are combination of concrete and wood.

The result is façade playfulness that shines through and provides a navigation system for the zoning of the building. Intriguing chequerboard goes way beyond the usual industrial park (non)aesthetic.

混凝土盒子（办公、仓库、购物）

整幢建筑规格为35X22.5X11.5（米），根据合同规定建筑公司负责整幢建筑的预制水泥结构（里面带有金属板结构）供应。

这一项目的主要目的即为在既定的结构内布置空间，同时重新设计原有立面。其中，建筑内一个客户公司的产品及生产设备需占据整个结构的三分之二空间。

立面上的金属板打破了外观的统一，同时更进一步突出了之间的水泥结构。办公区外观采用双层玻璃材质打造，仓库部分采用半透明聚碳酸酯材料建成，背面的开口用于装卸物品。屋顶由带有坡度的预制结构打造，四周镶嵌在外立面上，使其成为完整的结构。

在材质运用上突出内外对比，外观采用水泥、玻璃、金属及聚碳酸酯打造，内部则凸显功能性、灵活性及简约性。

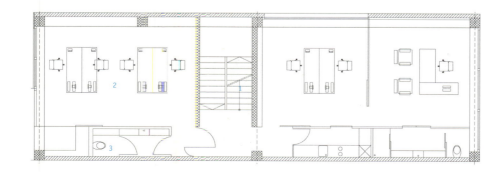

一层
1.楼梯
2.办公室
3.洗手间
level-1
1. stairs
2. office
3. washing room

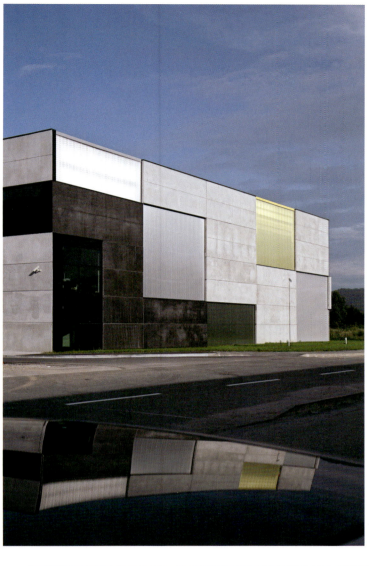

Photo: Tomaz Gregoric

Completion Date: 2009

Architect: OFIS Arhitekti

477

Side view 建筑侧面

Hotel Sotelia

Wellness Hotel Sotelia fills the gap between two existing hotels, neither of them hiding their different architectural origins. The new hotel is not trying to summarise samples from nearby structures but rather clearly distances itself from the built environment and connects, instead, with its natural surroundings.

In design process primary concern was to avoid immense building mass, like the one suggested in the client's brief, which would have blocked the last remaining view of the forest. The volume is broken up into small units arranged in landscape-hugging tiers. As a result, the four-storey 150-room building appears much lower and smaller than this description would suggest.

The specific shape of the hotel was dictated by the folds in the landscape. The unique structure offers passers-by some strong spatial experiences: from the front, the building is perceived as a two-dimensional set composed of parallel planes placed one behind the other; a walk around the hotel reveals entirely different views of the timber façade, from a plane vertical wooden slats to a rhythmic arrangement of balconies and wooden terraces.

索特利亚酒店

索特利亚酒店位于两家极具特色的酒店中间。新建的酒店并没有模仿其周边的建筑，而是将自身和周边的自然环境紧密结合在一起。

在设计过程，正如委托人所要求的那样，设计师极力避免以一个体量巨大的建筑遮挡住森林的景观。酒店被分割成若干个小块，沿着地势一次排列，这让这座拥有四层楼、150个房间的酒店看起来更加低矮。

酒店特殊的造型来源于自然地势，独特的结构给人以强烈的空间体验感。从正面看，建筑是二维平行面板的叠加；环绕酒店一周，就可以看到木板外墙的全景：从普通区域的垂直木板条到阳台和平台上独具韵律的不均匀板条。

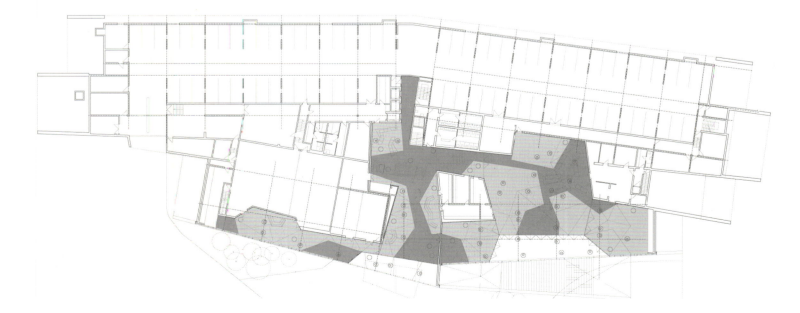

中庭 Atrium

大堂酒吧 Lobby bar

入口 Entrance

Photo: Miran Kambic

Hotel

Completion Date: 2006

Architect: Enota

Farewell Chapel

A farewell chapel is located in a village close to Ljubljana. The site plot is next to the existing graveyard. The chapel is cut into the rising landscape. The shape is following the lines of the landscape trajectories around the graveyard. Three curved walls are embracing and dividing the programmes. External curve is dividing the surrounding hill from chapel plateau, and also reinstates the main supporting wall. Services such as storages, wardrobe restrooms and kitchenette are on the inner side along the wall. Internal curve is embracing the main farewell space. It is partly glazed and it is opening towards outside plateau for summer gatherings. Roof is following its own curvature and forming an external porch. The cross as catholic sign is featured as laying feature positioned on the rooftop above the main farewell space. It also functions as a luminous dynamic element across the space during the daytime and lighting spark at a night time.

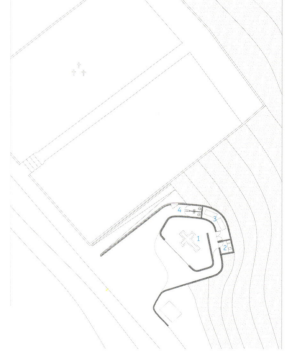

告别礼拜堂

这个告别礼拜堂位于卢布尔雅那附近的小村庄里，临近一片墓地。礼拜堂嵌入了上升的自然景观之中，顺应着景观的轨迹而建。三面弧形的墙壁分割着项目的布局，其中外侧的墙壁作为主承重墙，将周围的山体和礼拜堂分隔开。例如仓储、更衣室、洗手间和小厨房则靠内侧的墙设置。室内的墙壁环绕着告别仪式的空间。墙面的一部分镶有玻璃，在夏天可以向外开放。屋顶也依着礼拜堂的弧度而建，最终形成了一个户外门廊。作为天主教的标志，屋顶的十字架是项目的标志性特征，凌驾于告别仪式空间之上。同时，十字架也是一个动态的发光元素，在白天和黑夜都闪闪发光。

1. main space
2. kitchen
3. storage
4. washing room

1. 主空间
2. 厨房
3. 仓库
4. 洗手间

Completion Date: 2009

Architect: OFIS Arhitekti; Project Leaders: Rok Oman, Spela Videcnik
Project Team: Andrej Gregoric, Janez Martincic, Magdalena Lacka, Katja Aljaz, Martina Lipicer

Universidade Agostinho Neto

阿戈什蒂纽·内图大学

Located on the outskirts of Luanda, Angola, the master plan for a 2,000-hectare campus for 17,000 students consists of a core of academic buildings with research and residential buildings to the south and north respectively. Phase I, currently under construction, includes four classroom buildings housing faculties of chemistry, mathematics, physics and computer sciences and the central library and plaza. A refectory, student union and conference centre are also included.

The guiding principle of the master plan is to create a low-maintenance sustainable urbanism. Development is concentrated on the semi-arid rolling site, leaving as much of the existing vegetation and river washes as possible untouched. The ring road is conceived as a pure circle, distorted into an ellipse to fit between the washes. Within the ellipse, which differentiates natural landscape from man-made pedestrian streets, quadrangles pinwheel from the central plaza. The orientation of the man-made grid is approximately nineteen degrees east of the north/south axis, a compromise between the ideal solar orientation and the need to be perpendicular to the prevailing southwest breezes. Landscaping within the site channels the wind to maximise natural ventilation and cooling.

阿戈什蒂纽·内图大学位于安哥拉首都卢安达城郊，校园总面积为2,000万平方米，在南北两个区域为17,000名学生提供了教学楼、研究楼和宿舍楼。一期工程包含四座教学楼（里面设置着化学、数学、物理学和计算机教室）、中心图书馆、食堂、学生餐厅和会议中心。

学校总体规划的设计指导原则是设计一个低维护率的可持续城市化校园。校园的建设工程集中在一个半干燥的场地，尽量保留了场地上原有的植被和河流。环形公路被改造成为一个椭圆形场地的边缘。椭圆形结构划分了步行街上的自然景观和中心广场上的四方院。校园建筑网格的设计在南北轴线上偏东19度，以获得最佳的日照效果，并且垂直于西南信风。校园内的景观设计调节了风向，优化了自然通风和建筑的被动式制冷。

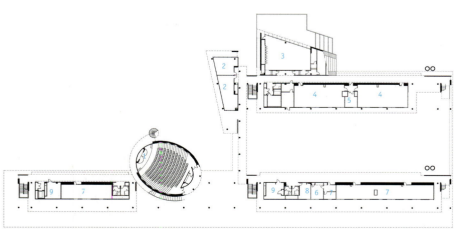

1. 礼堂
2. 小组讨论室
3. 集会处
4. 教学实验室
5. 辅助实验室
6. 主管办公室
7. 办公室
8. 接待处
9. 机械室

1. auditorium
2. seminar
3. assembly
4. teaching lab
5. lab support
6. department head office
7. office
8. receiving
9. mechanical

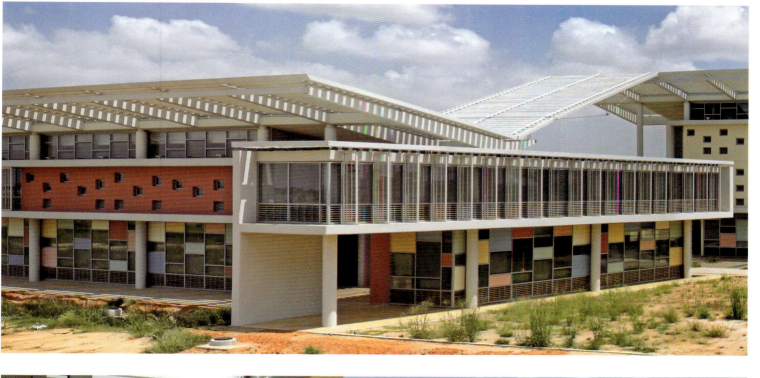

Main entrance, night view – a source of pride for the nation it represents 入口夜景，体现了利比亚的国家特色

Tripoli Convention Hall

Tripoli Convention Hall signifies the spirit of its homeland and claims to establish a physical relation with the global community. As an outcome of intensive urban and architectural movement in Libya, new innovative and prestigious buildings in Tripoli started to signify the power and contemporary style.

Open to the world cultures and where diverse languages meet up, the Convention Hall is a strong element of high representation and welcomes presidents of the world in the texture of the natural environment and as a source of pride for the nation it represents. Surrounded by the woods, the rectangular two-storey "block" is nestled in a metal envelop that opens up to the external landscape with a wide portico that defines the main entrance.

A semi-transparent perimeter "shield" of designed bronze mesh application flows around the building, protecting the inner glass walls; an eight-metre corridor encircles all three sides; the main building is flanked by a four-metre-wide reflection pool and another four-metre is left as a semi-open shady circulation area. The metal mesh walls carry incise patterns that are inspired by the trees that surround the site, permitting controlled daylight to diffuse into the central space.

的黎波里会议中心

的黎波里会议中心既具利比亚本土精神，又与国际社会相接轨。作为利比亚城市建筑运动的成果，的黎波里新建的创意建筑充满了现代风格。

作为各国语言文化交汇的中心，的黎波里会议中心以自豪的姿态代表自己的国家迎接着世界各地的领导人。会议中心的四周绿树环绕，两层高的建筑被包裹在一个金属外壳里，宽大的柱廊入口向外部的景观开放。

镂空的铜网保护层环绕着建筑，保护着里面的玻璃墙。8米宽的走廊包围了建筑的三面。主楼的两侧分别是一个4米宽的倒影池和一个半开放的绿荫休息区。金属网上图案的设计灵感来源于环绕场地的树木，控制了进入到大厅的日光直射。

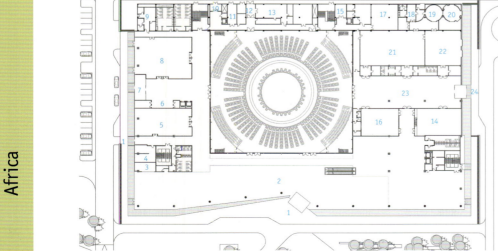

1. presidential entrance	1. 总统入口
2. main foyer	2. 主门厅
3. store	3. 商店
4. security office	4. 保安室
5. minister lounge	5. 部长休息室
6. front office/cloak room	6. 前厅/衣帽间
7. minister entrance	7. 部长入口
8. minister lounge	8. 部长休息室
9. cold storage	9. 冷库
10. staff entrance	10. 员工入口
11. technical area	11. 机械区
12. service entrance	12. 服务入口
13. service kitchen	13. 服务厨房
14. press hall	14. 新闻大厅
15. cold storage	15. 冷库
16. meeting room	16. 会议室
17. storage	17. 仓库
18. VIP security	18. VIP保安室
19. VIP lounge	19. VIP休息室
20. VIP suite	20. VIP套房
21. president lounge	21. 总统休息室
22. president salon	22. 总统沙龙
23. VIP meeting room	23. VIP会议室
24. VIP entrance	24. VIP入口

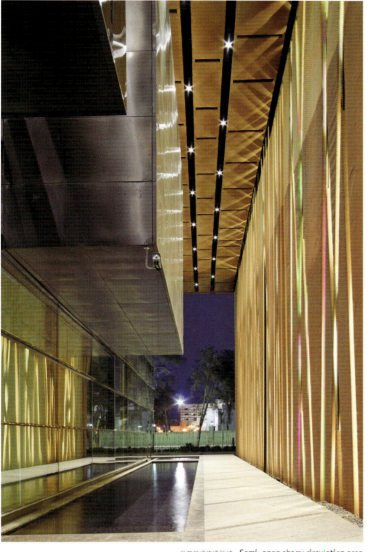

半开放式流通区域　Semi-open shady circulation area

可控的日光弥散在中心区域　Controlled daylight to diffuse into the central space

会议室　Meeting room with transitivity

Elandra Beach Houses

The project is a collection of fifteen stylish beach houses set in a pristine bushland setting. The site itself is just back from the beach and has stunning views of Port Hacking and Hastings Beach. What is unique about this development is that each house has been designed as if it were a stand-alone luxury designer home. The houses were designed for a sophisticated style-conscious market, for the person who would live in a luxury contemporary apartment in the city. The idea was to provide holiday housing for the sophisticated inner city dweller.

Stage 1 consists of a type called the "cross-over house". In this house, the top floor, which contains the living areas, is oriented east/west at and the lower level bedrooms run north/south. This allows the house to sit into the slope with minimal impact on the landscape. The top floor maximises exposure to the northern sun, whilst the lower level is directed toward the beach. These houses were designed specifically with environmental issues in mind, and have been orientated to maximise passive solar design. Design for water conservation is always important — all the plumbing relies on recycled water.

休息室 Lounge

兰德拉海滩度假屋

兰德拉海滩度假屋选址在原始森林地带，距哈金港及黑斯廷斯海滨仅有数步之遥，四周风景宜人，共由15栋现代风格住宅构成。更为吸引眼球的一点是，每一栋住宅都奢华至极，旨在为那些城市居住者们提供一个度假胜地。

一期工程以"交叉外形"为主要特色——顶层（生活区）为东西朝向，一层则为南北朝向。独特的设计方式使得住宅巧妙地运用了倾斜的地势，与周围环境相吻合。此外，这一设计充分运用了自然光线以及海滨景致。住宅设计格外注重环保因素：其一便是大量运用自然光线；其二便是充分运用循环水。

Front view 建筑正面

建筑侧面和树木 Side with trees

建筑正面 Front side

室内全景 Interior full view

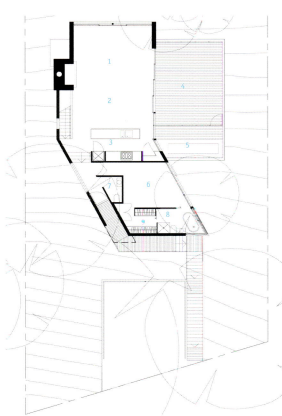

1. living
2. dining
3. kitchen
4. terrace
5. plunge pool
6. bed
7. WC
8. ensuite
9. robe

1. 客厅
2. 餐厅
3. 厨房
4. 平台
5. 水池
6. 卧室
7. 洗手间
8. 套房
9. 衣帽间

Photo: Brett Boardman

Residential

Completion Date: 2007

Architect: Tony Owen Partners

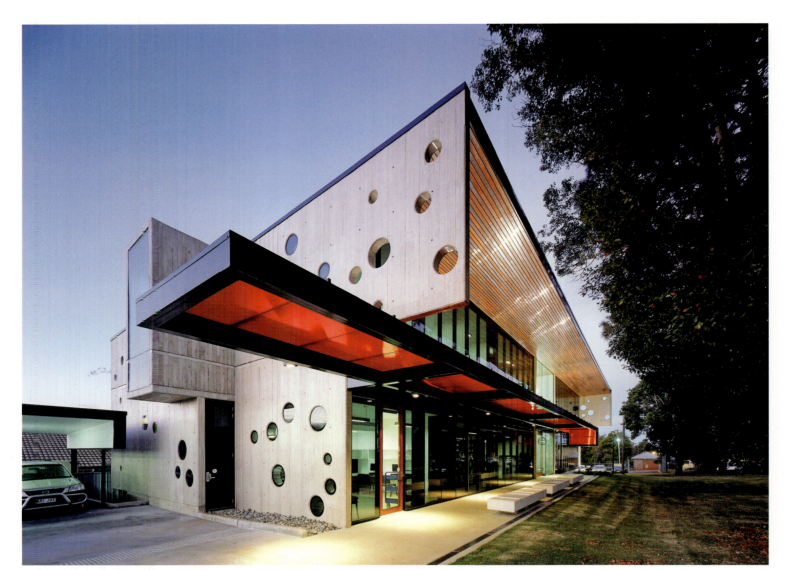

Night view 建筑夜景

University of Queensland Rural Clinical School

The principles of functionalism and rationalism are manifested in the UQ RSC project, which in turn, reflect to the programme. Reductive detailing, robust material selection and expressive volumes pay homage to these principles. These aspirations, not often associated with medical training facilities, offer an alternative to the cold institutional paradigm of past. The building's plan form is simple and rational, largely driven by the limitation of the site and programme. Entrance into the building itself is via a centrally located, double volume, addressing the two programmes contained within the building–training and administration.

The contemporary use of materials and the large expanses of glass brought the building in line with the expectations of a modern institutional building while delivering a warm, tactile and inspirational internal volume within, in which to learn. The east and west façades are formed up, textured in natural, white concrete walls. The walls have a series of perforations which increase in density and size across the façade to emphasise the entrance. The north and south façades are clad in full–height, structurally glazed curtain wall. Winter sun is allowed to penetrate the building while allowing views to the north over the suburban district and south over the hospital grounds.

昆士兰大学医学院

功能主义及理性主义原则在这一项目中得到充分体现。精致的细节、实用的材质以及表现性极强的结构更是将这一原则完美诠释。建筑格局造型简约而合理，在很大程度上受到地块及工程本身的限制。入口设在一个两层中心结构内，突出了建筑内部的空间分配——培训区及行政区。

材质运用的现代化方式以及大面积玻璃的使用使建筑在外观上完全符合现代化研究机构的标准，同时在内部营造温馨融合、触感十足的氛围，易于灵感的迸发。东、西立面采用白色水泥墙封闭，上面的小孔在视觉上增添了建筑的规格，同时将入口突显出来。南、北立面完全采用玻璃幕墙结构，使得冬季的阳光源源不断地涌进来，同时带来一个开阔的视野。

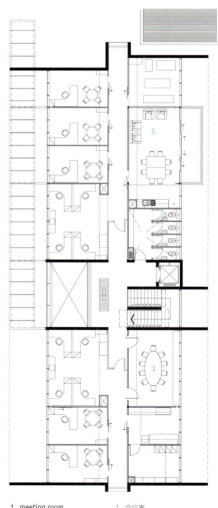

1. meeting room　　1. 会议室
2. washing room　　2. 洗手间
3. lecture room　　3. 教室

全景图 General view

楼梯 Stairway

外墙 Outside of the wall

General view 全景图

Adelaide Central Bus Station

The Adelaide Central Bus Station is the culmination of four years of master planning and design. This new station is a quantum leap in quality from the old bus station and provides travellers with modern facilities, and sets the standard for terminal design in Australia. The design strove to provide an architectural landmark offering an immediate sense of place and orientation to the traveller, and an environment that is light, airy, welcoming with a certain sensory impact.

The bus station's distinctive Mandarin Orange palette references its location, in the environs of Adelaide's lively Chinatown. The strong colour also reinforces the pedestrian access to the terminal from Grote Street.

A series of aluminium grids make up part of the bus station façades – these create a contemporary imagery that is drawn from its transportation–based uses–bus station and car parking. These curved and folded aluminium arcs are perforated, creating visual interest and emphasising the best features of both old and new transport facilities. The curved canopies on the street frontages project out from the line of the glazing, offering a degree of protection for pedestrians and continuing the notion of the verandah. In addition to the bus station the project provides 550 car parks over five levels and thirty-nine residential units. The bus station will be home to fifteen coaches, and approximately 300,000 people per year will pass through its doors on both interstate and intrastate travel.

阿德莱德中心汽车站

经过四年的建设，阿德莱德中心汽车站终于得以建成。新汽车站的设计在传统汽车站的基础上取得了巨大飞跃，是澳大利亚车站设计的典范。阿德莱德中心汽车站的设计为当地提供了一个地标，也为乘客指明了方向。建筑的整体造型轻盈、明亮，具有一定的感官效果。

汽车站采用了独特的橙黄色色调，与周边活跃的唐人街相得益彰。强烈的色彩还装饰了格罗特街通往汽车站的人行道。

汽车站的外墙由一系列的铝窗格组成，为建筑营造了现代化的形象。弯曲、折叠的铝板上密布着小孔，打造出奇妙的视觉感，强调了交通设施的功能与特性。街面上弯曲的华盖向外突出，为行人提供了屋檐保护，也延续了建筑的走廊。

1. 汽车站
2. 西侧中心广场
1. bus station plaza
2. west central plaza

建筑正面 Front view

入口 Entrance

接待处 Reception

Photo: Drew Lenman

Completion Date: 2007

Architect: Woodhead + Denton Corker Marshall

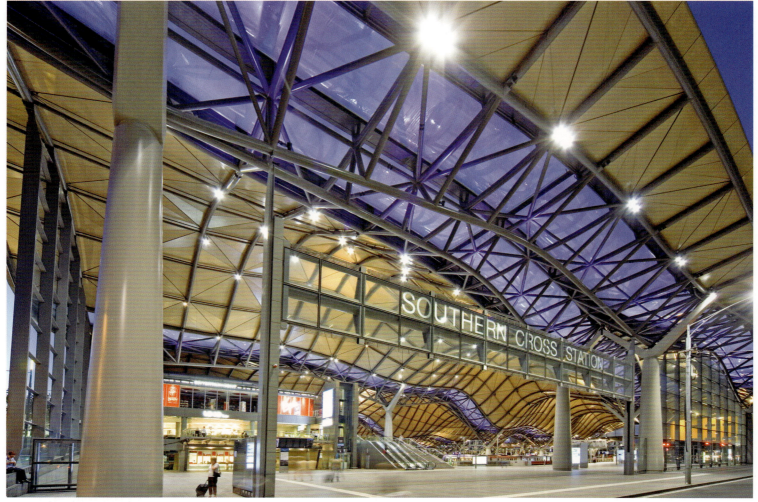

Entrance area 入口区域

Southern Cross Station

The commission for the redevelopment of Melbourne's Southern Cross Station (previously known as Spencer Street Station) arose out of the need for an upgraded terminus to accommodate the anticipated rise in demand for public rail services in the future. The brief required a fully-integrated transport interchange, which would also provide essential public transport upgrades as well as pedestrian connections between Melbourne's Central Business District (CBD) and the developing Docklands area. This increased connectivity would encourage regeneration and improve commercial growth locally.

While the design approach endeavoured to create a visually inspiring structure, the key generators for the station were always practical performance, ease of passenger circulation and an improved working environment for staff. The newly refurbished Southern Cross Station provides fifteen million users per annum with fully sheltered, high-quality ticketing, baggage-handling, waiting and retail services, all equipped with comfortable seating, lighting and passenger information display systems. Internally it is a vast hall, with uninterrupted vistas in every direction so that the interconnection of different streets surrounding the station can be easily understood. Pods of accommodation beneath the roof, house administrative functions as well as providing a defined retail space below. By their nature, stations must work across many different levels to enable passengers to access the various train lines. At Southern Cross Station, the ground plane itself changes, with Bourke and Collins Street rising in parallel to either side of the building. The concourses rise in response to the street plan so that level change with the station is almost imperceptible.

南十字星车站

南十字星车站原名为"斯潘塞街车站",为满足公众不断增长的出行需求,决定将车站重新修复。设计要求打造一个完整的公共运输交汇处,升级必要的公共运输设备以及建设连通墨尔本商业中心同港口住宅开发区之间的人行道,从而促进当地经济的发展。

设计师在努力打造一个极具视觉冲击力建筑结构的同时,更加注重功能性需求,确保旅客通道顺畅流通以及完善工作人员办公氛围。新建车站内设有便利而高档的售票区、行李存放处、等候区及零售区,舒适的座椅、明亮的光线以及旅客信息展板一应俱全,每年发送150万旅客。大厅内,每个方向都设有明确的指示标识,让人将周围的环境一目了然。尤为重要的一点是,车站必须保持不同楼层之间的连通性,以满足乘坐不同车次的旅客的需求。在这里,这一点得到完美体现——设计师通过车站本身地平面高度的变化,与两侧街道取得平衡,大厅提升到街道平面高度处,所以基本上感觉不到楼层之间的变化。

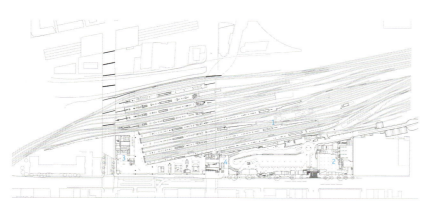

1. 车站
2. 控制中心
3. 休息区
4. 活动空间

1. terminal areas
2. control centre
3. lounge area
4. activity space

鸟瞰图 Aerial view

自动扶梯 Escalator

Transportation

Completion Date: 2006

站台 Platforms

Architect: Grimshaw Architects

View of main lodge 旅馆全景

Southern Ocean Lodge

Southern Ocean Lodge, perched on a forty-metre-high cliff with panoramic views over the wild Southern Ocean. The Lodge houses twenty-one spacious guest suites and restaurant/bar/lounge, primarily for international visitors seeking a unique Australian experience.

The architecture has a close relationship with the dramatic site. The Main Lodge is tucked back into the cliff top, with large sweeping window walls capturing the expansive views of the ocean, rugged coastal cliffs and pristine bush. A strong sculptural element is the 100-metre-long curving Kangaroo Island Limestone wall, which weaves from a covered entrance, through the largely untouched bush and into the Main Lodge/restaurant. It provides a textured backdrop for the refined details of the guest areas with recesses accommodating desks, seating and reception facilities for guests and staff.

Guest suites cascade down the slope from the Main Lodge, with access from a breezeway ramp. Roofs follow the slope of the land, but with a gentle upward, wave-like curve every fourth suite. The curves define the rainwater collection system with gutters extending out to galvanised iron rainwater tanks. Such tanks are ubiquitous iconic structures in dry rural Australia, and here they emphasise the sustainability principles of the project, with all rainwater collected for use within the Lodge.

南冰洋旅馆

南冰洋旅馆位于40米高的峭壁上，可以俯瞰辽阔的南冰洋。旅馆共有21套宽敞的客房以及餐厅、酒吧、休息厅等设施，为国际游客提供独特的澳洲风情体验。

建筑与优美的自然环境联系密切，旅馆的门房建在峭壁顶上，透过巨大的玻璃可以将大海和海岸的风景尽收眼底。100米长的袋鼠岛石灰岩墙壁从建筑的入口处穿过天然灌木丛，一直延伸到门房和餐厅的门口，极具雕塑特征。墙壁为设计精美的客房区域提供了纹理清晰的背景，使客房区的休息区和前台接待设施的设计更加突出。

客房从门房所在的坡上向下蜿蜒，可以通过带有屋顶的坡道到达。屋顶依随着场地的坡度，但是每隔四件套房便会微微上翘。这些弧度是雨水收集系统的一部分，屋顶的水槽延伸到下方的镀锌雨水收集槽里。这些水槽在澳大利亚的干燥区域随处可见，体现了项目的可持续设计特征。所收集的雨水将被用于旅馆的日常用水。

1. staff village	1. 员工村
2. service yard	2. 服务庭院
3. reception	3. 前台
4. guests office	4. 来宾办公室
5. disabled	5. 无障碍通道
6. departure lounge	6. 出发大厅
7. family suite	7. 家庭套房
8. deluxe	8. 豪华房
9. standard	9. 标准房
10. family suite	10. 家庭套房
11. deluxe	11. 豪华房
12. standard	12. 标准房
13. standard	13. 标准房
14. standard	14. 标准房
15. premium suite	15. 额外付费套房
16. rocky cliffs	16. 岩石峭壁
17. beach	17. 海滩
18. sub premium	18. 额外付费套房附属间
19. spa retreat	19. SPA

旅馆依斜坡而建 View of main lodge and suites stepping down the slope

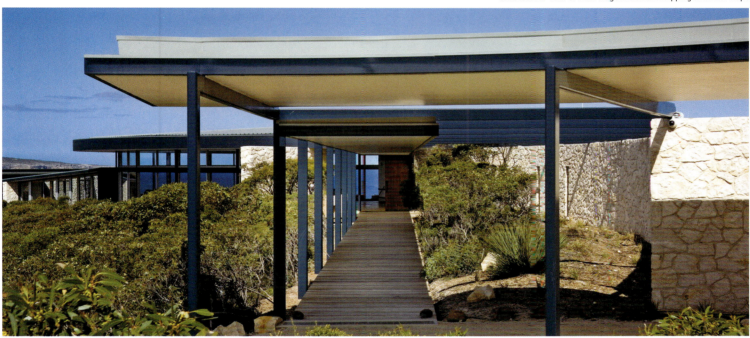

旅馆入口 Entrance to the Lodge

大堂 The Great Room

Photo: Sam Noonan, George Apostolidis

Completion Date: 2008

Architect: Max Pritchard Architect

Little Creatures Brewery

Little Creatures Brewery is a production brewery and multi-facetted hospitality venue located alongside the picturesque Fishing Boat Harbour in Fremantle Western Australia. In 2008 Little Creatures commenced a significant expansion and now occupies five different interlinked buildings comprising a mixture of nealy built and substantially converted existing buildings.

The completed facility now boasts a large purpose-made brewery operating twenty-four hours a day, six different bars, outdoor harbourside dining, a Bocce court, live music venues and even a gallery. The total capacity allows for up to 1,200 patrons interspersed around the brewing and fermenting tanks of a busy working industrial space.

The architectural expression consistently reflects the industrial nature of the brewing environment. All brewery processes, from the grain delivery to the chilled serving tanks in the main bar take precedent in planning and functionality and remain open, visible and accessible. The hospitality and associated services have all literally been fitted around the available remaining space.

The architecture makes few concessions to decorative interior spaces normally associated with hospitality venues. The customers are actually immersed in an active working industrial environment, which is reflected in the built form, the open kitchen, hard-edge finishes and furniture.

小生物啤酒厂

小生物啤酒厂集啤酒制造和餐饮娱乐于一身，位于西澳大利亚风景如画的弗里曼特尔渔港。2008年，小生物公司对啤酒厂进行了一次扩建，形成了五座新旧建筑连在一起的局面。

现在啤酒厂有一间24小时运作的工厂、六个各具特色的酒吧、一个露天港口餐厅、一个地掷球场、一个现场音乐表演场和一间画廊。整个场地最多可以同时容纳1,200人，场地四周还点缀着工厂的酿造发酵槽。

啤酒厂的整体建筑设计反映了整体环境的工业特征。从收购粮食到酒吧里的冰啤酒，消费者可以观看到这一系列的啤酒酿造过程。所有的餐饮娱乐设施都被合理地安排在啤酒制作过程中间。

建筑很少采用普通餐饮业所使用的室内装饰。消费者完全融入了啤酒厂的工作环境之中。建筑的造型、开放式厨房、硬装饰和家具都极具工业特征。

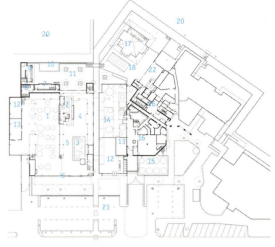

1. storage tanks	1. 储酒桶
2. main kitchen	2. 主厨房
3. beer serving tanks	3. 酒桶
4. restaurant	4. 餐厅
5. bar	5. 酒吧
6. front verandah	6. 正面长廊
7. male WC's	7. 男洗手间
8. female WC's	8. 女洗手间
9. courtyard kitchen	9. 庭院厨房
10. bocce court	10. 地掷球场
11. rear courtyard	11. 后院
12. retail bar	12. 零售吧
13. boiler	13. 锅炉
14. brewhouse	14. 酿酒区
15. storage tanks	15. 储酒桶
16. services	16. 服务区
17. deck bar	17. 平台吧
18. rear deck	18. 后部平台
19. boardwalk	19. 木板路
20. harbour	20. 码头
21. carpark	21. 停车场
22. burger kitchen	22. 汉堡厨房
23. mezzanine	23. 夹层
24. brewhouse mezzanine	24. 酿酒区夹层
25. music room	25. 音乐屋
26. loft	26. 阁楼
27. deck	27. 平台
28. verandah	28. 长廊

Front elevation 建筑正面

酿酒酒吧 Brewhouse bar

酿酒区里的沙发 Sofas in the brewhouse

Photo: Jody D'Arcy

Completion Date: 2009

Architect: Paul Burnham Architect Pty Ltd

Anglesea House

The Northern addition replaces an old timber deck that previously divided the two storeys and radically reduced sunlight to the living area on the lower level, making the space beneath damp, dark and disconnected from the rest of the house. The trafficable roof of this addition is now extruded down to the earth, creating a three–metre–thick deck and grounding the entire house to the site, while extending the top floor living spaces out into the treetops.

The glazing to the new northern box addition has been located in such a way as to allow winter sun to penetrate deep within its interior, warming the concrete slab provided for thermal mass and block out the high summer sun. Besides, carefully located, timber boxes appear on the southern and eastern edges of the existing structure. The former is a glass roofed & walled shower. Its transparent material, pushing the privacy boundary to create a shower experience immersed in gum trees and sky – something that cannot be easily achieved in the city – reminds the user of the natural beauty of their coastal environment. To the eastern side of the house, other newly introduced structures nestle under the existing carport providing much needed external storage space and a children's bunk retreat.

安格尔西别墅

别墅的旧观景木台分割了两层楼，使得底层起居区域的空间阴暗又潮湿，与别墅的其他部分相分离。北侧的扩建工程移除了这个平台。扩建结构的屋顶一直延伸到地面，形成了一个3米厚（从地面到屋顶）的平台，使整个别墅与地面相连，也让二楼的起居空间得以延伸到树顶。

北侧扩建结构的玻璃拉门让冬日的阳光渗透到屋内，温暖了起到隔热保暖作用的混凝土墙面。另外，木盒结构的扩建工程分别位于原有结构的南侧和东侧。前者是一个玻璃淋浴室。浴室的透明材料让人的隐私完全暴露在桉树树林和天地之间，这种体验是城市中难以获得的，它唤起了人们对海岸环境自然美欣赏之情。东侧车棚下方的区域提供了户外仓储空间和孩子们的游艺场所。

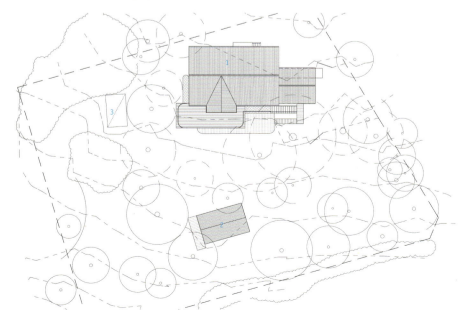

1. 原有的住宅
2. 舢板棚
3. 蹦床

1. Existing house
2. Boat Shed
3. Trampoline

Architect: AMA Team, Andrew Maynard, Mark Austin, Matthew McClurg

Completion Date: 2009

Photo: Peter Bennetts

Residential

Open room 开放式房间

Essex Street House

It is a residential alteration and extension to an existing double-fronted weatherboard house. The brief required two bathrooms, a bedroom, living area, kitchen and increased connection with outside areas.

The context is typical of inner suburban Melbourne. The site is double-fronted with a deeper than usual block running east-west. The initial brief asked for an extension along the full width of the existing house. The response to the brief was that any addition should run along a southern boundary to maximise solar access to new and existing spaces and to bring external space into the middle of the living areas.

The original house has been restored to its simple four-room square plan. The new structure sits lightly beside it like a loyal companion. Rather than build a hard-edged or strongly defined object, the new structure has a blurred or vague edge. The recycled grey iron bark portal frames are of a larger, non-domestic scale. They were envisaged as an old relic of a pre-industrial age, an old, wise element to a new and vibrant addition. Within the robust portals is the delicate layered box. The use of screening and the glazed garage doors creates a soft edge that allows the internal spaces to spill into the outdoor spaces. Within this structure are the small, colourful boxes of the bedroom and kitchen. These objects separate functions and act as a bridge between the original house and the extension.

亚瑟斯街住宅

项目是对一座双面挡风板住宅的扩建和改建，包含两间浴室、一间卧室、起居区、厨房和室外连接区。

住宅所在的环境具有典型的墨尔本近郊特点。沿东西向延伸的建筑比寻常的建筑更长。最初的设计大纲要求沿着原有住宅实行全面加宽。最终，设计把所有的扩建区域都设在了建筑南侧，以最大化室内采光、将户外空间引入住宅内部。

原有的住宅布局极为简单，只有四间房间。新的扩建结构坐落在它旁边，像一个忠实的伴侣。新建结构并没有过多的棱角，边缘十分模糊，再利用的灰铁门架异常巨大。门架仿佛前工业时代的老爷车一样，是充满活力的新建结构上一个古老而具有哲理的元素。粗犷的门架内部是一个精致的叠层盒结构。屏风和玻璃门的使用营造了一个柔化的边缘，将内部空间和外部环境联系起来。盒结构再往内是一个将原有建筑和扩建结构连接起来的廊桥。

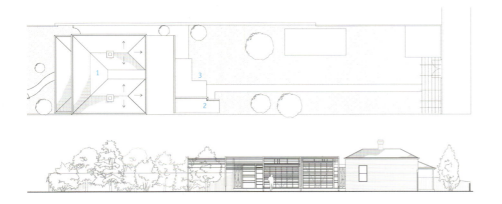

1. 传统圆形屋顶
2. 清树脂顶板
3. 浴室/洗衣房/洗手间

1. custom orb roofing
2. clear polycarbonate roof sheeting
3. bathroom & laundry & WC

浴室 Bathroom

建筑外观 Outside view

室内 Interior

Photo: External photography by Peter Bennetts
Internal photography by Dan Mahon

Residential

Completion Date: 2006

Architect: Andrew Maynard Architects

Front exterior 建筑正面

Beresford Hotel

Thomas Jacobsen, has had a hand in crafting every aspect of the heritage hotel's transformation: from the new architecture's sweeping curves finished in bespoke green Spanish tiles to the sculpted olive oil vessels in the 130-seat bistro.

Phase one of The Beresford hotel, now completed, is the ground floor public bar wrapped in tile and Tasmanian oak, the bluestone-clad bistro and a spacious garden where films can be projected under the stars. Upstairs is a 1,000-square-metre ballroom, a live music venue and function room that will open soon. Glamorous balconies lined with custom-made stainless steel tiles and faced with rusted steel panels lend the new external elements a tough elegance.

Little expense has been spared on this beautifully-finished project which Jacobsen calls "New Deco" for the way it takes traditional values and craftsmanship and give them a funky twist that is undoubtedly of our time. Even the hidden elements to the design have integrity. In the front bar, for instance, the lighting concealed behind new polished stucco deco cornices is neon rather than the usual fluoro to ensure just the right light colour tone. There are sixty individual pieces of neon subtly intalled into the back bar alone.

伯莱斯弗德酒店

设计师托马斯·雅各布森亲手操刀了这个具有历史意义的酒店的改造工程，从横扫建筑弧形外墙的绿色西班牙瓷砖，到小酒馆里橄榄油雕刻油罐，每一个细节他都没有放过。

已完成的伯莱斯弗德酒店一期改造工程包括一楼由瓷砖和塔斯马尼亚橡木包裹的酒吧、青石外墙的小酒馆和晚上可以放映电影的室外花园。二楼的1,000平方米的舞厅、现场音乐表演台和宴会厅即将开放。华丽的阳台上，特别定制的不锈钢砖板与生锈的铁板相对比，形成了一种粗犷的优雅。

雅各布森在这个装饰优美的项目上并没有花费过多的费用，他将具有传统价值的新装饰主义和手工艺与现代设计完美地结合在一起。甚至连设计的隐藏元素也十分完整。例如，酒吧里隐藏在抛光飞檐后面的灯光是霓虹灯，而不是常见的荧光灯，保证了色调的纯正，仅仅吧台后区就使用了60盏独立的霓虹灯。

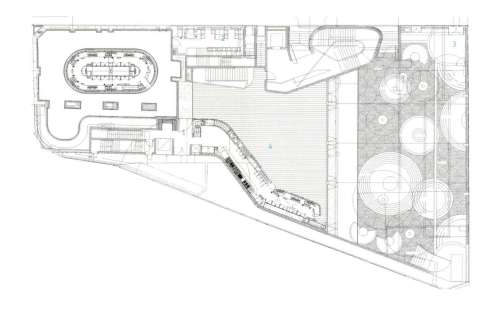

1. 员工入口
2. 游艺区
3. 消防设施
4. 厨房

1. staff access
2. gaming
3. fire service
4. kitchen

建筑外部和庭院 External with courtyard

夜景 Exterior at night

餐厅 Restaurant

Photo: Ross Honeysett

Completion Date: 2008

Architect: Thomas Jacobsen

Hotel

The Village at Yeronga

The Village at Yeronga is the master planned and integrated retirement village development located in the inner city Brisbane suburb of Yeronga on a 28,000-square-metre site previously occupied for light industrial building use. The first two stages of the village have now been completed comprising 91 independent living apartments and associated community facilities. On completion of all stages this medium density project will house 240 independent living apartments, 60 assisted living apartments and a 110-bed full aged care nursing home facility. This integrated retirement project is designed to respond to the Australian Government's "Ageing in Place" policy which encourages developments to cater for all three levels of care (i.e. independent, assisted and aged care) within the one facility.

The buildings were constructed predominantly in precast concrete which was chosen for its long-term low maintenance properties. The building uses different panel colours to assist the elderly residents to find their way within the project, such as the strong ochre colour which is used to accentuate the building entrance points.

The feature of the building is undoubtedly the resort-style communal facilities provided within the development. These facilities include a fully-equipped commercial kitchen and restaurant for residents and their visitors, cinema, club lounge and bar, library, lobby lounge and café, indoor swimming pool, gymnasium, beauty salon and wellness centre, billiards room, medical centre and administration. These facilities also assist the residents in building valuable social networks within the retirement village community.

Yeronga养老村

Yeronga住宅村是一项综合养老村开发工程，位于布里斯班近郊的Yeronga，总占地面积28,000平方米，前身是轻工业工厂厂房。目前，养老村的前两期工程已顺利完工，共有91套独立公寓和配套的社区设施。在所有阶段的施工结束后，养老村计划拥有240套独立公寓、60套辅助公寓和110套家庭护理公寓。这个综合养老项目是配合澳大利亚政府的"就地养老"政策而进行的。"就地养老"政策鼓励在同一地点开发具有三个层次（独立、辅助、护理）的养老住宅。

建筑采用预制混凝土结构，使用寿命长、维护费用低。建筑的外墙采用了不同颜色的面板，以帮助老人们找到自己的路。例如，设计师利用土黄色着重渲染了建筑的入口。

具有度假村风格的公共设施无疑是设计的亮点。这些设施包括配备齐全的商业厨房和餐厅、电影院、俱乐部、酒吧、图书馆、休息室、咖啡厅、室内游泳池、健身房、美容沙龙、桌球室、医疗中心和行政区。这些设施为住户在养老社区内提供了宝贵的社交网络。

Entrance of the village 养老村入口

Panorama of the village 全景图

建筑外观 Buildings

室内细部 Interior details

大厅 Lobby

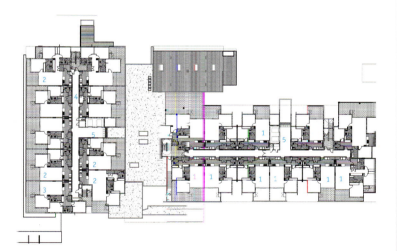

1. bedroom
2. bedroom
3. bedroom
4. corridor
5. lobby

1. 卧室
2. 卧室
3. 卧室
4. 走廊
5. 大厅

Photo: Scott Burrows (Aperture Photography)

Residential

Completion Date: 2009

Architect: Arkhefield (www.arkhefield.com.au)

Berry Sports and Recreation Hall

Set on sixty hectares of rolling countryside in Berry, three hours' ride south of Sydney in Australia, the site was originally an experimental dairy farm and has made way for a magical and innovative multipurpose hall for basketball, netball, rock climbing, dance and theatre.

Reminiscent of a modern farm shed, the building comprises two long sides of precast concrete panels, each pierced by 500 shards of glass in amoeba-like windows, allowing natural light to flood the halls in the day and interior lights to shine through at night, illuminating the building and making it "disappear" into the night sky.

The building also features environmentally sustainable design (ESD), with a dozen wind turbines combining with panels of louvers to create a natural ventilation system which cools the structure in summer and creates an insulation blanket in winter. Roof water is tracked back from the 3.5-metre cantilevered composite roof via a steel beam to provide water for irrigation tanks.

贝里运动休闲馆

项目位于悉尼以南3小时车程、总面积为60万平方米的贝里山区，原址是一个试验奶牛场。新建的贝里运动休闲馆可进行篮球、无板篮、攀岩、舞蹈和表演等活动。

建筑造型模仿了现代农场小屋，两侧由长方形的预制混凝土板构成，每块板上都有500个彩色玻璃碎片构成的不规则形状小窗。白天，自然光可以透过这些小窗射入室内；夜晚，建筑透出的点点灯光让它仿佛融入了星空一样。

建筑采用了可持续性环保设计，12个风轮机和一些百叶窗墙板共同构成了自然通风系统，使建筑冬暖夏凉。3.5米高的悬臂复合屋顶可以通过一个钢管将雨水导入灌溉水槽里。

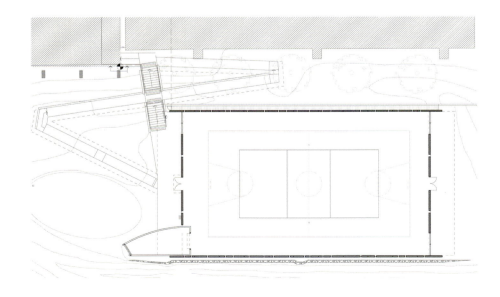

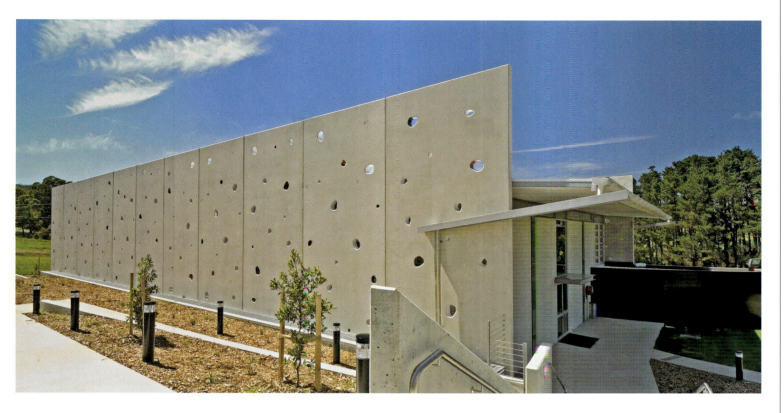

Architect: Allen Jack & Cottier

Photo: Nic Bailey

Completion Date: 2008

Recreational